Darbyshire

D1601658

Applications of Physiological Ecology to Forest Management

This is a volume in the
PHYSIOLOGICAL ECOLOGY series
Edited by Harold A. Mooney

A complete list of books in this series appears at the end of the volume.

Applications of Physiological Ecology to Forest Management

J. J. Landsberg
CSIRO Australia
Institute of Natural Resources
and Environment
Canberra, Australia

S. T. Gower
University of Wisconsin
Department of Forestry
Madison, Wisconsin

Academic Press

San Diego   London   Boston   New York   Sydney   Tokyo   Toronto

Front cover photograph: Canopy of a jack pine stand in Sayner, Wisconsin. Photo courtesy of Jeff Martin/JMar Photo Werks.

This book is printed on acid-free paper. ∞

Copyright © 1997 by ACADEMIC PRESS

All Rights Reserved.
No part of this publication may be reproduced or transmitted in any form or by any means, electronic or mechanical, including photocopy, recording, or any information storage and retrieval system, without permission in writing from the publisher.

Academic Press, Inc.
525 B Street, Suite 1900, San Diego, California 92101-4495, USA
http://www.apnet.com

Academic Press Limited
24-28 Oval Road, London NW1 7DX, UK
http://www.hbuk.co.uk/ap/

Library of Congress Cataloging-in-Publication Data

Applications of physiological ecology of forest management / edited by
   J.J. Landsberg. S.T. Gower.
         p.   cm. -- (Physiological ecology)
   Includes bibliographical references and index.
   ISBN 0-12-435955-8 (alk. paper)
   1. Trees--Ecophysiology. 2. Forest ecology. 3. Forest management. I. Landsberg, J. J. II. Gower, S. T. III. Series.
SD395.A66 1996
574.5'2642--dc20   96-27735
                                                                               CIP

PRINTED IN THE UNITED STATES OF AMERICA
96 97 98 99 00 01 BB 9 8 7 6 5 4 3 2 1

Contents

Preface ix

1. Introduction: Forests in the Modern World
  I. Forest Management: Levels, Decisions, and Influences 3
  II. Overview 9

2. Forest Biomes of the World
  I. Management 20
  II. Plantation Forestry 21
  III. Species Adaptations and Climatic Conditions 22
  IV. Forest Biomes of the World 33
  V. Future Distribution and Extent of Forest Biomes 48
  VI. Concluding Remarks 49
      Recommended Reading 50

3. Canopy Architecture and Microclimate
  I. Canopy Architecture 53
  II. Energy Balance and Interception of Visible (Photosynthetically Active) Radiation 63
  III. Heat and Mass Transport 72
  IV. Effects of Topography on Microclimate 83
  V. Concluding Remarks 87
     Recommended Reading 88

4. Forest Hydrology and Tree–Water Relations
  I. Hydrologic Balance 92
  II. Catchment Hydrology 111
  III. Tree–Water Relations and Their Effects on Growth 116

IV. Concluding Remarks   122
   Recommended Reading   124

5. Carbon Balance of Forests

I. Leaf Photosynthesis   128
II. Canopy Photosynthesis   133
III. Autotrophic Respiration   139
IV. Net Primary Production   144
V. Growth Efficiency   154
VI. Net Ecosystem Production   155
VII. Forests in the Global Carbon Budget   157
VIII. Concluding Remarks   158
   Recommended Reading   160

6. Soil Organic Matter and Decomposition

I. Soil Carbon Content and Accumulation   163
II. Sources of Soil Organic Matter   166
III. Litter Decomposition   170
IV. Carbon Losses from Forest Ecosystems   176
V. Influence of Forest Management on Soil Carbon Dynamics   180
VI. Role of Forest Soils in the Global Carbon Budget   183
VII. Concluding Remarks   184
   Recommended Reading   184

7. Nutrient Distribution and Cycling

I. The Essential Plant Nutrients and Ion-Exchange Capacity of Soils   187
II. Nutrient Distribution   191
III. Nutrient Cycling   194
IV. Impacts of Natural and Anthropogenic "Disturbances" on Nutrient Cycles   214
V. Concluding Remarks   226
   Recommended Reading   228

8. Changes in Ecosystem Structure and Function during Stand Development

I. General Succession Theory   230
II. Changes in Species Composition   233

     III. Stand Functional Characteristics   234
     IV. Forest and Ecosystem Productivity   236
     V. Nutrient Cycling   242
     VI. Concluding Remarks   244
           Recommended Reading   245

9. Ecosystem Process Models

     I. Forestry Models   249
     II. Current Process-Based Models   251
     III. Concluding Remarks   272
           Recommended Reading   276

10. Applications of Modern Technology and Ecophysiology to Forest Management

     I. Geographical Information Systems   278
     II. Remote Sensing   282
     III. The Use of GIS, Remote Sensing, and Models as Management Tools   289
     IV. Concluding Remarks   297
     V. Peroration   299
           Recommended Reading   300

Symbols and Definitions   301

References   304

Subject Index   351

Preface

An important objective in writing this book was to bridge the gap between foresters—whether they be managers, decision-makers at various levels, or policymakers—and ecologists and ecophysiologists. Unfortunately, there is a dichotomy between the teaching, the practice and profession of forestry, and the more basic science areas of forest ecology and ecophysiology. Undergraduate forestry students learn a great deal about forest types and forest management—genetics, silviculture and mensuration, forest engineering, and forest economics—but are given only a brief introduction to soils and plant sciences. The emphasis varies from school to school and from country to country, but in general there is not much space in forestry curricula for the basic disciplines. The dichotomy carries through to the world of forest managers and those who make decisions about the use of forested lands.

There are undoubtedly various reasons for this problem, but among them is the perception that ecology and ecophysiology are not of much relevance or importance in the practical world of management and decision-making. The reasons for this perception may lie as much in the failure of ecologists and physiologists to make a convincing case for the relevance of their disciplines as in the lack of interest among foresters in what they perceive as esoteric "background stuff." Management of any system, in any sphere of life, is a matter of making decisions and taking actions that are intended to have particular effects—the manager is trying to control or manipulate the system. In commercial forestry, the end results are usually economic, with ecology and sustainability often barely considered; decisions tend to be made by accountants, not scientists. This is unavoidable, but it is important that the consequences of those decisions are predictable. It is no less important when forests are managed for other uses, for example, as recreational areas, for the preservation of biodiversity, or as water-producing areas.

To predict the results of particular actions the manager must understand how the system works. This is what ecology and ecophysiology have to offer. We have tried, throughout this book, to show how the understanding and insights into the functioning of ecosystems offered by these disciplines are relevant and indeed essential to ensure that the decisions that will be made by people responsible for the management of forested lands are the best possible. We do not delude ourselves by thinking that we can bridge the gap with one book, but we believe that it will make a significant contribution and that it will stimulate forestry students to address the interesting ecophysiological questions that are relevant to the long-term sustainable management of forests.

This book is aimed at undergraduate students in the final courses for degrees in forestry and forest ecology and at postgraduate students in the same fields. We also hope that the synthesis contains enough of interest to be of value to our professional colleagues in research and that practicing foresters will find it useful in explaining observations made in the forests.

We have covered a wide range of topics, from the distribution of forest ecosystems, with some indication of the reasons for that distribution, to treatment of the basic disciplines of micrometeorology and hydrology, the factors affecting the carbon balance, organic matter decomposition, and nutrition. We have provided an introduction to the methods of modern process-based modeling, illustrated by a review of some of the better known extant models, and in the last chapter we have illustrated how all this can be used by managers in conjunction with the technologies of remote sensing and geographical information systems. Specialized textbooks are available for the areas covered by each chapter. We are aware that, by addressing such a wide field, our coverage will inevitably be patchy, and in parts inadequate, so that specialists in particular areas will find gaps and shortcomings. For these we apologize in advance, but we are prepared to gamble that the benefits of the overall synthesis will more than compensate for our shortcomings in specialist disciplinary areas.

It is a pleasure to acknowledge the help of friends and colleagues, experts in the fields covered by this book, who have generously given their time and expertise to read and criticize drafts of the various chapters and to keep us from straying too far from the paths of established scientific wisdom. As always, in these cases, we exonerate them from blame for omissions, mistakes, and failures—we did not always take their advice—and thank them most sincerely for helping to keep such errors to (we hope) a reasonable minimum. We list them in no particular order: Ross McMurtrie and Roddy Dewar (School of Biological Sciences, University of New South Wales); Peter Sands (CSIRO Division of Forestry, Hobart); Tim Fahey (Cornell University); Dale Johnson (Desert Research Institute, Reno, NV); Craig Lorimer (Department of Forestry, University of Wisconsin, Madison); Neal Scott (Landcare Research, Palmerston North, New Zealand); John Raison (CSIRO Division of Forestry, Canberra); Almut Arneth (Landcare Research, Lincoln, New Zealand); Peter Briggs (CSIRO Centre for Environmental Mechanics, Canberra); Ian Woodward (School of Biological Sciences, University of Sheffield); Frank Kelliher

(Landcare Research, Lincoln, NZ); David Whitehead (Landcare Research, Lincoln, NZ); Rob Vertessy (CSIRO Division of Water Resources, Canberra); Jim Vose (Coweeta Hydrologic Laboratory). Thank you all very much.

Jerry Melillo (Ecosystems Research Center, Woods Hole, MA) kindly gave permission for us to use a version of the world vegetation map (Fig. 2.2 in this book) that he and his colleagues developed as the basis for their analysis of terrestrial net primary productivity, published in *Nature* in 1993. David Kicklighter, part of that team, transmitted the (then) latest version of the map to us in digital form. Sune Linder (Swedish University of Agricultural Sciences, Uppsala), Douglas Malcolm (Institute of Ecology and Resource Management, University of Edinburgh), and John Bartle (Conservation and Land Management, Western Australia) provided comments about what managers in their parts of the world see as their needs and priorities.

On the production side, Greg Heath (CSIRO Centre for Environmental Mechanics) was patient and skillful in drawing and redrawing the figures—an invaluable contribution—and Suzie Bubb not only typed the tables and entered a great many of the references into a data base but also taught one of us (J.J.L.) to use the thing with reasonable proficiency. Tara Stow (Department of Forestry, University of Wisconsin, Madison) skillfully and cheerfully assisted with numerous facets of the book, as well as with some of the day-to-day tasks that S.T.G. did not have time for while working on this project. Thanks Tara! Connie Gower patiently proofed the Reference section, located missing references, and typed the Subject Index.

We are grateful to John Finnigan, Head of the Centre for Environmental Mechanics, for allowing J.J.L. to spend a great deal of his time on this project and to enlist the help of lab support staff. We also acknowledge the support provided by Landcare Research, New Zealand, and by University of Wisconsin College of Agriculture and Life Sciences to S.T.G. for allowing him to put a great deal of time into preparing this book while on sabbatical in Australia and New Zealand.

We acknowledge the support of NASA Grant NAGW-4181, which enabled J.J.L. to do some preliminary library work in the United States and S.T.G. to spend some time in Australia.

Finally, I (S.T.G.) thank my family (Connie, Kristen, and Cathy) for their support and love during the preparation of this book.

Joe Landsberg
Stith T. Gower

1

Introduction: Forestry in the Modern World

The objective of this book is to provide an up-to-date treatment of current scientific knowledge about the effects of the environment, humans, and their interactions on the growth of forests, and to show how this knowledge can be used in making decisions about the management of forests. To understand how environmental factors, such as incoming solar radiation, air temperature and humidity, precipitation, nutrients, and water balance, affect forest growth, we need to understand the way the physiological processes that determine the growth of trees are influenced by these factors. This requires an understanding of plant–environment interactions as physical processes as well as of the influence of environmental conditions on the physiological processes.

The general approach and philosophy are encapsulated in Fig. 1.1, which is a very simplified schematic outline of the major components of the forest/environment interactive matrix: the key components are water, nitrogen[1] and carbon. The diagram shows climate acting on the plant community through the foliage (radiation interception, temperature, and humidity), where CO_2 is absorbed and water vapor lost. The CO_2 absorbed is converted to carbohydrates through the process of photosynthesis; these are allocated to maintenance and the growth of foliage, stems, and roots. Precipitation is, of course, a major determinant of the soil water balance that, with temperature, affects the rate of organic matter decomposition and nitrogen release and uptake by the roots. Any management action in relation to forests will, directly or indirectly, influence all components of the system.

[1] Note that the focus on nitrogen as the most important nutrient may be arguable, particularly in relation to areas of the world where phosphorus is known to be the major nutrient limiting plant growth. However, a great deal of modern research has focused on nitrogen that, because of its fundamental role in the carbon cycle and because it is inextricably linked to growth processes through that cycle, appears to be the key nutrient. Phosphorus availability is much more strongly controlled by geochemical processes. We provide a more general treatment of tree nutrition in Chapter 7.

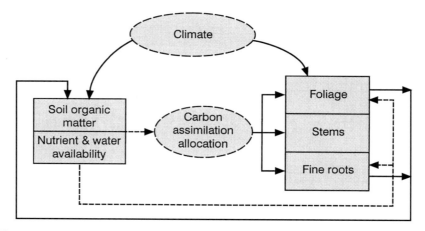

Figure 1.1 Schematic diagram showing the interactions between climate and forest ecosystems. The solid lines denote flows of energy or matter, the dotted line indicates the effects of environmental conditions on process—in this case, the influence of nutrient and water availability on carbohydrate allocation within the plants—and the boxes are pools of material (soil organic matter, foliage, stems, etc.). These interactions form the theme of this book.

The aim of the book is to describe the various processes that are involved in this system and, as far as possible, to provide a treatment that leads to, or at least provides the basis for, quantitative models that allow us to calculate the consequences of particular events or actions. This is not always possible because knowledge of various processes is not adequate or the interactions involved are too complex for anything more than estimates of the likely effects of change or disturbance. However, the aim of quantitative treatment is important and should be pursued; without it the current gap between environmental physiologists and the people who manage forests is likely to remain unbridged.

At various points in the book we use the word "sustainability," or sustainable management. We define sustainable management of forest ecosystems as management in a way that is likely to maintain or improve species diversity and structure, and the functioning and biological productivity of the ecosystem, for the foreseeable future. We do not accept that economic considerations provide a legitimate basis for evaluating ecosystem sustainability *per se*, although they may have to be taken into account when evaluating whether the (human) community considers the cost of managing for sustainability to be worthwhile. We note also, for use in later discussions in this chapter, the distinction between "conservation" and "preservation." Conservation of forests is taken to mean that some utilization of the resource, under a suitable management plan, is acceptable. Preservation does not allow for utilization.

I. Forest Management: Levels, Decisions, and Influences

Management involves making decisions, and taking actions, with the intention of achieving some specified result or goal. The decisions taken are more likely to lead to the desired results, or achievement of the goal, if they are based on sound knowledge about the (eco)sytem being managed and the way it will respond to change. Management actions taken without the ability to predict the way the system is likely to react to the action are nothing more than guesswork. A decision not to intervene in any way is as much a part of management (e.g., in wilderness management) as a decision to do something to the forest. The "no action" option will often occur by default, but it may also be based on some assessment of the future pattern of forest growth, assuming environmental conditions remain, on average, about the same as in the past.

A simple example of management action in forestry is the practice of thinning. Stands are thinned because the forest manager believes that thinning will lead to some desirable results, such as faster growth of the trees left behind or better quality wood from those trees. There will undoubtedly be all sorts of consequences as a result of thinning. These will include effects on future wood production, with economic implications and consequences for other stands, and a range of effects on the ecology of the thinned stand and its hydrology. The implications may be wide ranging if the stand is in a water-producing catchment. Therefore, it is important that the manager's assessment of the likely results of the decision to thin should be as accurate as possible, and that there should be a sound appreciation of the probable consequences of the action. In this case, and in many other cases that we will outline, sound understanding of the effects of management actions on forest/environment interactions at the physical level, and on the physiological processes and processes that affect physiology such as water relations and nutrient release and uptake, must be of immense value.

Forest management is carried out at a number of levels. We may regard the stand as the lowest level or unit subject to decision making. The block, or group of contiguous but not necessarily similar stands, is the next level. A region containing various forest types is the next, the national forest estate the next, and the forests of the globe are the highest level. Clearly, different people are responsible for decisions about forests at these various levels, and the types of decision they make will vary from operational decisions at the stand and block level to policy decisions at regional and national level. Forest blocks may be owned by, or leased to, forestry companies, or managed by the state.[2] In the case of companies

---

[2]The word "state" is used here in a generic sense, meaning the institution(s) of government. It does not mean state as opposed to federal, as in the American or Australian contexts.

concerned with wood production, the primary considerations will tend to be economic, although these will be modified by government policy and the extent to which the companies are inclined to behave in a responsible manner in terms of ecological sustainability. This will vary from country to country and is often influenced by whether the companies are local or international. Management policies at regional or national levels are likely to be determined by legislation or regulation arising from social and political considerations. At the global level, we can only hope for agreements that forests will be managed on a sustainable basis in as many countries as possible.

A person responsible for the management of stands, which we define as moderately homogeneous areas of forest of the same age and general characteristics, may be required to make or advise on a series of practical decisions that include matters such as thinning, fertilization, clear-cutting and reestablishment procedures, and weed or pest control. Areas of forest that should not be harvested must be identified and access roads constructed and maintained. Plannning wood supplies and cutting schedules requires information about how much timber is in the forests and how fast it can be expected to regrow after harvesting. Decisions about the use of various parts of the forest obviously require knowledge about the type of vegetation in the forest, including the dominant tree species, condition and stage of stand development, knowledge about the soils and their characteristics, and about the way various plant communities and forest types are likely to respond to disturbance. In the case of those who manage forests at the stand and block level, knowledge about what is in the forest tends to come from direct observation and experience, translated into local maps and charts; at higher levels, good data bases containing information about forest types and condition become essential management tools (see Chapter 10).

Those who manage a large number of forests for companies or governments (state forests) make different kinds of decision. In a wood-producing context, they are concerned with wood flows—the steady supply of wood to mills or contracted customers that must be maintained over long periods. This requires knowledge of the standing crop in all the areas under their control and the development of harvesting schedules that will maintain supplies both in the short and long term. Ensuring long-term supply requires knowledge of projected growth rates on a range of sites, a central issue and problem in production forestry and one to which environmental physiology and the modeling procedures arising from it have much to offer. We discuss this question in more detail in later chapters (see, particularly, Chapter 9).

In many parts of the world, management for wood production has consisted of deciding on areas to be clear-cut or selectively logged in the most economically efficient manner. Arguments about clear-cutting and its consequences, and about the sustainable management systems for natural forests, rage in countries such as the United States, Canada and Australia. Increased environmental awareness within large sectors of society in the so-called developed countries has led to increased pressures to reduce destructive clear-cutting and manage natural forests in ways that maintain aesthetic and recreational values as well as biodiversity. Many natural forests are now managed for multiple use. This may include access to forests for recreational purposes—for hiking, riding, camping, and, in some cases, hunting. Those who use the forests in this way demand that they are (or at least appear to be) unspoilt. People seeking recreation tend to react unfavorably to the ordered rows of plantations or to the spectacular damage associated with clear-cutting—at least as it appears a short time after the event. They want to see large old trees, diverse understory growth, and a range of animal and bird life. It is also recognized that forested catchments are sources of clean water, although the water yield may not be as high as if the catchments were cleared or thinned (see Chapter 4). Obviously, the simplest method of management to meet these requirements is to leave the forests alone—to accept that they should not be utilized for wood production. This may have significant economic implications, at least locally, and decisions to preserve forests in this way have been, and are, the cause of much debate and controversy in various parts of the world. Even where the decision is made to preserve forests, controversy is likely to arise about their management in relation to matters such as burning policy. In Australia, there are great debates about whether to burn deliberately, at intervals, to avoid the massive wildfires that occur if fuel is left to accumulate for long periods, or whether to accept the natural cycle of fire. The United States is not immune to similar debates; the consequences of "no burn" policies were evident in the massive fires in Yellowstone National Park in 1988.

The resolution of debates about management systems for natural forests depends on a multitude of biophysical, ecological, sociological, political, and economic factors, which vary from place to place. We cannot even begin to consider them here. However, we note that it is possible to manage forests for wood production and still conserve most of the characteristics demanded by those who see their primary values as aesthetic and ecological. Practices that tend to minimize the impact of wood extraction from forests include careful selective logging, restricting clear-cutting to small areas, avoiding logging on steep hillsides, and the development of suitable, "environmentally friendly" machinery. It is also essential to preserve adequate protective strips along stream banks and leave untouched the rich, biologically diverse areas that occur in gulleys. Careful use of seed trees, sympathetic road engineering, and careful consideration of the structure and diversity of the forest(s) as a whole are also important. Clearly, multiple-use management is likely to be more complex than if wood production were the sole objective, but the forest management procedures followed in Swedish and Finnish forests, which have a long history of being manipulated by humans, show that ecologically sympathetic and economically viable multiple-use management is perfectly feasible. Many of the principles discussed in this book are relevant to multiple-use forest management and need to be understood if it is to be practiced successfully.

In some countries, plantation forestry is being promoted as a substitute for logging natural forests. This may not be an advantage if native forests are being cleared in order to establish the plantations, and it may also be unpopular with long-established forest industries, because it is not only necessary to invest considerable capital to establish plantations but, in the case of hardwoods, plantation-grown trees tend to have wood properties different from those of the wood of old trees. This leads to the need for changes in the technologies used for wood processing—whether sawing, pulping, or the production of products such as chip and fiber boards.

In many undeveloped regions and countries, social pressures are different. The demand for high-quality hardwood is a major factor driving the exploitation of forests in southeast Asia and the Pacific region. There, and in South America and Africa, high and rapidly growing human populations lead to strong pressures for land so that forests are cleared piecemeal by peasants practicing slash and burn agriculture. This practice may not adversely affect forest ecosystems where the areas affected are relatively small and they are allowed adequate time to recover. However, rapidly increasing human populations in these regions have increased both the areas affected and the frequency with which they are revisited. Access to the forests is often facilitated by roads cut by logging companies that extract the large, high-value tropical hardwoods. The number of trees removed may be relatively small—perhaps 10–20 per hectare—but there is a high potential for serious damage to the ecosystems unless considerable care is exercised, including care in road construction, the disposal of residues, and the use of machinery in the forests.

Unfortunately, for a variety of reasons, ranging from official corruption and the dominance of the economic motive to the lack of adequate, or adequately enforced, regulations, this care is lacking in many cases. The voracious apetite of the modern world for wood products, particularly paper and packaging and disposable products such as paper napkins, tissues, diapers, and geriatric hygeine products, as well as the more "traditional" uses such as wood for building and furniture making, ensures that the economic pressures for forest products will remain high, and increase, in the foreseeable future. These demands for wood products also ensure that clashes will continue between those concerned primarily with goals such as short-term profit and the current economic welfare of a company or even a society, and those concerned with the longer-term goals of maintaining the integrity of forests and their noncommercial (in the short term) values. Such clashes are important in the political arena, where forest policy and regulation are, eventually, determined.

At political levels, forest policy and the regulations governing forest management are determined by all the usual pressures brought on politicians by a range of interest groups. Commercial interests have, particularly in the past, tended to favor exploitive management—wood extraction with very little effort to ensure that regeneration was successful or that cutting rates were sustainable. This approach was encouraged by the illusion that natural forests were so extensive that areas damaged or even destroyed would not make any significant difference or cause shortages of wood. It was also encouraged by the lack of knowledge of forest ecosystems and lack of appreciation of the consequences of major disturbances to forests. These views are now changing rapidly, although there is still resistance in the commercial world to expensive management practices designed to minimize damage and sustain and conserve the forest estate. This translates into political lobbying, made more effective by the difficulty of obtaining good quality data on forests: What area of particular types exists; what is the volume of wood and what species are present in any particular area; and how will the forest respond to logging, fire, fertilization, etc.? It is always possible to interpret poor-quality data in different ways and use them to support particular arguments. Therefore, if lobby groups are so inclined they can often present a strong case for the commercial exploitation of forests, while arguing that the management practices they intend to follow will not cause any long-term damage.

Political pressure in the other direction comes from conservationist groups and those who are aware of the long-term damage and loss of biodiversity brought about by practices such as clear-cutting large areas, cutting riparian forest, burning the residues from logging, using heavy, steel-tracked machinery under wet conditions, and so on. The responses of governments to such lobby groups depend on the effectiveness of their arguments and the degree of public support the groups can muster. Support for conservation and preservation is invariably higher in developed than developing countries: Such matters tend to be of more concern and interest to people for whom the essentials for a comfortable life are readily available. In countries where high populations and poorly structured and poorly developed economies make it difficult for large proportions of the population to reach and maintain acceptable standards of living, forest conservation and preservation tend to be low on the agenda of most people, unless they happen to be dependent on the forests for their livelihood, as is the case with some groups of native peoples in South America and southeast Asia. It is this lack of public concern, coupled with economic incentives, that leads to corruption and exploitation by commercial interests, often based outside the countries that contain the forests. Politicians can be induced, by bribery and payments of doubtful legality, to provide licenses and concessions for the extraction of high-value wood from large tracts of forest without concern for the manner of extraction and the consequences of the resulting damage. This is happening in a number of countries of southeast Asia, the Pacific island nations, the Commonwealth of Independent States (previously the Soviet Union), and in South America.

One of the major sources of argument and concern relating to forests, as well as to a range of other facets of human life, is the question of global climate change. Increased emissions of greenhouse gases (CO_2, NO_x, CH_4, etc.) are predicted to increase the average global air temperature by 1.5–4.5°C, although the increase is not expected to be uniform. High-latitude regions are expected to experience the greatest change in temperature, whereas the tropics may experience little or no change (IPCC, 1995). Precipitation patterns are also expected to change, but predictions about these are less certain than those relating to temperature. Forests are a significant component of the global carbon balance (see Chapter 5), both because they store large quantities of carbon and because actively growing forests are significant sinks for carbon—they absorb and store more carbon than they emit, and therefore are considered, by some, to be important as a means of mitigating or even halting the rise in atmospheric CO_2. In fact, the area of new forests that would be needed to make a significant difference to the rising atmospheric CO_2 concentrations would be enormous and completely impractical as a solution to the problem. However, the increasing CO_2 concentrations may result in more rapid growth of young trees if other resources, particularly nutrients, are not limiting.

We do not claim that knowledge of the effects of the environment on the physiological processes that govern the growth of trees, and hence the growth and responses of forests, and of the interactions and feedbacks between forests and the environment is likely to be used directly in developing policies about forest management or in making decisions about how to manage forests. The previous discussion leads to the conclusion that social and economic considerations are the primary drivers in these areas. However, we are convinced that better policy and decision making, and improved ability to develop sustainable management systems and to evaluate the sustainability of existing sytems, will come from an improved, and steadily improving, capability to predict the productivity of forests in contrasting environments. Progress will come from the improved ability of scientists to provide reliable, supportable, and relevant information to bureaucrats and politicians, and from improved ability to model and predict the results of particular strategies or to assess the realism of management goals. We recognize that judgments about the relative value of short-term economic benefits and long-term forest sustainability and unquantifiable aesthetic values will, unavoidably, remain subjective. However, if scientists can evaluate in quantitative terms the likely impacts of disputed practices, they can provide a rational rather than an emotional basis for settling arguments about those practices.

The remainder of this chapter provides an overview of the subject matter covered in this book, with comments on the relevance to management of the various subject areas. The subsections are not chapter summaries but are intended to indicate the main ideas underlying the detailed treatments in the chapters.

II. Overview

A. Chapter 2: Forest Biomes of the World

The broadest vegetation classification system involves vegetation units of similar physiognomy or appearance known as biomes. Common biomes include deserts, grasslands, tropical forests, etc. The number of forest biomes in the world depends on the criteria used to identify ecosystems. Within any particular biome, there is likely to be considerable variation in ecosystems, brought about by differences in topography, soils, and local differences in precipitation amounts and patterns, so that any "lumped," or large-scale grouping will encompass a number of smaller ecosystems that are coherent and readily identifiable both in terms of their structure and function. Nevertheless, it is necessary to decide on some sort of classification that allows general statements about the properties of forest ecosystems, the functioning of important processes in different forest ecosystems, and the way they are likely to react to disturbance. We have therefore used a classification based on leaf habit (evergreen or deciduous), leaf morphology (needle leaved or broad leaved), and climate (boreal, temperate, or tropical). The general characteristics of these forest biomes, and the climatic patterns characteristic of each region, are outlined in Chapter 2, which is intended to provide background and a framework for the rest of the book, where principles and processes are discussed in general terms. The principles and processes apply to all forests, but their importance will vary among forest biomes and ecosystems within a biome. Figure 2.2 is a map of the global distribution of biomes and Table 2.1 contains estimates of annual net primary productivity (see Chapter 5) for the major forest biomes. Clearly, the application of the principles of environmental physiological analysis to management problems in any forest ecosystem will depend on more detailed knowledge about the area of concern than can be provided here, but these data give an indication of what we can expect, and some basis for considering problems at the level of the global carbon cycle, and global estimates of forest productivity.

B. Chapter 3: Canopy Architecture and Microclimate

This chapter on architecture and microclimate is essential background for the rest of the book. The way a stand interacts with its environment is determined by the amount of foliage (leaf area) it carries and by the stand architecture—the way the branches and foliage are arranged and displayed. These parameters determine the absorption of radiant energy and momentum. The physical principles and the procedures discussed in Chapter 3 provide the basis for calculations of CO_2 uptake by canopies, and hence for all the carbon balance calculations, and for calculation of the rates of water loss from canopies, and hence for all the calculations necessary to evaluate the hydrological implications of management actions or climatic conditions. The carbon balance models presented later in the book (Chapter 9) all depend on some form of radiation absorption calculation, discussed in this chapter. Knowledge of the energy balance and exchange processes is essential for calculating water use rates by stands, and hence stand water balance and the hydrological implications of stand structure. Knowledge of radiation penetration and air flow patterns is valuable for assessing the effects of practices such as thinning and the probable influence of gaps on the ecology of forest stands, including understorey dynamics. Few managers will use the principles and equations in this chapter directly, but it is important that professional foresters and forest ecologists be familiar with them because they provide the underpinning for so many models that managers may use or that may be used in the generation of management advice.

C. Chapter 4: Forest Hydrology and Tree–Water Relations

Water is essential for tree growth and it is also among the important commodities obtained from forests. Anyone concerned with forest management on large scales is likely to be concerned with the water yields of catchments, which are often a source of debate and dissension between competing interests. The character of the debate will depend on the priorities placed on adequate water and the cleanliness of the water supply. The idea of crystal streams from forested catchments is embedded in the folklore of western peoples and tends to be synonymous with the concept of unspoiled environments. It is indeed true that streams and rivers from forested catchments are usually clean, but it is less generally appreciated that the water yields from such catchments are, in most cases, lower than the water yields from, for example, catchments under grassland for the same amount of precipitation. The hydrological consequences of thinning or clear-cutting tree stands on catchments are therefore likely to be increased water yield, but this may be accompanied by increased soil erosion and increased quantities of nutrients in the streams. The actual effects in any particular case will depend on the biogeochemical characteristics of the area and the rates of plant regrowth.

The principles of catchment hydrology and stand water balance calculations are developed, in this chapter, from the basic hydrological (mass balance) equation, with relatively detailed treatments of canopy interception, soil water-holding characteristics, and transpiration losses. Results from catchment hydrology experiments are reviewed, and the last part of the chapter deals with tree water relations and their effects on growth.

Forest managers at the stand/block level, concerned with wood production, will be aware of the importance of the soil water balance for two reasons: It is generally inadvisable to carry out forest operations when soil is wet, and prolonged periods of drought will reduce the growth of trees. In neither case, it might be argued, are managers likely to calculate water balances; they will generally be guided by experience and, in any case, there is nothing that can be done about prolonged dry periods. In fact, research in at least one Australian state (and probably elsewhere) has resulted in guidelines for the soil wetness safe for forest operations. These are issued in terms of rainfall within a period for particular soil types: If there is more that a specified amount then the soil will be too wet for safe (in terms of damage, compaction, etc.) operations and work with machines in the forest is prohibited. Although expressed in crude "rule of thumb" terms, these guidelines were derived from studies in which soil water balances were properly calculated.

There is nothing that can be done about drought in relation to established trees, but it is important to have some appreciation of drought probabilities when new plantations are being established and in relation to the probability of success of reestablishment of stands after harvest or fire. Few managers (or scientists) calculate drought probabilities, which depend on precipitation patterns and probabilities, soil water-holding characteristics, and rates of water loss by evapotranspiration. The procedures can be derived directly from the principles described in Chapter 4. Very briefly, if the soil can hold $(\theta \cdot z)$ mm of water within the root zone (where θ is volumetric water content and z is the rooting depth), and the average rate of evapotranspiration over a period Δt is E_t mm/day, then the time scale for adequate water for growth is about $(\theta \cdot z)/E_t$ days. The probability of drought is derived from analysis of long-term precipitation records, from which the cumulative probability of periods longer than $(\theta \cdot z)/E_t$, without significant precipitation, can be calculated. Rooting depths will vary with age and type of stand as well as soil type; clearly, the rooting depth used for calculations relating to seedling establishment will be very different from that used for mature trees. Decisions about what constitutes "significant" rainfall will be made by the analyst. The probability of drought will (presumably) influence management decisions about plantation establishment and the timing of intervention in stand reestablishment.

D. Chapter 5: Carbon Balance of Forests

The carbon balance of forests at any time is the net result of carbon uptake by photosynthesis and losses by respiration. Standing biomass is the net result of photosynthesis, autotrophic respiration, litterfall, fine root turnover, and coarse root mortality (if any). It is important to distinguish between net primary productivity (NPP)—generally taken as the difference between photosynthesis and autotrophic respiration—and net ecosystem productivity, which includes heterotrophic respiration. Chapter 5 deals with the process of photosynthesis at leaf and canopy (stand) level, with respiration, carbon balance, and carbon allocation, and includes modern data on net carbon fluxes to (and from) a range of forest ecosystems.

Forest management for wood production is essentially a matter of manipulating stands to optimize the harvestable yield, which entails maximizing the production of carbon and its allocation to useful product, i.e., tree stems. Our ability to calculate the carbon balance of stands is improving rapidly (see Chapter 9), but our ability to predict carbon allocation within trees (to stems, branches, leaves, and roots) is still hampered by incomplete understanding of physiological controls on carbohydrate allocation.

One of the areas in which most progress has been made in recent years has been in our understanding of the interactions between carbon fixation and nutrient dynamics. Laboratory measurements of leaf photosynthesis rates generally indicate that maximum leaf photosynthesis rates are strongly related to leaf nitrogen (N) concentrations. This has led to the assumption—supported by model analysis—that growth, in terms of dry mass production, will be related to leaf N concentrations. (In fact, this has proved difficult to demonstrate in field experiments on conifer forests.) Nitrogen availability has a major effect on foliage mass and area and influences carbon fixed because increased foliage area results in greater radiation absorption and hence greater carbon fixation. However, it also results in more transpiration and hence tends to lead to adverse water balance, which appears to cause greater allocation of carbohydrates to roots. There is strong feedback between the nitrogen in foliage, the rate of decomposition of litter, and the release and availability of N in the soil for uptake by the trees. This may be a matter of considerable importance in relation to the long-term carbon balance of forests, particularly in relation to rising atmospheric CO_2 concentrations; some model analyses indicate that the long-term effect of these rises on growth may be small because of limitations imposed by the availability of N.

From the management point of view, the need to be able to predict the carbon balance of stands, and the pattern of allocation of the carbon, remains a central problem, whether the manager is concerned with stand productivity or the carbon balance of ecosystems or is a policy maker, politician, or bureacrat concerned with carbon fixation by forests and prediction of the global carbon balance. The problem can only be solved by ecophysiological research.

E. Chapter 6: Soil Organic Matter and Decomposition

Soil organic matter, its structure and rate of decomposition, exerts strong control over the availability of some nutrients for uptake by plants that, in turn, affects the health and sustainability of the forest ecosystem. Sustainable productivity depends on the maintenance of the soil organic matter and the balance between rates of decomposition and nutrient supply and forest growth. The quality and quantity of organic matter depend on the type of litterfall—which is affected by numerous ecological factors—and the management of the forest. The rate of decomposition of the material depends on temperature and moisture conditions. As a general rule, large material decomposes more slowly than fine material and, quite obviously, organic material does not decompose when it is frozen or very dry. In some systems, such as boreal forests, the rate of organic matter accumulation in the soil exceeds the rate of decomposition so that over long periods there is accumulation in the soil: The organic matter layer in the boreal forest regions is estimated to contain billions of tons of carbon. In other forest biomes, notably lowland tropical forests, where high temperatures and continuously wet soil cause high rates of decomposition, there is very little, if any, organic matter accumulation. The rate at which nutrients become available to plant roots depends on the rate of organic matter decomposition.

Management can have direct effects on soil organic matter. Any harvesting operation, whether thinning or clear-cutting, results in the addition of large amounts of organic matter to the soil, including material that may take a long time to decompose. The disturbance of the soil may cause increased aeration or compaction. Temperature and soil water relations are altered. There have been cases in which site-preparation practices have had such massive effects on soil organic matter that the sustainability of the forestry enterprise was threatened: one of these was in the state of South Australia where, for many years through the 1950s, 1960s, and 1970s, the standard practice, after harvesting *Pinus radiata* plantations, was to push the debris into windrows and burn it. The soils in the region are sandy and low in organic matter and the result was declining growth and yield (Keeves, 1966; Woods, 1976) until the problem was recognized and rectified. Where fire is used as a forest management tool, it can affect organic matter content and distribution.

F. Chapter 7: Nutrient Distribution and Cycling

The objective of this chapter is to present an outline of current knowledge about forest nutrition. It deals with the macro- and micronutrients and their occurrence, availability, and role in the growth of trees. Manipulation of nutrients has been, for many years, a major target of forest managers at the stand level, particularly in the management of plantation forests. There was a strongly held view, despite the paucity of convincing evidence to support it, that the nutrient status of trees could be determined and fertilizer regimes developed on the basis of the concentration of nutrients in foliage. A great deal of research was carried out to try to determine the optimum sampling and analysis procedures and to try to establish correlations between foliage nutrient concentrations and growth. The problem with this now largely discredited approach was that it ignored the fact that nutrients and the nutrient cycle are highly dynamic, both in terms of the rates of release of nutrients from organic matter and the soil (by weathering) and the uptake and utilization of nutrients by the trees. There is significant internal cycling of nutrients, which vary in their mobility within the plant; the rates of movement are also dependent on the stage of stand development, the condition of the trees, and the season.

Detritus production and decomposition represent a major pathway of nutrient supply for trees. The quality and quantity of detritus production are affected by forest type, climate, soil fertility, and events such as storms and drought. There is accumulating information demonstrating that changes in species composition, even removal of the understory by herbicides, alter detritus production and nutrient cycling.

Humans can have tremendous impact on a number of the nutrient cycling processes. Nutrients are lost by direct removal when forests are harvested as well as by leaching and in fires, and the combination of unusual climatic events with such actions can lead to major changes in the nutrient balance of stands. It is important to be able to calculate nutrient budgets and to have some knowledge of rates of uptake and replacement so that management practices are not only geared toward optimum productivity but also result in sustainable forest systems. Drastic changes in the nutritional status of a stand are likely to lead not only to changes in the growth rates and productivity of species of interest for wood production but may also lead to changes in species composition and shifts in dominance patterns.

Typically, when we discuss nutrient cycling we are concerned about nutrient depletion. Atmospheric deposition of nitrogen may initially cause increased forest growth rates, as we note in the outline of Chapter 9. However, there is increasing evidence that excessive amounts of nutrients, in the form of atmospheric deposition, can saturate the demand by microorganisms and vegetation, causing imbalances and system malfunction. Relatively high concentrations of nitrates, in particular, are produced from industrial effluents and deposited in precipitation: The adaptation of species to high soil nitrogen may vary considerably, and if the balance between available nitrogen and other nutrients is heavily biased then we can expect to see changes in the composition, structure, and function of forest ecosystems. Abiotic mineralization of nitrogen is likely to be important in forests subject to large additions of nitrogen in the form of fertilizer or as a result of atmospheric deposition.

G. Chapter 8: Changes in Ecosystem Structure and Function during Stand Development

The species composition and structure of forests are subject to continual change. Catastrophic natural disturbances, such as wildfires or hurricanes, or human-induced disturbances such as clear-cutting, destroy forests so the process of development must start from the beginning. During the development cycle there may be changes in species composition and stand functional characteristics. The patterns of carbon and nutrient flow change with the changes in forest structure.

There is no guarantee that regrowth forest will have the same species composition as a forest that has been destroyed. For example, the seeds of some Australian *Eucalyptus* species will not germinate unless they have been subjected to fire, so it may happen that, if there is a large reservoir of such seeds in the soil of an area that has not been burnt for many years and is then burnt, there will be massive germination of the species in question. This can result in a virtually monospecific stand that is much less diverse than its predecessor. Serious insect damage, or disease, will also lead to alterations in forest succession cycles.

In addition to exogenous factors, there are endogenous factors associated with stand development and succession that alter species composition. Changes in stand structure affect forest microlimate, which in turn can affect nutrient and carbon cycling and allocation. Most notable is the well-established decline in above ground net primary production with stand age. This may cause forest managers to harvest forests on shorter

16   *1. Introduction*

rotations than might otherwise be the case, leading to more frequent site disturbance and possibly greater depletion of nutrients from the soil.

Successional patterns depend on stand structure and the life cycles and growth patterns of the species in stands, as well as on such matters as nutrient cycling. These factors are reviewed in Chapter 8.

H. Chapter 9: Ecosystem Process Models

The complexity of forest ecosystems and the interactions with which we are concerned make it essential that we use models to describe, analyze, and understand them. It is possible to envisage qualitatively the flows and interactions of mass and energy in a forest ecosystem, but it is not possible to describe them with any precision, to predict responses not observed, or to analyze observed responses to disturbance or change without using models. We are concerned primarily with process-based models, i.e., models that describe the system(s) under study in terms of the biophysical processes involved and the way those processes are affected by and interact with external conditions. Most of the work on such models in recent years has been concerned with water and carbon balance, but there is now an increasing trend toward the development of models in which water, carbon, nutrients, and their interactions are central to the models.

The focus in forest modeling of the type with which we are concerned has been on the development of models as research tools, aimed at explaining experimental observations or measurements of forest processes such as the CO_2 and water vapor fluxes that can be measured with modern equipment and that provide estimates of the net uptake of carbon, total water loss by forests, or nutrient removal. There have been few, if any, serious efforts to use process-based models as management tools; they are usually too complex and too highly parameterized, the scientists who develop them never have the time to provide the backup necessary to deal with the problems that will arise in practical use, and forest managers, if they are aware of them at all, consider them too esoteric and impractical. In Chapter 9, we consider the empirical models managers use, based on the development of growth curves from site indices, and review them briefly.

Present-day empirical growth models will undoubtedly become less accurate as the conditions under which they were developed change with the climate, with atmospheric pollution, and with the impact of forest management practices on soil nutrient status. There are already cases in which forest managers have noted that growth and yield tables routinely overestimate forest growth of young plantations; the discrepancy has been attributed to ozone. In Europe, the opposite seems to be happening, apparently as a result of nitrogen "fertilization" by atmospheric deposition and increasing atmospheric CO_2 concentrations. We consider that the concept of site index could be supplanted by modern models utilizing the principles of environmental physiology. We review in more detail some of the more important and widely used process-based models, grouping them in terms of the time scales across which they operate and the number and type of parameters they need. Several examples are provided demonstrating how process-based models can and should be used in ecosystem management.

I. Chapter 10: Applications of Modern Technology and Ecophysiology to Forest Management

This chapter focuses on three areas of technology: models (discussed in Chapter 9), remote sensing, and Geographical Information Systems (GIS). All are already used in forestry to some extent and in various ways. The chapter includes a basic outline of remote sensing, with emphasis on satellite measurements and the information that can be obtained from these measurements. This includes mapping and stand classification, with information about canopy architecture and leaf area index. It also includes the possibility of obtaining estimates of the radiant energy absorbed by forest canopies and using these with models of the type we advocate to make estimates of forest growth rates. GIS is obviously an essential tool in this exercise—superimposing radiation absorption on stands mapped by a combination of remote sensing and ground mapping or aerial photography, and using weather data to drive and modify the model calculations, will yield spatial estimates of productivity and water relations that can be varied on the basis of sequential series of satellite measurements. The idea is not new, having been pioneered by Steven Running (Running *et al.*, 1989) who has emphasized global carbon balance and NPP modeling, but the approach has not been used at the level of stand and block management and there remains great scope for rapid progress and further developments. We explore the possibilities and implications of the use of these techniques.

2

Forest Biomes of the World

In this chapter, we provide an overview of the main forest biomes in the world, their characteristics and distribution, as background and context for the more detailed treatments of various aspects of forest/environment interaction in the following chapters.

Forest ecosystems are usually considered in terms of natural forests, and there is a tendency to assume that these are undisturbed systems in equilibrium with their environments (Melillo *et al.*, 1993). In fact, there are now few such forests in the world. Most have been disturbed by humans—some very severely—and the area of "pristine" forest of any type is relatively small. The largest areas of such forests probably exist in the boreal regions, although even there logging and mining have affected significant sections of the forest ecosystems. Tropical forests—the largest biome in the world—have been, and are being, grossly disturbed.

The reasons for forest disturbance vary. Wood extraction has always been a primary factor, but the major cause of deforestation in the tropics is population pressure and growth. Myers (1996) provides data to show that farmers are responsible for about half the forest felling in Third World countries, whereas the lumber industry accounts for about one-quarter of forest destruction or degradation. Halting these processes is therefore largely a socioeconomic problem rather than a technical one. Clearing for farming has always been a major cause of forest destruction in the temperate areas; the forests of Europe were largely cleared in the Middle Ages. In North America, vast areas of temperate mixed and deciduous forests have been cleared for farmland (Delcourt and Harris, 1980) and there are few areas of temperate coniferous forest that have not been exploited for wood production. Tropical deciduous forests, as we note later, have all but disappeared from much of their range because of burning, grazing—with its accompanying effects of seedling and understory destruction and associated effects—and clearing for farming. Any consideration of forest ecophysiology must, therefore, recognize the fact of disturbance and the fact that many forests are managed in some sense. The principles that we discuss in later chapters are applicable to any forest ecosystem, but because we are concerned with their application in management we provide here some general ideas about management and the consequences of manipulation and note the major differences between natural forests and plantations in terms of the actions taken in the course of managing them.

I. Management

Management implies manipulation in some sense, so a managed forest may be defined as any forest that is, or has been, subject to some form of management action, i.e., manipulation aimed at achieving some specified objective. Managed forests cover the range from plantations, where management includes all silvicultural activities, through natural forests managed for sustainable wood production, to natural forests exploited for timber with little or no long-term perspective (Landsberg et al., 1995). Exploitation for timber without any attempt to manage the system in a sustainable way is arguably not management, but the line between such exploitation and constructive management is often very fine. For example, natural forests in Australia have been logged for timber since settlement by Europeans in the 18th century. The practice did not initially include clear-cutting, but with the advent of the chain saw and mechanized heavy equipment clear-cutting has become widely practiced. It is opposed by conservationists but defended by the forestry industry on the grounds that the procedure results in the regeneration of "overmature" or partially degenerate forests—which means forests of mixed age, with old trees that are degenerating from a timber production point of view. Clear-cutting results in the regeneration of even-aged stands, often with a species composition different from that of the original stand. Regenerating, even-aged stands, whether they result from clear-cutting or fire (which may follow clear-cutting), may involve loss of biodiversity, compared to old stands, because the seed stores in the soil, or transport of seeds from adjoining areas, can allow particular species well suited to take advantage of the changed conditions to become dominant in the stand. In any event, the result is a changed ecosystem (see Chapter 8).

An even-aged, regenerating stand is similar, in many ways, to a plantation, and could be managed in the same way, particularly if early stage thinning was carried out to produce stem populations lower than those that might develop naturally. A major series of studies was carried out to examine the feasibility of intensively managing regrowth *Eucalyptus* stands in Australia (see Kerruish and Rawlins, 1991). It was concluded that regrowth can be profitably managed, mainly by appropriate thinning, to produce greater yields of commercially usable wood than unthinned stands. Regrowth can be used to substitute for, and therefore conserve, older growth forests of high conservation value or to offset transfers of native forests to reserve status.

Many countries and organizations now have forest management plans to ensure the sustainability of natural forests: an excellent example of this is the publication *A Richer Forest* (1992), produced by the National Board of Forestry of Sweden. This book incorporates, in a clear and simple way, a great many of the principles of forest ecology that are needed for the successful management of natural forests for multiple use. The U.S. Forest Service is moving away from the idea that the primary purpose of forest management is wood production to the view that the complexity of forest ecosystems must be recognized and the systems managed so that the nonwood values of the forests are not lost. These values range from biodiversity to aesthetic values to the need to ensure that disturbance of forests in catchments does not cause serious disruption of the catchment's hydrological cycle (see Chapter 4).

II. Plantation Forestry

A large proportion of the timber products used in the world today, particularly pulpwood, but, increasingly, sawn timber products, come from plantations rather than from natural forests. For these purposes, plantations have many advantages over natural forests: they can be planted on prepared land, using genetically improved and uniform seedlings at standardized spacings that allow optimum growth rates of the individual trees; and it is economically feasible to control weeds and use fertilizers to ameliorate problems of soil nutrition. Overall, the management of plantations is much more intensive than that of natural forests—the objective is to achieve a specified product in the minimum possible time, and the procedures are essentially those used in crop production. The effects and implications of some of these procedures will be considered elsewhere in this book.

With the exception of the widely grown *Eucalyptus* plantations, most plantations are softwoods, although the idea of introducing extensive hardwood plantations in the tropics is gaining ground. Wood from these plantations should substitute for wood from native forests and so reduce the impact of logging in the native forests. However, a great deal of research is necessary before the problems of managing tropical plantations to produce acceptable timber growth rates are solved. *Eucalyptus* are grown for pulp in Portugal, in Brazil—where they apparently achieve some of the highest forest growth rates in the world—in South Africa and other countries in Africa, and increasingly in countries such as China and India. Major softwood plantations around the world include large areas of Sitka spruce (*Picea sitchensis*) in Britain, extensive plantations of loblolly and slash pine (*P. taeda* and *P. elliottii*,[1] respectively) in the southern United States, Douglas fir (*Pseudotsuga menziesii*) in the Pacific Northwestern United States and Canada, and *P. radiata* as the major softwood plantation species in Australia, New Zealand, Chile, and South Africa (although South Africa also uses several other softwood species). There are few plantations in northern Europe, but many stands of Scots pine (*P. sylvestris*) and Norway spruce (*Picea abies*) are intensively managed.

It is interesting to note that plantation forestry is dominated by exotics—trees that are not native to the area where they are cultivated as plantations. Sitka spruce is native to the Pacific Northwest of the United States, where some of the trees grow to great size. *Pinus elliottii* is native to the southeastern United States, but *P. radiata,* now so widespread as a plantation species in the southern hemisphere, is native to the Monterey Peninsula in southeastern California.

Australia is the only continent where *Eucalyptus* species are native, yet they are now grown in about 130 countries around the world. In general, new species are introduced to areas where the climate is similar to that of the region of origin, but it is a remarkable fact that many successful introduced species grow much faster in areas where they have been introduced than they do in their natural habitat. Some of the success of exotic trees as plantation species is attributed to the lack of their native pests in the countries where they have been introduced. This may contribute to the high production rates, but it is not the only reason; factors such as better soils and cultural practices such as weed control, fertilization, and general silvicultural management also contribute. It also does not follow that the range of climatic conditions bounding the natural distribution of a species in its native habitat will necessarily restrict the areas where it can be grown when transported to other regions. This has been neatly demonstrated by Mitchell and Williams (1996), who found that the areas where *Eucalyptus regnans* grows and thrives in New Zealand do not coincide with the climatic regions to which the species is restricted in its native Australia.

III. Species Adaptations and Climatic Conditions

A. Physiological Adaptations to Climate

Recognition of the importance of climate in controlling the distribution of vegetation dates back almost two centuries to early plant geographers (Humboldt and Bonpland, 1805; Humboldt, 1807; Schouw, 1823). Many of the early plant geographers understood that similar climates produced similar vegetation, although cause and effect were not established. It was not until the late 19th century that Schimper (1898) recognized

---

[1] *Pinus* is, throughout the book, abbreviated to *P.* and *Eucalyptus* to *E.* Other generic names are usually written out.

that the climatic control of the distribution of vegetation can only be explained in terms of basic physiological processes—arguably marking the founding of the discipline of ecology now known as physiological ecology. Holdridge (1947, 1967) and Box (1981) demonstrated strong correlations between major physiognomic, or life form groups (i.e., desert, grasslands, deciduous or coniferous forests, etc.) and two climatic characterisitics—temperature and water availability.

The distribution of the major forest biomes, and terrestrial biomes in general, is strongly influenced by climatic, geologic, ecologic, and anthropogenic factors, varying in importance across timescales ranging from historic (10^1–10^4 years) to geologic ($>10^4$ years). Discussion of the physiological basis for the geographic distribution of the major forest biomes can be found in Box (1981), Walter (1979), and Woodward (1987). The importance of nutrient availability and environmental variables favorable for photosynthesis in determining the distribution of evergreen and deciduous forests is discussed in Chapters 7 and 8, respectively. There are, however, several critical minimum temperature thresholds that strongly affect key physiological processes, which in turn define environmental limits for the major forests biomes. These are discussed briefly here.

Plants have evolved a number of mechanisms to survive minimum temperatures less than +10°C, all of which are energy-demanding processes. Therefore, in many instances the gradual disappearance of a species near its altitudinal or latitudinal limit may be attributable more to the increasingly noncompetitive carbon balance than to damage directly related to low temperatures (Woodward, 1987, 1995). The first critical minimum temperature threshold ranges from −1 to +12°C and is related to chilling tolerance (Larcher and Bauer, 1981). Trees can survive chilling by maintaining a fluid membrane; failure to do so results in leaky membranes and, commonly, death. Raison *et al.* (1979) described a striking relationship between the minimum temperature at which the cell membranes change from fluid to gel state and the geographic range for a number of species, and Sakai and Weiser (1973) explained the distribution of North American tree species based on leaf and bud frost resistance.

The next critical temperature is −15°C and roughly corresponds to frost resistance. The frost resistance of leaves is much lower for broad-leaved than coniferous species, with a noticeable minimum threshold of −15°C (Woodward, 1987). The minimum temperature threshold for buds is similar for conifers and broad-leaved trees, but broad-leaved angiosperms are more likely to experience ice formation in the xylem during winter. This commonly results in the formation of air bubbles in the spring when the ice melts, thereby rendering the vessels nonfunctional (Becwar and Burke, 1982). The next critical temperature threshold is −39 to −40°C, which coincides with the temperature at which supercooled, pure water nucleates spontaneously to form ice (Woodward, 1987). However, the buds of boreal tree species, such as *Picea glauca, Larix laricina,* and *Larix sibirica,* have been reported to survive temperatures below $-70°C$ (Sakai, 1979; 1983) and Rasmussen and Mackenzie (1972) demonstrated that by accumulating solutes in intercellular-free water, plants can lower the critical temperature of spontaneous nucleation below $-40°C$. Boreal tree species (i.e, *Picea, Abies,* and *Larix*) at the extreme northern distribution often have a parenchyma pith cavity beneath the crown of the primordial bud that prevents ice nucleation from spreading from the xylem to the bud (Sakai, 1983; Richards and Bliss, 1986).

The inherent properties of the water-conducting architecture of trees may also influence their geographic distribution. The tracheid xylem of conifers is more resistant to, and has a greater potential for recovery from, ice-induced embolism than the conducting system of angiosperms (Borghetti *et al.,* 1991, Sperry and Sullivan, 1992). Pit membranes, unique to the tracheid vessels of conifers, were once thought to provide greater resistance to widespread embolism, but it is now believed that the substantially smaller lumen diameter of tracheids compared to vessel xylem is an important factor (Pallardy *et al.,* 1995). It is interesting to note that the few deciduous genera that occur in boreal forests have either tracheids (i.e., *Larix* spp.) or diffuse-porous xylem (Gower and Richards, 1990; Pallardy *et al.,* 1995). Species with diffuse porous xylem have water-conducting characteristics more comparable to tracheids than species that have ring-porous xylem. These latter species are noticeably absent from boreal forests (Woodward, 1995). Havranek and Tranquillini (1995) provide a thorough review of the physiological processes operating in boreal and cold-temperate conifers during winter dormancy and their ecological significance.

B. Species Composition of Forest Ecosystems

Ecologists have had difficulty explaining the wide range in species diversity and species distribution in forest ecosystems. Plant family diversity is positively correlated to absolute minimum temperature but even within tropical forests there is great variability, with species diversity inversely related to length of the dry season (Kira, 1983; Whitmore, 1984). It appears that there is a negative correlation between species richness and resource supply (soil fertility, water, etc.), so an explanation for the high species diversity in tropical forests is that the constancy of climate produces a constantly poor supply of mineral nutrients in the soils, suppressing the opportunities for competition among species. Connell (1978) and Doyle (1981) concluded that the highest diversity of tropical forests is achieved under moderate disturbance frequency over small areas, thereby creating a myriad of niches. Iwasa *et al.* (1993) reviewed all the explanations for forest species diversity and used models to examine various hypotheses. They concluded that narrow niche width tends to enhance diversity when niche width (the time for which conditions are suitable for species regeneration) is shorter than the time period suitable for regeneration. In tropical forests, the period suitable for regeneration can be several months and there is great seasonal variation in germination and regeneration rates, thereby providing, in effect, a storage mechanism. If the niche width is greater than the period of high regeneration opportunity, then the number of coexisting species increases with niche overlap, as first proposed by Huston (1979). Runkle (1989) speculated that seed storage is the basic mechanism that enables many similar tree species to coexist in a forest. Although the temporal patterns of gap formation and gap size are similar for tropical and temperate forests (Denslow, 1987), gaps created during the unfavorable period (i.e., winter in temperate regions or dry periods in tropical regions) remain vacant and increase in number until the next favorable growing period. Synchronized regeneration provides a competitive advantage to species having their maximum regeneration at the start of the peak. Therefore, as the length of the unfavorable growing period increases (see Fig. 2.1), the peak rate of supply of gaps at the beginning of the favorable season becomes more important and species diversity decreases.

It has been suggested that the evergreen habit provides a more favorable carbon balance in harsh climates (Mooney, 1972; Schulze *et al.*, 1977a; see also Chapter 5), an argument supported by the dominance of evergreen trees in cold montane and boreal forests. Kikuzawa (1991), using a cost–benefit analysis, demonstrated that "evergreeness" should have a bimodal distribution with peaks at low latitudes (tropics) and high latitudes (boreal). His general biogeographical distribution pattern agrees well with observed vegetation patterns, except in two regions. First, although temperate forests occur predominantly from 30° to 55° latitude, broad-leaved evergreen species dominate forested landscapes in the southern hemisphere, whereas broad-leaved deciduous tree species predominate in the northern hemisphere. Axelrod (1966) concluded that the temperate climates, ample rainfall evenly distributed throughout the year, and rarity of frost favored the evolution of broad-leaved evergreen rather than deciduous forests in the temperate regions of the southern hemisphere (Fig. 2.1e vs 2.1b). Also, the minor amplitude of Quaternary change, the absence of large ice sheets and the lack of full-glacial environments persisting through the interglacial periods in the southern hemisphere—compared to the northern hemisphere—lead to the dominance of broad-leaved evergreen forests in the temperate regions of the southern hemisphere (Markgraf *et al.*, 1995). The second area where Kikuzawa's prediction is incorrect relates to the widespread distribution of the deciduous conifer, *Larix* species, in the cold montane and boreal environments, especially in Eurasia (Gower and Richards, 1990), where evergreen species would be expected because of the short growing season, infertile soils, and snow loading on the canopy of broad-leaved angiosperms. At the more local scale, soil fertility, disturbance frequency, and edaphic conditions affect the distribution of trees species. Temperate deciduous forests are often restricted to the more fertile sites, whereas conifers occur on the more infertile sites because deciduous species have a greater annual nutrient requirement (Cole and Rapp, 1981; Son and Gower, 1991; see also Chapter 7).

C. Climates of the Forest Regions

In the second half of this chapter (Section IV) we discuss the distribution and some of the properties of the major forest biomes of the world. The environments of these forest types are illustrated by the diagrams in Fig. 2.1.

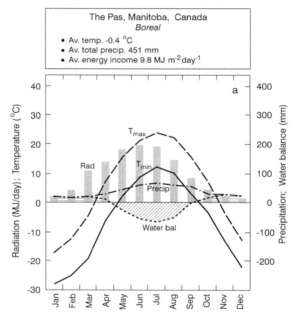

Figure 2.1 Climate diagrams for representative locations illustrating the conditions in which we can expect to find (a) boreal forests, (b) temperate deciduous forests, (c) temperate coniferous forests, (d) temperate mixed forests, (e and f) temperate broad-leaved evergreen forests, (g) tropical evergreen forests, and (h) tropical broad-leaved deciduous forests. The diagrams show long-term monthly averages of maximum (T_{max}) and minimum (T_{min}) air temperatures (°C), precipitation (mm), radiation (MJ/day), and the water balance, calculated as the difference between precipitation and evaporation using the Thornthwaite (1948) equation. The diagrams were produced from data presented by Müller (1982). Radiation data were not available for every station (they were missing for The Pas, Hobart, Manaus, and Jamshedpur); where this was the case, data from other stations at similar latitudes, which closely matched these in terms of sunshine hours, temperature, and rainfall patterns, were used.

III. Species Adaptations and Climatic Conditions 27

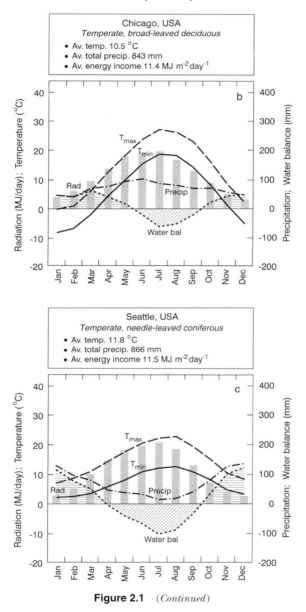

Figure 2.1 (*Continued*)

Each panel of the figure provides information about the climate of an area representative of those where the specified forest biomes occur.

The water balance data (precipitation–evaporation) were derived from the monthly potential evaporation figures provided by Müller (1982),

28   2. Forest Biomes of the World

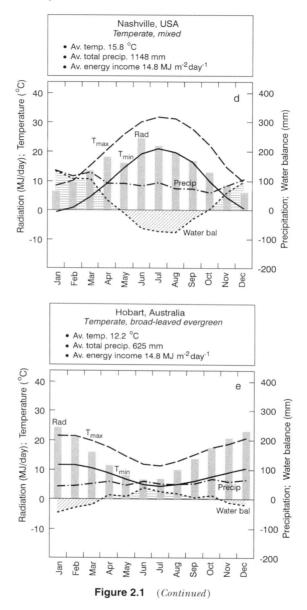

Figure 2.1   (*Continued*)

which were calculated from the evapotranspiration formula derived by Thornthwaite (1948). This formula is based on temperature and is unlikely to provide accurate values for the water use of forests—or indeed any type of vegetation or even open water. However, as Müller points

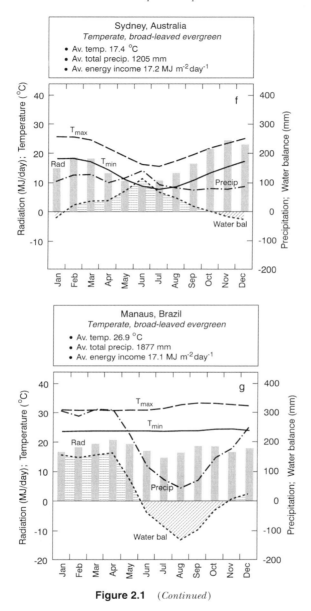

Figure 2.1 (*Continued*)

out, the Thornthwaite equation is the only one that gave comparable values for every station, and it does provide a reasonable indication of evapotranspiration regimes and hence the overall water balance. Methods more soundly based in physics, from which much better estimates of

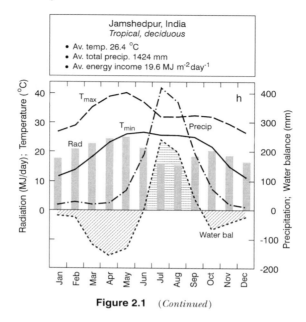

Figure 2.1 (*Continued*)

forest water use rates can be obtained, are presented and discussed in Chapter 3.

Forest growth rates are determined primarily by average climatic conditions, but ecosystems are not generally altered by normal conditions. They are changed, often abruptly, by extreme events: Destructive storms may flatten areas of forest, causing large gaps in which regeneration takes place and which alter the microclimate and growth patterns in adjacent areas; extraordinarily high or low temperatures may kill whole cohorts of plants, making available niches in which different species establish themselves; drought may have the same effect, and the community that develops after the breaking of a killing drought may be different in species composition from the original. Drought may also lead to conditions conducive to fire, which can destroy plant communities. Again, the regenerating community may not be the same as the original (see Chapter 8). It is arguable that most ecosystems are in some phase of recovery from disturbance of some sort, and that ecosystem heterogeneity is largely a function of historical disturbances and edaphic conditions that have affected patches of varying size. It is therefore worth considering briefly, in relation to the climatic areas in which the major forest types occur, the likelihood of extreme events and their probable effects.

The boreal forest regions, represented by climatic data from The Pas, in Manitoba, Canada (Fig. 2.1a), are characterized by long cold winters.

The mean daily minimum temperature at The Pas is below 0°C for more than 7 months of the year, and during the period when temperatures remain above zero there are significant water deficits. This is clearly an extremely difficult environment for plant growth, a fact reflected in the very slow growth of trees in the boreal forests. Above-ground net primary production ranges from 1 or 2 t ha^{-1} year^{-1} for poorly drained black spruce forests to 4–6 t ha^{-1} year^{-1} for early successional balsam poplar, aspen, and birch stands (Van Cleve et al., 1983; Gower et al., 1997).

The low temperatures of the boreal forest regions lead to slow soil development so that the soils tend to be nutrient poor; the fact that the soils tend to be very young—derived from parent material left by retreating ice sheets—contributes to this. Permafrost, which may not retreat more than a meter below the surface, causes restricted root zones, and the poor drainage results in extensive fens and bogs when the soil is not frozen. Damaging summer storms are rare, although lightening strikes may cause fires that can burn unchecked over large areas. Severe winter weather is unlikely to cause damage to trees.

In the temperate deciduous forest zone of the northern United States (represented by Chicago, IL; Fig. 2.1b), minimum temperatures are well below freezing for at least 4 months of the year. The best period for growth is in the spring, when temperatures rise and water is adequate; high evaporation during the summer more than balances the precipitation, and water deficits are likely to restrict growth.

The climate of temperate evergreen forests cannot be encompassed by one climate diagram; these forest ecosystems can occur in environments ranging from cold subalpine to near subtropical. However, an area notable for such forests is the northwest coast of the United States, represented by Seattle (Fig. 2.1c), which has cooler summers than the temperate deciduous zone, warmer winters—the average minimum temperature for any month is never below zero—and a different precipitation pattern. Highest precipitation is in the winter months, with very little during the period of highest evaporation, so significant water deficits and reduced tree growth during summer are the norm. The fact that the trees in this area tend to grow to great sizes indicates that severe storms that cause physical damage to trees are relatively rare. A more probable cause of ecosystem disturbance is fire, which may occur in exceptionally dry summers, when the normal summer drought (Fig. 2.1c) is extended and exacerbated by unusually hot weather and lack of precipitation.

In the temperate mixed (evergreen, needle-leaved conifers and broad-leaved deciduous) region, represented by the climate at Nashville, Tennessee (Fig. 2.1d), higher rainfall is not enough to prevent summer water deficits, but early season temperatures are significantly higher than those in the deciduous and coniferous areas. The winters are cold enough to cause leaf fall, but the period when temperatures are low enough to prevent growth is relatively short.

Severe weather in the temperate mixed forest regions may take the form of drought or, in the southern United States, tornadoes (highly localized, high-energy rotating winds, which may flatten small areas of forest), or occasional larger-scale damage from hurricanes that penetrate further north than usual. In the "old world" of Europe and the Mediterranean region, most of the forests have long since been destroyed by man.

The climate at Hobart, Tasmania, Australia (Fig. 2.1e) is not dissimilar to that of the west coast of South Island, New Zealand, and both areas are characterized by broad-leaved evergreen forests. The rainfall in the Hobart area is not high (it is high along the west coast of South Island, New Zealand) but is evenly distributed through the year; evaporation is low and only mild water deficits develop during the summer. Average minimum monthly temperatures are never below zero; therefore, some tree growth can be expected throughout the year—a fact that helps explain the evergreen growth habit in broad-leaved trees. Further north, the climate of Sydney, Australia (Fig. 2.1f) shows similar characteristics, although temperatures are higher (the annual average is 17.4°C compared to 12.2°C at Hobart). Higher rainfall (1200 mm) compensates for the higher temperatures and evaporation; therefore, environmental conditions are good for growth throughout the year.

The tropical evergreen forests are generally considered to be wet at all times, but Fig. 2.1g indicates that in the Amazon there are quite long periods when evaporation may exceed rainfall. However, the water deficits estimated using the Thornthwaite equation may be particularly misleading (longer and more intense than in fact occur) in this region. The reason is that the Thornthwaite equation is temperature based; it does not include a humidity term. Evaporation is strongly dependent on air humidity (vapor pressure deficit; see Chapter 3), and the vapor pressure deficits in the tropical rainforest regions are, at all times, low (mean relative humidity is about 80%). Wind speeds are also low. We would therefore expect evaporation to be overestimated by the Thornthwaite equation.

Nevertheless, we can accept that, for several months of the year, the amount of water lost by evaporation exceeds the amount of rainfall in the South American tropical broad-leaved evergreen forest regions. This pattern also occurs in the African tropical rainforests, but in many of the southeast Asian areas rainfall exceeds evaporation in every month of the year. The remarkably stable temperatures are characteristic of the lowland tropical forests.

Jamshedpur, India (Fig. 2.1h), is a monsoon area characterized by heavy rains for 4–6 months of the year, with high water deficits developing after the monsoon season. The drought leads to leaf fall and the deciduous growth habit in trees. Temperatures are high throughout the year. A very similar climate is found in the Central and South American tropical deciduous areas; Managua, Nicaragua has a mean annual temperature of 27.3°C (cf. 26.4°C at Jamshedpur) and a total annual precipitation of 1142 mm, most of which falls in 5 months.

Natural disturbances in tropical forests include cyclonic storms, wildfires, and volcanic eruptions. Inland forests, in the Amazon and Congo basins, for example, are not subject to hurricanes and appear to be at very little risk from environmental hazards, but the hazard is significant through much of the southeast Asian area, particularly in the Indonesian archipelago, and Malesia. The destruction of an area of forest by hurricane is the classic cause of large gaps. The weight of epiphytes has also been suggested as a factor contributing to individual tree fall in tropical forests (Strong, 1977). We have little information about weather hazards to the tropical deciduous forests. Fire is an important ecological factor in these forest ecosystems (Murphy and Lugo, 1986).

IV. Forest Biomes of the World

Whatever the mechanisms determining the distribution of forest ecosystems, we need some sort of classification scheme as a framework for discussing them. Numerous land cover classification schemes exist based on environmental factors, physiognomy of the vegetation, soil types, or a combination of these variables. The number of land cover types recognized in each scheme largely reflects the objective for developing the classification system. For example, Holdridge's (1947) Life Zone classification system includes 22 vegetation cover types and was developed to illustrate the climatic influences on vegetation distribution, whereas Running *et al.* (1994) proposed a scheme including only 6 cover types; the critical features of this classification are that the structural characteristics can be identified by remote sensing and used as the inputs to drive global vegetation models.

For the purposes of this book, we have adopted the vegetation classification system used by Melillo *et al.* (1993) because it is a reasonable compromise between complex and simple schemes, and they have compiled one of the most recent sets of information about the global distribution of (potential) vegetation types. The forest types are based on major climatic zones (tropical, temperate, and boreal) and the physiognomy (broad-leaved evergreen, broad-leaved deciduous, and needle-leaved evergreen conifer) of the vegetation.

Figure 2.2 is derived from Melillo *et al.* (1993). The map (Fig. 2.2) shows the regions where particular vegetation types could occur, although it is unlikely that the areas concerned are completely covered by those vegetation types. The impact of humans has resulted in vegetation loss and change across very large areas of the globe; to determine actual areas of particular vegetation types will take a vast effort involving ground surveys and satellites. The simple land cover classification scheme of Running *et al.* (1994) was designed with the objective of making this possible. Nevertheless, the data in Table 2.1 probably provide a good guide to the relative areas of the different forest biomes. The net primary production data in the table represent estimates updated from currently available literature.

We note from Table 2.1 that two categories—boreal forests and tropical evergreen forests—dominate all other forest types in terms of area. Of these, we know that the rates of deforestation in tropical regions are by far the highest because it is these regions that have the highest human population densities. Because of the inhospitable nature of the climate in the regions where they occur, the boreal forests have been subject to far less disturbance by humans than tropical forests. Human populations are low and vast tracts are uninhabited. However, there are areas, for example, along the southern edges of the forests in Canada, and in Siberia, where there has been extensive commercial logging without ecological safeguards or concern for the sustainability of the forests (Shvidenko and Nilsson, 1994). The problem with the boreal forests is that the rate of recovery is so low. There is no growth during the extremely cold winters, and even in summer temperatures restrict growth rates. Soils are also generally infertile. Consequently, it may be several hundreds of years before logged or burned forests recover, in stark contrast to tropical forests where, if the forest is not deliberately completely destroyed and converted to (usually poor-quality) pasture, but is allowed to regenerate, recovery is rapid. Full canopy and fine roots of tropical forests may be restored within 1–8 years (Raich 1980, Berish, 1982) and there are tall trees, with multilayered canopies, within 10–20 years.

The following sections contain descriptive outlines of the major forest biomes we have chosen to discuss. As noted previously, the classification scheme we have chosen to follow is only one of a number of possibilities; within that scheme, there will be overlaps between categories as well as gaps in information—the amount of information readily available about each category varies. The information that follows is necessarily sketchy;

Figure 2.2 Global vegetation map showing the distribution of the forest types discussed in this book. The map is a modified version of the one published by Melillo *et al.* (1993) (with acknowledgments to D. Kicklighter and J. Melillo, who provided this version in digital form). Reprinted from *Nature,* Melillo et al. **363**, 234 © 1993 Macmillan Magazines Ltd.

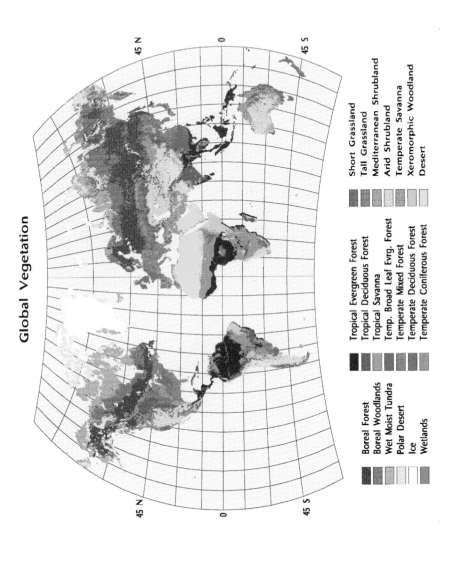

Table 2.1 Area (ha $\times 10^{-8}$), and Maximum, Minimum, and Mean Net Primary Production (NPP, t ha^{-1} year^{-1}), of Selected Forest Biomes[a]

Forest biome	Area	NPP		
		Maximum	Minimum	Mean
Boreal	12.2	4.3	1.2	2.4
Temperate coniferous	2.4	7.0	2.1	4.7
Temperate mixed	5.1	10.7	2.3	6.7
Temperate deciduous	3.5	9.8	0.8	0.2
Temperate broad-leaved evergreen	3.2	10.0	3.2	7.4
Tropical deciduous	4.6	14.0	3.2	8.7
Tropical evergreen	17.4	14.2	4.1	11.0

[a]Data derived from a table by Melillo *et al.* (1993).

it is intended to provide a framework for thinking about the processes dealt with in the other chapters, and their implications for forest ecology and management, rather than detail about the forest types considered. For such detail, reference must be made to books and papers dealing exclusively with particular forest biomes.

A. Boreal forests

Boreal forests cover about 12×10^8 ha and occur exclusively in the northern hemisphere. The greatest single area of boreal forests is in Eurasia, where they extend from Scandinavia to eastern Siberia. The second largest boreal forest region occurs as a 500 to 600-km wide band from eastern Canada and the northeastern United Staes westward into northern British Columbia and Alaska. The northern and southern boundaries of the boreal forests in North America correspond, roughly, to the summer and winter position, respectively, of the arctic front. The length of growing season must be sufficient to ensure that evergreen conifers develop an adequate cuticle needed to minimize winter desiccation (Tranquillini, 1979) and mycorrhizal associations to facilitate nutrient and water uptake (Read, 1991). Woodward (1995) suggests that the northern limit of boreal forests may be crudely defined as the number of months in which the air temperature is greater than 10°C.

The climate of the boreal forests is one of the harshest in which trees can occur. There may be less than 50 frost-free days in summer (Fig. 2.1a) and permafrost is characteristic of these regions. Fire is the most important natural disturbance in boreal forests. Fires, ignited by lightning, tend to cover large areas and may burn as much as 25,000–50,000 ha (Dyrness *et al.*, 1986). Fire frequency in the boreal forests in North America ranges from 30 to 200 years, depending on species composition and topographic position. Fire strongly influences species composition, nutrient availability, and forest productivity (Larsen, 1982; Dyrness et al., 1986).

There are few major soil taxonomic groups[2] found in boreal forests. Histisols, or organic soils, are common to poorly drained forests. The upland soils are relatively young compared to temperate and tropical forest soils. Entisols have little or no horizon development and are typically associated with early successional riparian forests (*Populus and Betula*) and coarse-textured, excessively drained pine forests (*P. banksiana* in Canada and *P. sylvestris* in Eurasia). Some Spodosols can be found, but in general the boreal forests are too dry for these soils to develop.

Tree species diversity is very poor in the boreal forests. There are only 9 dominant tree species in North America (Payette, 1992) and 14 in Fennoscandia and former Russia (Nikolov and Helmisaari, 1992). The low species diversity is attributed to the short-term geologic history and the harsh climate of this biome (Woodward, 1995). In general, boreal species arrived in this region less than 2000 years ago. Therefore, the ecology of this large forest biome is younger than any other (Takhtajan, 1986). The distribution of species, and the species composition of stands are strongly influenced by topography and soils. Important genera include *Abies, Betula, Larix, Picea, Populus,* and *Salix. Picea* and *Larix* commonly occur on poorly drained lowland soils. Pines commonly occupy well-drained upland soils, whereas *Populus, Abies, Salix,* and certain species of *Picea* occur on the finer-textured upland soils. In North America and Europe, needle-leaved evergreen conifers tend to dominate the boreal landscape, especially at northern latitudes. However, *Larix*, a deciduous conifer, increases in importance in Eurasia and often dominates the boreal treeline in Siberia (Gower and Richards, 1990). Ericaceous shrubs commonly dominate the understory of boreal forests and the soil is covered by lichens on the drier (xeric) sites or mosses and sphagnums on the mesic to wet (hydric) sites.

Unique structural characteristics of boreal forests include low leaf area index (L^*, See Chapter 3, Section I) and spiral canopies. The spiral canopies help shed the snow and maximize light interception when the sun is low in the horizon (Oker-Blom, 1989). Above-ground net primary production (ANPP) rates are typically low but the production rates of upland boreal forests on southern slopes can approach those of cold temperate forests (Van Cleve et al., 1981; Gower et al., 1995). Typically, only a small fraction of ANPP is allocated to stem wood production. Bryophytes comprise less than 1% of the total above-ground biomass of boreal forests, yet they play numerous important roles. Bryophytes insulate the soil, which strongly affects the thermal regime and hence overall

---

[2] U.S. 7th Approximation Soil classification system.

nutrient cycling and productivity patterns of boreal forests. Despite their low biomass relative to vascular plants, the productivity of bryophytes can comprise 18% of total ANPP (Van Cleve et al., 1983).

Because of the low stemwood productivity, the inhospitable climate, and inaccessability, boreal forests are one of the least managed forest ecosystems of the world. Nonetheless, as we noted earlier, there is great interest in them, especially in Siberia, because of the large areas (see Table 2.1) of mature forests. The political instability and poor infrastructure in Siberia may be the only factor stopping large-scale harvesting of this fragile ecosystem. Plantation forests are scarce in the boreal regions, although forest management programs in Scandinavia can approach the intensity level of plantations.

B. Temperate Deciduous Forests

Temperate broad-leaved deciduous forests cover 3.5×10^8 ha (Melillo et al., 1993; Table 2.1) and occur primarily between 30° to 50° N latitude (Rohrig and Ulrich, 1991). The major areas of these forests occur in the eastern United States, Europe, the western parts of Turkey, the eastern border areas of Iran, western China, and Japan (Rohrig and Uhlich, 1991). With the exception of the western coast of southern Chile, deciduous forests are noticeably absent in the southern hemisphere (Axelrod, 1966; Schmaltz, 1991).

Temperate forest soils are highly variable so we can provide only a cursory treatment of this topic. A more detailed treatment is provided by Pritchett (1979). Many of the riparian and swamp forests in the southeastern United States occur on Histisols and it is interesting to note that the use of these soils is a topic of intense debate in the southeastern United States today (Richardson, 1981). When drained and fertilized with nitrogen and phosphorus, they can be extremely productive and are highly valued for agriculture and forestry. However, for these soils to be productive their hydrology must be altered by drainage, which results in large freshwater intrusions into adjacent saltwater estuaries, adversely affecting the delicate food chain. Many of the mountain soils in temperate regions that are covered by deciduous and some coniferous forests are Entisols, Inceptisols, or Alfisols, with the former being young and infertile and the latter being moderately weathered but fertile. Spodosols are restricted to cool- to cold-temperate conifer forests that receive abundant rainfall. In warmer climates (e.g., southeastern United States and southern Europe), the soils have undergone extensive weathering and the dominant soil order is Ultisols: These soils can be productive, especially if nitrogen and phosphorus fertilizer are applied.

Both tree and understory diversity are greater in temperate deciduous than temperate conifer and boreal forests; approximately 30 plant families and 65 genera occur in the overstory canopy of temperate deciduous forests (Rohrig and Ulrich, 1991). Species diversity is higher in the deciduous forests in North America, China, and Japan, where numerous refugia are believed to have existed during glacial periods, than in Europe where the predominantly east–west mountain ranges prevented retreat to warmer climates. Species composition varies according to topography, soil fertility, and successional status. A few important temperate deciduous tree genera include *Acer, Ailanthus, Albizzia, Betula, Carya, Castanopsis, Fagus, Fraxinus, Juglans, Liriodendron, Magnolia, Nothofagus* (endemic to Chile and Tasmania), *Populus, Quercus, Tilia,* and *Zelkova.*

Because of large-scale clearing and conversion to agriculture, pasture, and urban areas, temperate deciduous forests do not usually occur in extensive tracts. Except for stands of *Populus,* it is also unusual to find pure stands of one species. The forests are being further altered by exchange of species, extirpation of species, and pathogen outbreaks. Collectively, all these factors have greatly altered the nature of temperate deciduous forests, especially in Europe. Management may range from periodic selective tree removal to the most intensive form of forest management—short-rotation plantations for fiber or fuel production. Depending on the species, life history, and ecophysiology, both even-aged and uneven-aged management practices can be successful. Even-aged management is most prevalent for shade-intolerant and coppicing species, whereas uneven-aged management is commonly used for shade tolerant species. Species commonly used in short-rotation plantations include *Populus, Liquidambar, Salix,* and *Platanus.*

C. Temperate Coniferous Forests

Temperate evergreen coniferous forests cover approximately 2.4×10^8 ha (Melillo *et al.*, 1993; Table 2.1) and are largely restricted to the northern hemisphere. Temperate evergreen conifers occur in a wide range of climates, perhaps the most diverse of all the major forest biomes, ranging from subtropical, open woodland to ecotonal temperate-boreal forests to temperate rainforests (Fig. 2.2). Conifers dominate the montane forests in North America, Europe, and China and smaller areas of temperate conifers are located in montane regions of Korea, Japan, Mexico, Nicaragua, and Guatemala. In addition, *Pinus* species have been planted extensively in the southern hemisphere (see comment in "Plantation forestry"). Temperate conifers tend to occur on xeric or infertile soils that cannot supply the greater water and nutrient demands of deciduous species (Son and Gower, 1991); perhaps the best example of this is the dominance of evergreen conifers in the Pacific Northwestern United States, where dry summers and mild winters provide a favorable environment for evergreens (Waring and Franklin, 1979).

Common genera in the temperate coniferous forests include *Abies, Picea, Pseudotsuga,* and *Thuja* in northern latitudes and montane/subalpine regions, whereas genera such as *Tsuga* occur over a much broader range of environmental conditions. *Pinus* species, an important genera from both an economic and ecological perspective, occurs in a wide variety of environments ranging from hot, arid southwestern United States to cold regions of Scandinavia and Eurasia (Knight *et al.*, 1994).

Given the diverse environments in which temperate conifers occur, it is not surprising that the ecophysiology and structure of these forests are varied. For example, needle longevity can range from less than 2 years for *P. taeda* to greater than 40 years for *P. longaeva*. Above-ground biomass of mature forests can range from a low of about 100 t ha^{-1} for *Pinus* forests in southwestern United States to 3300 t ha^{-1} for giant redwood (*Sequoia sempervirens*) forests in northern California (Cannell, 1982). Some of the lowest leaf area indices (<1) occur in temperate coniferous forests, whereas the highest measured leaf area index (20) was for a western hemlock (*Tsuga heterophylla*) forest in coastal Oregon (Fujimori *et al.*, 1976; Gholz, 1982;[3] but see footnotes in Chapter 3, Section I). Above-ground net production is also quite variable, ranging from about 2 to more than 20 Mg ha^{-1} year^{-1}.

Management practices in temperate coniferous forests vary greatly; the intensity of management tends to be strongly correlated to the suitability of environmental conditions for tree growth. At one extreme, these forests are allowed to regenerate naturally following disturbance such as fire or harvesting. Despite the slow growth rates on low-quality sites, biomass accumulation can be substantial over several centuries. The harvesting of mature, slow growing forests is controversial, because these forests serve other valuable functions, such as providing wilderness areas extensively used for recreation, wildlife refuges, and stable water catchments. They also have great aesthetic appeal. Because of the long growth cycle and the destruction of many of these values, harvesting these "old-growth" forests more closely resembles resource mining than sustainable forest management.

Management practices of intermediate intensity are becoming more common and there is increasing pressure from society to manage mature forests on an uneven-aged basis because the resulting stands retain strong structural similarities to old-growth or pristine forests (Goodburn, 1996). At the other extreme, temperate conifers are managed at a level of intensity that rivals or exceeds agriculture. Intensive management includes mechanized site preparation and planting, use of genetically superior

---

[3] The leaf area index value of 20 is unrealistically high. Maximum achievable values are about 12 (Marshall and Waring, 1986).

seedlings, application of herbicides and fertilizers during the rotation, pruning (the removal of low branches to increase wood quality), and mechanized harvesting, which can include the complete removal of all above-ground biomass.

D. Temperate Mixed Forests

Mixed (deciduous plus evergreen) forests occur throughout the temperate regions—particularly the southeastern United States, Europe through northern Iraq and Iran, and China—where the climates are the same as those described for temperate deciduous and coniferous forests. Mixed forests have been studied less than pure deciduous or evergreen forests. Their occurrence reflects past land use change, successional status and local variations in edaphic conditions. For example, in the southeastern United States, early successional forests are dominated by conifers, mid-successional forests are dominated by mixed forests, and late successional forests are dominated by broad-leaved hardwoods (Oosting, 1956). In the Lake States, needle-leaved evergreen forests dominate the xeric, infertile soils, mixed forests are most common on the soils of intermediate edaphic conditions and broad-leaved hardwood forests occur on the mesic fine-textured soils (Kotar *et al.,* 1988). There may be a wide range of leaf morphology combinations in natural mixed forests: needle-leaved evergreen and broad-leaved deciduous (pine and oak, Douglas fir and alder); broad-leaved evergreen and deciduous (eucalyptus and acacias); and needle-leaved evergreen and deciduous (Douglas fir, true fir or pine, and larch).

One of the major, and most interesting, questions arising in relation to mixed species forests relates to their long-term stability and the way they regenerate after disturbance. There is no straightforward answer to the question. Stand dynamics are dependent on the almost infinite possible range of environmental conditions interacting with seed stores and species responses—such as shade and drought tolerance, relative rates of species growth under different conditions (water, temperature, fertility, etc.), and the presence and impact of pests that may affect some species but not others. Shade tolerance and initial growth rates are often invoked as important factors and are used as the mechanistic basis in models of the FORET type (see Shugart, 1984) for predicting growth and survival rates. There are arguments to be made for the assertion that mixed stands will be more stable than pure stands because of the capacity of different species to exploit a range of niches and conditions, and there are also arguments to be made to the contrary on the grounds that a single species, well-suited to its environment, will dominate the system and eliminate competitors. Examples that support both arguments can be found.

There appears to be remarkably little information on the growth rates of species in mixtures. A useful study was carried out by Smith and Long (1992), who found that stemwood production of mixed stands of *P. contorta* and *Abies lasiocarpa* was no different than that of pure stands. Differences could be explained on the basis of leaf area index and leaf area per stem (see Chapters 3, 5, and 9 for a discussion of the biophysical reasons for this). However, Binkley (1992), reviewing some of the literature and considering the mechanisms involved, points out that the "productivity of trees on nitrogen-limited sites generally increases in the presence of nitrogen-fixing species, in some cases doubling relative to pure stands of non-nitrogen-fixing species. On nitrogen-rich sites productivity remains unaffected or even decreases in the presence of nitrogen-fixing species."

Because mixed forests consist of species with very different ecophysiological characteristics (i.e., shade tolerance, growth rates, etc.), it is more difficult to determine management principles for these forests. There is a long tradition of mixed species management in parts of Europe, particularly Germany, but the principles underlying the practices followed appear to owe more to empiricism and accumulated experience than to clear understanding of the likely consequences of particular actions, such as thinning regimes.

E. Temperate Broad-Leaved Evergreen Forests

Two categories of temperate broad-leaved evergreen forests are usually recognized: broad-leaved sclerophyll and broad-leaved rain forests (Ovington, 1983). The broad-leaved sclerophyll forests occur in areas with a Mediterranean-type climate, generally characterized by winter rain and summer drought, whereas the rain forests are found in areas with humid, frost-free climates. The sclerophyll forests occur in scattered areas of the United States, around the Mediterranean, and over large areas of Asia from northern India through southern China. The greatest continuous areas still existing are the *Eucalyptus* forests of Australia. Temperate broad-leaved evergreen rain forests occur in Japan, Chile, New Zealand, Australia (Tasmania), and in scattered, remnant patches in Asia. Melillo *et al.* (1993) give a figure of 3.2×10^8 ha for the potential area of temperate broad-leaved evergreen forests (see Table 2.1).

The sclerophyll forests in the Mediterranean area and the United States are dominated by *Quercus* species, whereas the broad-leaved temperate forests of Australia are dominated by *Eucalyptus*. In Tasmania, Victoria, New Caledonia, New Guinea, and southern Chile there are relatively small areas of evergreen *Nothofagus* forests; these are extensive in New Zealand. In both these areas, the distribution of forest (sub) types is determined, to a large extent, by the topography (Wardle, 1984), which creates strong precipitation and temperature gradients, leading to differences in forest type and structure. Topography also plays a dominant role in relation to the temperate evergreen forests of India, which occur mainly in the Himalayan mountains. There appears to be greater species variation in these areas than in the southern forests.

The natural structure of the *Quercus*-dominated forests is a dense, often continuous canopy, less than 20 m tall, but there are in fact few examples of undisturbed forest in the "old-world" Mediterranean areas, or indeed in the United States, and stand structure will generally be a consequence of the treatment the forest has had and the way it has been exploited. Forests in the Mediterranean area have been utilized by humans for thousands of years, and in the United States cutting and clearance for agriculture and various other forms of development have been rapid.

Eucalyptus canopies may vary from tall closed forest (up to 60 m in height), with relatively few individual trees, to shorter closed forest and woodland. The term "closed forest" does not imply, in the case of *Eucalyptus* forests, dense, high leaf area stands. In the tall stands, trees tend to be straight stemmed, with relatively sparse leaf area: L^* is generally low (Anderson, 1981). The forests have relatively dense shrub understories, presumably because there is adequate light and reduced evaporative demand (see Chapter 3). Australia has had a forestry industry, based on native forests, for more than 100 years, during which large areas of the so-called old-growth forests have been either clear-cut or selectively logged. Natural *Eucalyptus* forests produce about 3–5 m^3 of timber (about 1 or 2 tons dry mass) per hecare per year. This is much lower than the 20–30 m^3 ha^{-1} year^{-1} that can be obtained from *Eucalyptus* plantations. Attiwill (1979) gives a peak rate of NPP of 14 t ha^{-1} year^{-1} for a natural E. *obliqua* forest, at about 40 years of age, falling to about 10 t ha^{-1} year^{-1} at age 80.

Most *Eucalyptus* species are tolerant of fire, and the forest ecosystems are adapted to it. Australia—including Tasmania—is subject to recurring droughts that, when combined with hot weather and windy conditions, periodically result in bushfires of considerable intensity. Under natural conditions, these may occur with sufficient frequency to avoid the build up of fuel (litter on the forest floor) to levels at which fire intensity is such that most of the trees are killed. The native people of Australia used fire for thousands of years in a manner that appeared to be consistent with natural regimes (fires caused by lightening strikes). Since European settlement forest management has been greatly changed, large areas have been cleared, others logged and regenerated, and because modern human communities are not well adapted to fire, it has been excluded, as far as possible, from the remaining forests. The result has been fuel build up so that, when fires (inevitably) do occur, they are likely to be much more intense than was normal historically (see Chapter 7).

Large areas of the native forests in Australia have been cleared and replaced by softwood (*P. radiata*) plantations, but the development of hardwood plantations has been slow for economic and sociological (traditional practice) reasons. Consequently, the harvest of native hardwoods for saw timber and wood chips, for pulp and paper, continues. A consequence of the forest harvesting activities is that there are large areas of regrowth forest, the structure of which varies considerably, depending on how it was treated (clear-cut and burned, selectively logged, burned, etc.). Species composition in regrowth forests may not be the same as in the original stands.

Temperate broad-leaved evergreen forests in New Zealand vary from multistoried, mixed-species coastal forests, with tall conifers (*Podocarpus, Dacrycarpus,* and *Agathis*), to the pure, dense-canopied montane and subalpine stands of beech (*Nothofagus*). The lowland Podocarp forests have been ruthlessly exploited and few stands now survive. There are large areas of beech forest. New Zealand has now virtually halted native forest logging; it is a major exporter of softwoods and softwood products, based on extensive plantations of *P. radiata.*

The South American temperate broad-leaved evergreen forests include a range of types, from lowland to Andean slopes. The dominant species is generally *Nothofagus.* These forests have all been heavily exploited, with the wood being used for construction or the forests cleared for agriculture. Timber extraction continues today and many of the forests have been destroyed or are seriously degraded. Like Australia and New Zealand, Chile now has extensive *P. radiata* plantations, which may serve to slow the destruction of native *Nothofagus* forests.

F. Tropical Evergreen Forests

Tropical evergreen forests (generally called rainforests) comprise the largest single forest biome in the world (see Table 2.1). They lie (by definition) within the tropics, in areas where air temperatures are relatively high, with little seasonal or diurnal fluctuation (20–30°C), rainfall is generally greater than about 1500 mm per year, and relative humidity is uniformly high. As in all the forest types we consider, there is a broad range of subtypes within tropical forests, ranging from lowland to montane types. The large differences in climate and parent material have a pronounced impact on the structure and function of tropical forests (Vitousek and Sanford, 1986), and it is dangerous to make generalizations about such a large forest biome.

Like temperate forest soils, tropical soils are highly variable. An excellent summary of tropical soil distribution, extent, and the key pedogenic processes that control their fertility is provided by Sanchez (1976). The two dominant soil orders in the tropics are Oxisols and Ultisols, although they are less common in Asia than in Africa or the "neotropics" (Sanchez, 1976, 1981). As a direct result of their high degree of weathering (Oxisols are the most highly weathered soils), both these soil types are typically infertile and cycle phosphorus, and to a lesser extent calcium, very efficiently (Vitousek, 1984; Chapter 7). They have extremely low cation-exchange capacity and base saturation. Alfisols, more fertile than Oxisols and Ultisols, are often found in regions with lower precipitation (e.g., tropical deciduous forest regions) than lowland wet tropical forests. Because of their higher fertility, the forests on these soils are often cleared and used for agriculture.

Lauer (1989) gives a thorough treatment of the climate in regions with tropical rainforests. The greatest single area of tropical evergreen forest is the Amazon Basin, in the northern half of South America. Similar forests are found through the isthmus of Panama and into southern Mexico, and they cover the Congo Basin in equatorial Africa and the southern fringe of West Africa. In Asia, tropical rainforests occur along the southeast coast of India, in Sri Lanka, the Malasian Peninsula, the Indonesian archipelago, Borneo, Sarawak, and Papua New Guinea (collectively called Malesia). There is a small, remnant strip of tropical rainforest along the northeast coast of Australia.

Tropical evergreen forests are the most diverse terrestrial ecosystems on earth, with the greatest number of species per unit area. The Amazonian forests contain more than 2500 different tree species, with thousands more in the African and Asian forests. Some species are widespread: Prance (1989) gives a diagram showing the distribution of the six most widespread tree species (*Brosimum guianense, Casearia commersoniana, Rhamnus sphaerosperma, Guarea guidonia, Hymenaea courbarii,* and *Trichilia quadrijuga*). Malesia is characterized by the Dipterocarps, and the eastern tropical rainforests differ from others in the abundance of conifers (Whitmore, 1984) such as *Dacrydium* and *Podocarpus, Agathis,* and *Araucaria.*

The African rainforests appear to be relatively poor in species compared to those of America and Asia. Characteristic species are *Lophira alata, Turraeanthus africana, Tarrieta utilis,* and species of *Uapaca*. The most important commercial tree species of the evergreen forests of Africa belong to the *Meliaceae* family (Borota, 1991).

Rainforest canopies are characterized by layered architecture with, very generally, an upper layer of emergent, usually tall, trees, a middle or main canopy layer, and a subcanopy of smaller trees and shrubs. Ecologists have used a range of models and concepts to describe the architecture of tropical evergreen forests, with some identifying a series of layers, classified on the basis of tree crown geometry and size, leaf size, and mass (see Brunig, 1983). The best approach seems to be that of Webb *et al.*

(1972), who concluded that tropical forests should be considered on two levels of organization: vegetation types provide the framework; within each framework there is a mosaic of patches composed of species occurring probabilistically, each species association leading to a somewhat different structure. The stochastic element in the approach arises from uncertainties in seed sources, weather, and a range of other factors. The approach also serves well for describing spatial and temporal change: gaps of different sizes, created at different times, will lead, at any particular time, to a series of patches with different structure and different canopy architecture.

The standing, above-ground biomass of tropical rainforests obviously varies enormously from place to place and will clearly depend on topography, soil type, stage of development, and other factors. Estimates range from 100 to 1500 t ha^{-1} (Brunig, 1983). Jordan (1983) evaluated rates of NPP by tropical rainforests: he analyzed 139 studies of wood and litter production, characterizing them in terms of the radiant energy received at the study sites, and concluded that NPP in tropical rainforests is, as is often asserted, among the highest recorded for forests but, because litterfall is very high, the rates of wood production by tropical forests are not significantly higher than those of other forests. Mean rates of wood and litter production by tropical forests were 7.5 and 9.6 t ha^{-1} year^{-1}, respectively. The ratio of litter to wood production (1:3) is substantially higher than that of forests in areas receiving less radiant energy: It appears that tropical forests—or at least those in high-energy income regions—produce far more litter per unit of wood produced than forests in temperate climates.

A point that may be worth considering, in relation to the productivity of tropical rainforests, is the amount of biomass that must be consumed by insects. It is well known (and documented; see Wilson, 1992) that wet tropical forests are the home of enormous numbers of insects: Estimates of the numbers of insect species in the canopies of tropical rainforests run to 20 million. These are all part of a complex ecosystem in which the patterns of herbivory and predator–prey relationships are extremely complex; nevertheless, vast numbers of these organisms must be consumers of foliage, bark, and wood. Their fecal matter and carcasses contribute to the detritus falling to the forest floor, but the biochemical form of that material is very different from that of the original plant material (see Chapter 7, Section II). Few estimates exist of the biomass consumed and turned over by such populations, but it is likely to be considerable and would not show up in conventional litterfall estimates. This aspect of the ecology and productivity of these forests clearly warrants study.

Tropical rainforests are under heavy pressure from human activities. A study by Skole and Tucker (1993) provides an indication of the magnitude of the problem. They used remotely sensed data to quantify rates of clearing in the Amazon Basin and noted that, because the adverse effects of clearing on some plant and animal species extend into the forests from the edges, spatial analysis of the geometry of deforestation is critical to the assessment of its effects. Assuming the edge effect of clearance extends one kilometer into the intact forest, a 10 × 10 km clear-cut would affect 143 km^2 of forest, whereas if 100 km^2 is deforested as 10 strips, each 1 × 10 km, the affected area would be about 350 km^2. Skoles and Tucker's estimate of current deforestation for the entire Amazon Basin was about 280,000 km^2 year^{-1} and, under the assumption outlined above, the area affected by clearing—the so-called rate of fragmentation—was estimated at 380,000 km^2 year^{-1}. Although the assumption made by Skoles and Tucker about the magnitude of the edge effect is controversial, the exercise clearly demonstrates the importance of harvesting patterns in managing forests.

In Africa, the greatest pressures on the tropical forests probably come from expanding populations and constant clearance for agriculture. Shifting cultivation is a major factor in South America, exacerbated there by pressures from ranchers and commercial interests moving into the Amazon. In Asia, the pressures are again from burgeoning human populations as well as from ruthless commercial logging, subject to very little control in most countries of the region. We noted in the introduction to this chapter Myers's (1996) statistics relating to the impact of farmers and the lumber industries on deforestation in the tropics. Solutions must be found for these problems—perhaps the most important will be control of human populations, but that is outside the scope of this book. Agroforestry appears to hold out the possibility of viable and sustainable systems for tropical areas, based on various combinations of arboreal and herbaceous crops, and logged areas of tropical forests can be replanted with timber trees, although research will often be required to determine the best methods for establishing those trees—in many cases, knowledge of their ecology and requirements for growth is sketchy or nonexistent. In all cases, good ecological and physiological information will greatly assist the empirical research that must be done to develop viable management systems and procedures for preserving the forests and utilizing them on a sustainable basis.

G. Tropical Deciduous Forests

As a criterion for classification, the distinction between tropical deciduous and tropical evergreen forests is somewhat artificial and not very helpful because deciduous trees are numerous in tropical rainforests in all parts of the world (Axelrod, 1966). The difference is one of water balance: Leaf shedding of tropical trees appears to be controlled mainly by drought (Borchert, 1980).

Murphy and Lugo (1986) calculated that tropical deciduous forests (potentially) cover 42% of tropical and subtropical regions. As annual rainfall decreases and interseasonal differences in precipitation increase, tropical broad-leaved evergreen trees are replaced by broad-leaved deciduous species. Deciduous tropical forests occur on the borders of evergreen forests in South America and Africa, where the mountain forests of central Africa may be included in the deciduous category. The largest areas of tropical deciduous forests are the monsoonal forests of southern and southeastern Asia, in India, the Himalayan countries, and Bangladesh, stretching through to Burma, Thailand, Laos, Cambodia, and Vietnam (Borota, 1991). Melillo *et al.* (1993) give the total potential area of tropical deciduous forests as 4.6×10^8 ha.

Species diversity is less in tropical deciduous than in tropical evergreen forests (Murphy and Lugo, 1986). Dominant species in Africa include *Antiarus africana, Ceiba pentandra, Triplochiton scleroxolon,* and others, whereas important species in America are *Andira, Dalbergia,* and *Tabebuia* genera, with conifers represented by *P. caribaea* and *P. oocarpa* (Borota, 1991). Hartshorn (1989) gives stand tables for subhumid (deciduous) forests in Central America that include *Calcophyllum candidissimum* and *Licania arborea* in lowland forests, *Luehea seemannii* and *Guarea excelsa* in lowland riparian areas, and an oak, *Q. oleoides,* which occurs in scattered populations over a wide area. Deciduous tropical forests in Asia include *Tectona grandis* (teak), *Shorea robusta,* and species of *Dalbergia* and *Terminalia* (Borota, 1991).

The canopies of deciduous tropical forests tend to be shorter and characterized by more open structure than those of tropical evergreen forests, with two layers rather than three. Dense shrubs often occur as the second layer, presumably because of the better light environment and reduced evaporative demand at lower levels in the canopy. Jordan (1983) cites two wood production values (4.2 and 9.3 t ha^{-1} year^{-1}) from Indian studies, with litterfall rates of 4.5 and 6.2 t ha^{-1} year^{-1}, respectively. These figures fall within the median range of the data collated by Jordan and indicate that we can expect the NPP of these forests to be about 10–15 t ha^{-1} year^{-1}. It will, presumably, be strongly controlled by the periods for which the trees have leaves and water relations are such that the trees can utilize radiant energy and grow relatively unchecked by water stress [see Chapters 4 (water relations) and 9 (ecosystem models)].

Tropical deciduous forests in all continents have been subject, over long periods, to burning and clearing for grazing and arable agriculture. Current human population pressures in these areas virtually ensure the continued destruction of the few remnants, with progression toward degraded forests, woodland, and savanna. Soil erosion is among the many serious effects of forest destruction and degradation, and even where this is not a great problem the disruption of the hydrological cycle (see Chapter 4) will result in significant change to the ecology of the areas concerned. We acknowledge that protection of these forests for their own sake is, in most parts of the world, an unattainable goal; the objective must always be to manage their utilization so that sustainable ecosystems are preserved. Where damaged forests can be protected, their recovery depends heavily on the state of the soils in terms of organic matter content, structure, and nutrient status.

V. Future Distribution and Extent of Forest Biomes

There is great interest in understanding how anticipated climate change will affect the distribution and extent of future forests. The most common approach to the prediction of future vegetation distribution is the use of empirical relationships between climatic variables and biome distribution (Holdridge, 1967; Box, 1981) with general circulation model (GCM) simulations of future climate parameters to estimate future vegetation distribution (Emanual *et al.*, 1985; Prentice and Fung, 1990; Smith *et al.*, 1992). These exercises all predict that the extent of boreal forests will decrease and tropical forests will increase, although the magnitude of the change differs among the models. Given the uncertainties of GCMs and the empirical models commonly used to predict future vegetation distribution, the predictions of future vegetation patterns should be interpreted with caution.

Physical factors known to influence vegetation distribution include the expected doubling of atmospheric CO_2 and the associated influence on temperature over the next 50 years. The increase in temperature is far faster than changes in climate in the geologic past and some forest tree species may experience future climates that do not currently occur in their present-day range (Solomon *et al.*, 1984), causing widespread dieback. Forest dieback attributed to stress induced from climatic change has already been suspected in Europe (Nilsson and Duinker, 1987) and in the northeastern United States (Cook *et al.*, 1987; Johnson *et al.*, 1988). Assuming that average annual temperatures increase by $0.7°C$ km^{-1} for regions poleward of the 40° latitude (Leemans and Cramer, 1990) and assuming warming of $0.3°C$ $decade^{-1}$, species would be required to migrate 5 km $year^{-1}$ toward the pole to remain in a climate comparable to present-day conditions.

Based on paleoecological records, species migration rates for the post-

Holocene Era range from 150 m year^{-1} for animal-dispersed seeds to 40 m year^{-1} for wind-dispersed seeds (Davis, 1981, 1986). Little is known about migration rates for future climates. Models for migration rates for invasive species based on diffusion (Williamson, 1989) or epidemic theory (Carter and Prince, 1991) have produced invasion patterns that were often unexpected; accurate predictions were rarely obtained. More sophisticated models that incorporate feedbacks between temperature, elevated CO_2, and ecosystem processes suggest that lags in population response to climate change could cause transient decreases in forest growth (Solomon and Bartlein, 1992; Pastor and Post, 1993). Highly evolved pollination and seed dispersal symbiotic relationships may also be disrupted. Paleoecological records demonstrate conclusively that our present-day forest ecosystems varied in species composition during and after the last glacial advance, therefore, there is no reason to expect current-day forest ecosystems to migrate as an "organism." Physical barriers such as mountain ranges, especially those that run east–west, may impede or prevent the migration of plant species north. A more serious problem has to do with soil development, especially in the far northern latitudes. It seems unlikely that boreal forests will be able to become established on substrates with little or no soil.

Because our (scientific community) understanding of the ecophysiology of woody plants, the complex organization of even the most simple forest ecosystems, and the interaction of multiple stresses on the composition, structure, and function of forest ecosystems is still incomplete, predictions about how global change will affect future forests are premature (Jarvis, 1987).

VI. Concluding Remarks

This chapter has provided a cursory account of some of the more important and interesting aspects of the forest biomes of the world, with a brief discussion of ideas about the physiological and ecological factors that govern species distributions. One of the points that emerges clearly, as we consider the various forest types, is the fact that we are repeatedly forced to recognize the dominant influence exerted by humans on forests, their distribution, and ecological functioning. This does not mean there is no benefit in studying undisturbed ecosystems, but it does mean that we also have to study disturbance, the forms it takes, and its effects, not only in terms of empirical observations about areas damaged or changed but also in terms of the processes affected by particular types of disturbance and the extent to which those processes influence responses to disturbance. This, as we noted in Chapter 1, is the essence of management: It is essential to understand the way managed systems will respond to particular actions or changes.

The aim of preserving large areas of (relatively) undisturbed forest ecosystems must be pursued wherever possible, but it is equally important to be realistic about the fact that disturbance does not necessarily destroy the value of forests, either aesthetically or in terms of other values (wood producing, biodiversity preservation, etc.). One of the most important things we need to do is learn how to manage systems sustainably, which means we have to know what is there initially and understand how the forests change naturally and in response to normal climatic fluctuations as well as in response to anthropogenic disturbance.

Recommended Reading

Box, E. O. (1981). "Macroclimate and Plant Forms: An Introduction to Predictive Modeling in Phytogeography." Junk, The Hague.

Goodall, D. W. (editor-in-chief). "Ecosystems of the World." Elsevier, Amsterdam. [Röhrig, E., and Ulrich, B. (1991). Temperate deciduous forest; J. D. Ovington (ed.) (1983). Temperate broad-leaved evergreen forests; Leith, H., and Werger, M. J. A. (1989). Tropical rain forest ecosystems.]

Murphy, P. G., and Lugo, A. E. (1986). Ecology of tropical dry forest. *Annu. Rev. Ecol. Syst.* **17,** 67–88.

Woodward, F. I. (1987). "Climate & Plant Distribution." Cambridge University Press, New York.

3

Canopy Architecture and Microclimate

The atmospheric environment acts on plant communities through their canopies. Therefore, if we are to understand and be able to describe quantitatively the way the atmosphere acts on forests, we must have good descriptions of forest canopies and methods of describing them that can be applied to all forests. (We should also have good information about root systems and their distribution in the soil. Root systems will be considered in Chapter 4.) The physiological processes that govern growth are driven—or strongly influenced—by the interactions between physical factors, such as radiation, air temperature and humidity, and wind, and ecological factors, such as nutrient availability, canopy architecture, and leaf longevity. Analysis of the effects of these factors must be in terms of the biophysical processes involved. This chapter provides an outline of the biophysical principles that we need for understanding the way the atmospheric environment acts on, and interacts with, forest canopies.

The most important component of the canopy is the foliage. Leaves are the organs that intercept radiation and convert it to carbohydrates by the process of photosynthesis. The CO_2 that provides the substrate for this process is absorbed through the stomata, the minute pores on leaf surfaces that are at the interface between the plant and the atmosphere. CO_2 is trapped on the wet cell surfaces in the substomatal cavities. Loss of water by evaporation from these surfaces (transpiration) is inevitable and could be regarded as a cost associated with CO_2 absorption. However, there are other aspects to the process of transpiration, such as leaf cooling and water uptake and movement through the plant, that are important in relation to the metabolism and functioning of plants.

The effectiveness with which canopies intercept radiation depends on leaf area density (leaf area per unit volume of the canopy) and arrangement in the canopy space (see Canopy Architecture). Analysis of radiation interception is a problem in physics and mathematics rather than biology, but the problem has been solved to the point where, we would argue, there is not much need for further research, although there would be considerable value in testing available formulations and solutions for particular applications. The solutions are available at a number of levels (shoot, tree, and forest) and degrees of complexity. For most applications in forest management, we do not need very complex solutions, the value of which may be limited by the availablity of the data required to implement them, but it is important that forest scientists appreciate the principles, applicability, and limitations of the available models. We provide some discussion of the principles and present the simplest useful model, under Section II.

The microclimate in forest stands is influenced, to a very large extent, by their available energy, determined by local solar radiation, the amount reflected from the canopy, and the longwave radiation balance. The book by Monteith and Unsworth (1990) provides an excellent treatment of the physics of radiation, with particular reference to terrestrial systems. Jones (1992) also provides a detailed and valuable treatment, including discussion of methods of measurement. Some basic information about radiation and its units is provided in the section on energy balance and interception.

During daylight hours most (forest) surfaces have a positive radiation balance: The canopy is absorbing energy faster than it is losing it by outgoing radiation. The excess energy goes to heating the plants, the air in the canopy space and the ground, or it is used to evaporate water, i.e., it is converted into latent heat. Because the processes of heat and mass transfer in natural systems are highly effective, the temperature differences between plants and the air seldom become large, although they may be important in some circumstances. Quantitative descriptions of the processes of heat and mass transfer provide the basis for the calculation of transpiration; these are coupled to stomatal behavior through leaf-level equations that have been extended to canopies. These processes are considered under Section II,B.

In principle, the energy balance of the forest floor can be treated in the same way as that of the forest canopy, although the amount of energy available there will always be lower than at the upper surface of the canopy. Clearly, the amount of energy reaching the ground will be much greater in open than in closed canopies. Wind speeds will be small so that transport processes are less effective, and humidity will tend to be high, particularly if the ground and litter layer are wet. We do not treat the forest floor explicitly in this chapter, but note here that, in some circumstances, the fluxes there—particularly of water vapor—will be a significant fraction of the total canopy flux. The possible contribution of forest floor to net canopy flux should be considered in analyses of particular situations.

All the physical processes discussed in this chapter apply at all scales and regardless of variation, but it must be remembered that they are all very strongly affected by the variability of forests. At leaf level, stomatal behavior is among the most variable of all physiological processes (see Watts, 1977; Beadle *et al.*, 1985a,b). The architecture of an even-aged plantation may be relatively simple and easily described in terms of stand density, leaf area per tree, and crown depth. However, even in such plantations there are always gaps caused by mortality and differences in growth as a result of differences in soil fertility, or perhaps soil depth and water-holding characteristics. In natural forests, the variability may be considerable, particularly in areas with uneven topography. There may be differences in species composition from point to point, in tree size, and in canopy architecture. There may be gaps of various sizes. All this variability gives rise to problems of scaling and averaging: What is the appropriate scale for measurements and how can these be averaged to represent the scale we are interested in? (see Ehleringer and Field, 1993). These questions have been brought into sharp focus, in recent years, by the increased use of remote sensing, in which questions of pixel size and the meaning of signals integrated over large areas in relation to the known variability of the land surfaces scanned are a matter of considerable importance. We make some comment about scaling at the end of this chapter, and it arises again in relation to remote sensing (Chapter 10). When we have to calculate canopy fluxes, or use models to predict the productivity of stands that we know to be heterogeneous in their structure and (functional) properties, the best we can do, in many cases, is to make estimates of parameter values on the basis of reasonable assumptions about what can be considered uniform. The information derived from such calculations is likely to be better and more reliable than guesswork or the use of empirical relationships derived from a limited range of conditions and not transportable to other situations.

I. Canopy Architecture

The canopy of a forest is the continuous, or almost continuous, cover of leaves and branches forming the upper layers of the forest. Canopy architecture is described by the vertical and horizontal distributions and arrangement of foliage through the canopy space.

Canopies are formed by the crowns of trees. At the stand level, stand structure—the number, spacing, height, and size distribution of the trees—is obviously a major determinant of canopy architecture. At the tree level, environmental and genetic factors influence canopy architecture. The shape of the crowns of individual tree species is under strong genetic control. The conical canopy architecture of many gymnosperms (*Abies, Picea, Larix,* etc.) and to a lesser extent some angiosperms (*Liriodendron* and *Liquidambar*) is refered to as excurrent canopy form and is the result of the terminal leader exerting dominance over all lateral branches. Conversely, the terminal leader does not exert dominance over lateral branches in some gymnosperms (*Pinus* and *Juniperus*) and many angiosperms (*Acer* and *Quercus*) resulting in more rounded canopies. However, extreme winds and blowing ice, common in subalpine and timberline conditions, can severely modify the canopy architecture, producing trees that are missing all the branches on the windward side of the tree (flag trees) or even kill upright portions of the tree, resulting in trees growing prostrate (krummholz).

As a general rule, we can expect to find trees with narrow, columnar crowns associated with high latitudes and broad or spherical crowns associated with low-latitude sites. This is because narrow, columnar crowns (large height to crown radius ratio) result in greater energy interception, especially diffuse radiation (Oker-Blom, 1989) in regions where the average sun angle above the horizon is small. Such crown shapes also reduce damage from snow and ice loading. Where sun angles are larger, low densities (small numbers of trees per unit area) of narrow-crowned trees with relatively dense foliage will use less water than trees with the same leaf area exposed in broad, spherical crowns (smaller height to radius ratio). Trees in tropical forests may cover the range from tall, large-leaved species with broad crowns to short, small-leaved species in drier areas. Many forests, most notably tropical forests, are multilayered (see Chapter 2, Section IV,E), with understories of shrub and herb species, whereas others are essentially single-layered.

Although forest canopies tend to be characteristic of particular forest types, there will generally be considerable variation in canopy architecture within a forest type, induced by differences in soil fertility, precipitation and the water balance, the stand structure, and the understory species present (Grier and Running, 1977; Gholz *et al.,* 1991; Fassnacht, 1996). If they are not destroyed by fire, wind, or human activities, forest canopies may be relatively stable over many years, but there is always some change, even if only on local scales as a result of trees falling and creating gaps. When forests are completely destroyed and either replanted or left to regenerate naturally, they move through phases of development during which canopies vary in their architecture. In deciduous forests, the foliage disappears each year and is replaced in the spring (or when the rains come, in the case of tropical deciduous forests). The pattern of leaf production varies with species, some producing virtually all the leaves for particular years simultaneously, others producing an initial "crop" of leaves on the existing branch and shoot structure, then producing new leaves on summer-growing shoots. Evergreen canopies retain their foliage, enabling them to take advantage of favorable growing conditions whenever they occur.

There are metabolic energy costs associated with both evergreen and deciduous habit. For example, leaf longevity is positively correlated to initial construction costs, specific leaf area (σ_f, m^2 kg^{-1}) and photosynthesis rates are negatively correlated to leaf longevity (Reich *et al.*, 1991, 1992), and species with greater longevity tend to support greater foliage mass (Schulze *et al.*, 1977a; Gower *et al.*, 1993b; Reich *et al.*, 1995). Leaf structure or σ_f is affected by numerous environmental factors and is strongly correlated to a suite of ecophysiological characteristics that regulate carbon assimilation. Maximum net photosynthetic rate under optimal conditions is positively correlated to σ_f; however, low nutrient availability, increased needle longevity, and low atmospheric humidity reduce specific leaf area (Linder and Rook, 1984; Gower *et al.*, 1993a). Therefore, the actual specific leaf area is a compromise between maximizing instantaneous carbon gain and long-term carbon, water, and nutrient use efficiency (see Box 3.1).

Equations relating foliage mass (w_f) to stem diameter at breast height (d_b) or to sapwood cross-sectional area (A_s) [see Eq. (3.2)] allow the calculation of leaf area if specific leaf area is known (see Box 3.1). σ_f varies with species, leaf age, position in the canopy, and nutrition (Lambers and Poorter, 1992; Reich *et al.*, 1995). It is positively correlated to leaf nitrogen concentration (weight basis) for a wide range of tree species, regardless of leaf habit (evergreen or deciduous), leaf morphology (needle-leaved or broad-leaved), or environment (Field and Mooney, 1986; Reich *et al.*, 1991, 1992; Gower *et al.*, 1993b; Schulze *et al.*, 1994). Cromer *et al.* (1993) noted that values for eucalypts varied from 20 m^2 kg^{-1} when trees were 3 months old to about 5 m^2 kg^{-1} when they were more than 12 months old. In a fertilizer experiment on eucalypts in Queensland, σ_f declined from 16 to 10 m^2 kg^{-1} between 6 and 16 months of age, with strong indications that fertilization contributed to high values (Cromer *et al.*, 1993). As the trees aged, the effects of fertilizer tended to disappear. Values of σ_f for *P. radiata* ranged from 4.0 to 5.5 m^2 kg^{-1} (Raison *et al.*, 1992), with the lowest values in low-nutrition plots. Kaufmann and Troendle (1981) gave σ_f values of 8–10 for total (not projected) leaf area of various conifers. In general, shoots receiving greater amounts of solar radiation (e.g., at the top of the crown or at the end of lateral branches) have higher σ_f values than leaves in shaded regions of the canopy (Schoettle and Smith, 1991). Figure 3.1 provides some interesting contrasts between the distribution of leaf area, and the variation in σ_f, on trees in different ecosystems. [Throughout this book, unless explicitly stated otherwise, leaf area and leaf area index (see below) refer to single-

> **Box 3.1 Specific Leaf Area**
>
> Specific leaf area (σ_f) is the ratio of fresh foliage surface area to unit dry foliage mass. From a physiological point of view, it is an important factor in canopy architecture because it describes the photosynthetic surface area that can be constructed from unit dry mass of organic matter. It is also an important parameter in ecosystem process models (McMurtrie et al., 1994) because it provides the number(s) needed to convert foliage mass to leaf area: direct estimates of stand $L*$ are commonly calculated as the product of foliage mass and σ_f for each age cohort and vertical strata in the canopy (see Figs. 3.2 and 3.3). One half of the total leaf area is the appropriate parameter for radiation transfer models and is useful when comparing different leaf shapes. It is therefore essential that σ_f is measured consistently. This has not happened in the past and estimates of σ_f have been based on projected, total, and one-half of leaf surface area. The inconsistencies can in part be attributed to poor theoretical understanding of radiation interception by nonflat objects such as conifer needles (Chen and Black, 1992) and by failure to distinguish between canopy-scale and leaf-scale geometry. For example, if the projected area of needles is calculated for randomly oriented needles then a conversion of π should be used to convert to total leaf area (Grace, 1987), but if the needles were oriented in a nonrandom pattern then a different conversion factor based on the geometry of the needle (commonly ranging from 2.2 to 2.6 for conifers) should be used. Published data, in many cases, are not accompanied by enough information for other scientists to determine how σ_f and $L*$ were calculated. To avoid further confusion, total needle surface areas should be given. It is then straightforward to calculate one half of the canopy surface area.

sided or projected foliage surface area—not to total surface area. The difference, in the case of flat leaves, is a factor of 2; in the case of conifer needles, it depends on the details of their structure (see Box 3.1)].

As a basis for any study of forest growth, or assessment of forest microclimate, it is important to have models, or at least descriptions, of canopy dynamics—changes in leaf area and leaf area density profiles through time. Examples are provided by the work of Kinerson et al. (1974), Beadle et al. (1982), and McMurtrie et al. (1986). These models have to take account of rates of leaf growth and litterfall. The complexity will vary with the purpose and with the availability of data from which to derive values for the parameters of equations.

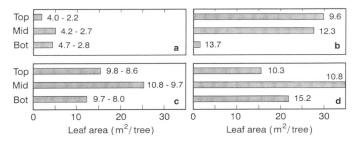

Figure 3.1 The distribution of leaf area (m^2) and specific leaf area (m^2 kg^{-1}) with height for (a) *Pinus ponderosa*, (b) *Quercus rubra*, (c) *Picea mariana*, and (d) *Populus tremuloides*. The specific leaf area values are given as a range in the upper, middle, and lower parts of the crown.

The distribution and arrangement of foliage through the canopy space may be formally described in terms of the vertical distribution of foliage area density (a_f) per unit volume of canopy space at height (z). We define $l^*(z)$, the cumulative leaf area, as the surface area of leaves per unit ground area above some level z in the canopy:

$$l^*(z) = \int_z^{h_c} a_f(z) \cdot dz,$$

where h_c is the average height of the canopy top. The most widely used measure of foliage amount in forest stands is leaf area index (L^*, m^2 m^{-2}), which is

$$L^* = \int_0^{h_c} a_f(z) \cdot dz. \tag{3.1}$$

There have been a number of studies of the distribution of a_f. Kinerson and Fritschen (1971) described the individual crowns of Douglas fir as triangular, although most studies on conifers have indicated that foliage area density distribution can be described by a normal curve (Whitehead, 1978; Beadle *et al.*, 1982; Halldin, 1985). Wang *et al.* (1990) fitted a β function to the vertical and horizontal distribution of needle area density for the four needle-age classes of *P. radiata*. The patterns for deciduous forests tend to be less regular, particularly if there is understory. Several different forms of canopy architecture are illustrated in Fig. 3.2, in which the normal distribution of foliage area density with height, commonly shown by conifers, is illustrated by data from Halldin (1985). The leaf area density profile of a *Nothofagus* forest in New Zealand and the complex profile of a deciduous forest in Tennessee, United States, are also shown in Fig. 3.2.

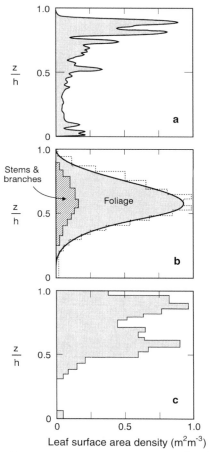

Figure 3.2 Vertical profiles of leaf surface area density (m² m⁻³) in (a) a pine forest, (b) a broad-leaved evergreen forest, and (c) a deciduous forest. (Redrawn from Halldin (1985), Hollinger (1989), and Hutchison *et al.* (1986), respectively.)

Similar L^* values do not necessarily imply similar canopy architecture; for example, mature, relatively widely spaced conifers, each carrying a large foliage mass but with relatively little overlap between trees, may have the same L^* as a young, dense stand of broad-leaved trees. The vertical and horizontal distribution of foliage—and hence the way the stands interact with the atmospheric environment—would be totally different in the two cases.

Foliage arrangement within the canopy space varies from approximately random distributions in space to organized arrangements based on clumps at shoot, branch, and tree level. There are large differences in the shapes of leaves, the most obvious being between the plane shapes of broad-leaved species and the needles of conifers. Leaf shape and foliage arrangement strongly influence leaf–air exchange processes (Schuepp, 1972, 1973) and radiation interception by canopies (Norman and Jarvis, 1975; Stenberg et al., 1995).

Foresters and forest ecologists tend to regard attainment of canopy closure as the most important point in the development of a stand. Canopy closure is the point where the branches of neighboring trees meet and overlap so that, viewed from above, the forest surface presents a complete cover. This is the point where increases in leaf mass tend to be matched by litterfall (see Chapter 6), and after canopy closure the radiation interception characteristics of the canopy are unlikely to change much, unless there is massive defoliation by insects or disease or the stand is thinned or subjected to catastrophic events like windthrow. There is no agreed criterion or method of identification; canopy closure is identified subjectively. If we take L^* as a criterion, we would expect a closed canopy to have an L^* of at least 3, although the converse does not necessarily hold—a canopy with $L^* \geq 3$ may not be closed because of spatial variation. Densely planted or regenerating stands may reach canopy closure within a few years of starting growth, whereas widely spaced trees may take many years. Tree communities that do not reach canopy closure are usually designated as woodlands, not forests.

In general, when discussing L^* in relation to forest stands it is assumed that we are dealing with reasonably homogeneous areas. When canopies are not closed, L^* can vary widely within the same stand, from quite high values in relation to the projected area of the tree crowns to zero in gaps. There is radiation penetration through gaps, to the forest floor (note earlier remarks about forest floors). The average value of L^* for the land area will depend on the size and number of gaps and on the size and foliage mass of the trees. Widely spaced trees may develop large foliage mass so that L^* varies enormously spatially (see Beets and Pollock, 1987). This is a matter of considerable importance in relation to spatial averaging of, for example, satellite data, and to the application of models to large areas. Even in less extreme situations, in which the canopy of a natural forest or plantation is closed, there is likely to be considerable spatial variation in L^*. However, for the purposes of discussion of the microclimatic processes considered in this chapter, we will assume that, for the most part, canopies are reasonably homogeneous.

Quite clearly, L^* depends on the leaf area per tree and the number of trees per unit area (stand density). There are many publications (e.g., Gholz et al., 1979; Gholz, 1982; Pearson et al., 1984) that present allometric relationships between leaf mass per tree and stem diameter—normally measured, in forestry, at "breast" height (1.37 m). The relationship is normally of the form

$$w_\mathrm{f} = c_\mathrm{f} d_b^n. \tag{3.2}$$

A simpler relationship emerged from a study by Grier and Waring (1974) in which they established a linear relationship between the foliage mass of conifers and sapwood cross-sectional area (A_s; see Whitehead, 1978; Kaufmann and Troendle, 1981). A_s can be estimated from stem cores, although detailed studies based on destructive harvests should be carried out wherever possible to increase our knowledge of how A_s varies with species and growing conditions (see Fig. 3.3)

In general, trees occuring in xeric environments have a smaller leaf area/sapwood cross-sectional area ratio (LA/SA) than those occuring in mesic environnments (Waring *et al.,* 1982; Gower *et al.,* 1993a; Mencuccini and Grace, 1995). The small LA/SA ratios of trees in dry climates are presumably an adaptation to minimize transpirational surface area and respiration losses. There is also evidence from fertilization studies (Brix and Mitchell, 1983) that nutrient availability influences the LA/SA ratio. In addition, it appears that sapwood can store significant amounts of water, thereby providing the tree with some buffer against soil water deficits. Waring and Running (1978) estimated that a *Pseudotsuga menziesii*

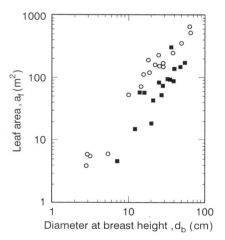

Figure 3.3 An example of the commonly observed relationship between leaf area per tree and diameter at breast height. A logarithmic plot of leaf area (a_f) against stem diameter for sugar maple (*Acer saccharum,* ○) and red oak (*Quercus rubra,* ■). Both data sets give a slope [n in Eq. (3.2)] of about 1.5; the average specific leaf area (σ_f) values were 17.7 and 17.1 m^2 kg^{-1} for the sugar maple and red oak, respectively (from Chapman and Gower, 1991).

I. Canopy Architecture   61

stand could store 270 m³ of water in the stem sapwood that, removed at 17 m³ day⁻¹, could sustain transpiration for several days. Unlike conifers, angiosperms, especially ring-porous species such as *Quercus* species, cannot refill their sapwood. In angiosperms, withdrawal of water often results in cavitation, rendering the xylem disfunctional (Siau, 1971).

It follows from the previous discussion that, if we know the stand density (n trees ha⁻¹) and have the appropriate information about allometric ratios [Eq. (3.2)] and σ_f for the species of interest, we can calculate L^* for a stand. Because trees in stands are not all the same size, we also need to know the tree size distribution. Statistical descriptions of tree size distribution are very much the domain of conventional mensuration studies; various equations are employed, among which the Weibull distribution (see Rennolls *et al.*, 1985; Vose and Allen, 1988) appears to be particularly useful.[1] Given the parameters for such a distribution, it is possible to calculate the number of trees in a stand with diameters within a particular range. L^* can then be calculated for a stand as follows: If the number of trees in size class m is x, then the leaf area per unit ground area attributable to those trees is $x_m A_{f \cdot m}$, where $A_{f \cdot m}$ is the leaf area per tree in that size class. L^* for the stand is, therefore, $\Sigma_m A_{f \cdot m} x_m$.

L^* can range from 1 to 3 for *Eucalyptus* (temperate broad-leaved evergreen) forests of Australia (Anderson, 1981), from 1 to 5.4 for boreal forests (Gower *et al.*, 1996), from 1.3 for pine stands to 8.4 for broadleaved deciduous forests of theLake States in the United States (Fassnacht 1996), and can reach values >10 in the evergreen conifer forests in the Pacific Northwest in the United States (Marshall and Waring, 1986; Runyon *et al.*, 1994).[2] The low L^* values in the boreal forests [<1 for jack pine to ca. 5 for black spruce (*Picea mariana*) in Canada (Gower *et al.*, 1996a)] are, presumably, a result of extremely cold winters and dry summers. When they are growing in similar climates and soils, evergreen conifers generally support a greater leaf area than deciduous trees (Gower *et al.*, 1993b) because of leaf longevity and shoot architecture. The factors affecting carbon production and allocation in trees, and therefore L^*, are discussed in Chapter 5, Section IV.

One more aspect of stand structure and canopy architecture should be noted: the relationship between tree size and stand density. Essentially,

---

[1] A simple, and possibly very useful, one-parameter probability distribution function is the Rayleigh distribution: $f d_b > d_b' = \exp[-d_b'/\bar{d}_b^2]$, where $f d_b > \bar{d}_b$ is the probability that a given stem diameter (d_b') will be exceeded and $\bar{d}_b$ is the average stem size. (We are indebted to Dr. Frank Kelliher for bringing this to our attention.)

[2] The L^* values for some forests in the Pacific Northwest in the United States were originally reported as $L^* > 20$ by Grier and Running (1977) and Gholz (1982), but these values were found to be overestimates.

there is a maximum size that any tree can attain at a particular population density. This is described by the so-called −3/2 power law, first identified by Japanese workers, and demonstrated to hold for trees by Drew and Flewelling (1977). Essentially, it describes the fact that high populations of young trees undergo self-thinning as they develop; the surviving trees increase in size but there is a limit to how large they can grow—the maximum stem mass W_{max} can increase at a rate given by the equation

$$W_{max} = k \cdot p^{-1.5}. \qquad (3.3)$$

(see Fig. 3.4). Smith and Hann (1984) have developed an analytical model based on Eq. (3.3), which they applied successfully to the calculation of mean stem volume of red pine (*P. resinosa*) stands. The power −1.5 is extraordinarily stable across plant communities (see White, 1981) of all species (in pure stands) and ages. The equation does not quantify plant mass if p is less than some limit value; W may then take any value less than that given by Eq. (3.3). Landsberg (1986) derived a general expression for L^* of a stand in terms of Eqs. (3.2) and (3.3), which indicated that, approximately

$$L^* = C_f p^{-0.2}. \qquad (3.4)$$

The "constant" C_f is a composite containing σ_f and k.

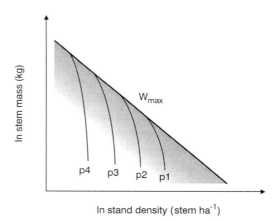

Figure 3.4 Graphical representation of the −3/2 power law describing the change in stem mass with population. The W_{max} line is the highest stem mass that will be achieved by a given species at any specified population. The p_i curves indicate that the mass of trees in a population can increase, independent of population, until significant competition begins in the zone of imminent competition (shaded), after which W tends to W_{max} (redrawn from Landsberg, 1986).

II. Energy Balance and Interception of Visible (Photosynthetically Active) Radiation

Shortwave radiant energy from the sun (φ_s) is normally measured on a horizontal surface and is the basic energy source driving plant growth, and hence, all biological processes. Solar energy reaches the earth as either direct (beam) or diffuse radiation in wavebands between about 300 and 3000 nanometers (nm). The visible part of solar energy (light) is the band from about 370 nm (blue light) to 700 nm (red light), commonly referred to as photosynthetically active radiation (PAR). The amount of shortwave radiation at the top of the atmosphere, and the depletion of energy as it passes through the atmosphere, are shown in Fig. 3.5.

When considering processes such as transpiration or the energy balance of bodies, we are concerned with the whole spectrum of radiation and use energy units (W m^{-2}, where 1 W = 1 J sec^{-1}). When considering photochemical processes such as photosynthesis, which is driven by the number of photons absorbed rather than by the energy of the photons, the unit used is photon flux density [(φ_p); mole m^{-2} sec^{-1} or, more usually, μmol m^{-2} sec^{-1}]. φ_p has the great advantage that photosynthetic efficiency can be expressed in terms of the number of moles CO_2 fixed per mole of photons, i.e., per mole equivalent of quanta (photons) in the visible wavebands. About half of the incoming shortwave radiation is in

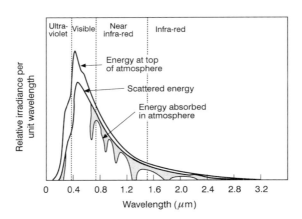

Figure 3.5 Incoming, reflected, and scattered solar radiation. The figure shows the shortwave radiation at the top of the atmosphere and the depletion of energy as it passes through the atmosphere. Radiation is scattered by clouds, water vapor and aeorosols and absorbed by ozone (in the ultraviolet region) and water vapor, at wavelengths greater than 1 μm. The lowest lines show the amount of radiation reaching the earth's surface under cloudless conditions (redrawn from Carleton, 1991).

the photosynthetically active waveband (Monteith and Unsworth, 1990). It is not strictly correct to convert directly from energy units to φ_p because the energy of a photon depends on its wavelength, but in normal sunshine 1 J PAR contains about 4.6 μmol, i.e., for solar radiation 1000 W m^{-2} ≈ 2200 μmol m^{-2} sec^{-1}. Jones (1992) provides a useful treatment of radiation in relation to the plant environment.

Because good radiation measurements are difficult to make and because the fundamental importance of radiant energy to all branches of biology has only recently been recognized by those responsible for meteorological networks, values of φ_s are often not available for forested areas. However, for periods of the order of a week or more, φ_s can be estimated from the Angstrom relationship with sunshine hours, measured by a Campbell–Stokes sunshine recorder:

$$\varphi_s = \varphi_{at} (a + b\, n_\varphi/t_\varphi), \tag{3.5}$$

where φ_{at} is the average value of φ at the outside of the atmosphere, n_φ is actual hours of bright sunshine, and t_φ is the maximum possible sunshine hours (obtainable from tables). Values of a and b vary with latitude and time of year, but values of $a = 0.23$ and $b = 0.5$ will give good estimates in most areas (see Stigter, 1980, for discussion). Linacre (1992) has provided a very useful treatment of the measurement and estimation of solar radiation at the surface of the earth, including a thorough discussion of Eq. (3.5). Values for particular periods and areas can now be obtained from satellite data (Eck and Dye, 1991).

In our consideration of forest ecosystems of the world (Chapter 2) we presented climate diagrams that included annual average energy income figures. These range from about 10 MJ m^{-2} day^{-1} at The Pas in Canada, to almost 20 MJ m^{-2} day^{-1} at Jamshedpur in India. On a daily basis, the highest figures for the places considered are summer values in Australia (23–25 MJ m^{-2} day^{-1}), whereas the lowest values—for boreal forests during the winter—are near zero. On individual days, in the high radiation areas, incoming solar radiation may reach 30 MJ m^{-2} day^{-1}.

A. Radiation Balance

The forest surface is composed of tree crowns exposed to the sky. In the case of a closed canopy of deciduous trees of about the same height, the surface will be relatively smooth compared to the surface of a coniferous forest which will generally be much rougher, with elements (tree crowns) projecting well above the average height, and with relatively deep "valleys." The roughness of the surface will have major effects on its radiation balance and on its momentum absorption properties (see Section III,A).

We consider the energy balance of a forest (or any surface) in terms of the net amount of radiant energy retained by the surface; conservation of energy requires that we account for this energy, of which a very small amount (<5%) goes to photosynthesis while the rest goes to warming the ground, the trees, or the air or to evaporating water from the ground or plants. The net radiation (φ_n) retained by a surface is

$$\varphi_n = (1 - \alpha) \varphi_s + (\varphi_L \downarrow - \varphi_L \uparrow), \qquad (3.6)$$

where α, the albedo of the surface, is the fraction of incident short-wave radiation that is reflected from the surface (α is also called the reflection coefficient, or reflectivity, of the surface), $\varphi_L \downarrow$ is the longwave downward flux of radiation from the sky, and $\varphi_L \uparrow$ is the longwave upward flux from the surface.

Any body with a temperature above absolute zero emits radiation at a rate proportional to the fourth power of its temperature (Stefan–Boltzman law): i.e., $\varphi_L = \varepsilon_L \sigma T^4$, where σ is the Stefan–Boltzman constant (= 5.67×10^8 W m^{-2} K^{-4}), T is the absolute temperature, and ε_L is the emissivity. For a perfect black body $\varepsilon_L = 1$. For most vegetative surfaces, ε_L is in the range 0.9–0.96. The flux of long-wave radiation from the earth's surface is in the wavelength range between about 3000 and 100,000 nm (long-wave radiation), with maximum energy output at about 10,000 nm. The downward flux of long-wave radiation depends on atmospheric conditions, such as cloud amount and type (e. g., cirrus, cumulus), and the amount of water vapor, and other gases in the atmosphere. Under cloudy skies, the outward and inward fluxes of long-wave radiation will be almost in balance, whereas the net loss of long-wave radiation is highest under clear skies—hence the increased likelihood of night-time frost under those conditions.

As a general rule, the reflectivity of surfaces decreases with increasing roughness, and because forest canopies are rough surfaces their reflectivity is low. Figure 3.6 is derived from Stanhill (1970) and illustrates the changes in albedo with the height of the elements of natural surfaces. As a general rule, shortwave albedo does not vary greatly among forest cover types, it decreases with canopy height and roughness and is typically low for forests relative to other cover types.

Representative mean daily values of α are about 0.10 for coniferous forests (Jarvis *et al.*, 1976) and 0.16 for deciduous forests (Rauner, 1976). There have been several thorough studies of the radiation balance of coniferous forests, for example, by Moore (1976) on *P. radiata* in South Australia and Stewart and Thom (1973) on *P. sylvestris* in England. Moore identified variations in albedo through the day but these were relatively trivial and he concluded that a mean albedo of 0.11 ± 0.01 could be applied to the canopy for all solar elevations. Pinker *et al.* (1980a) found the average albedo of a tropical evergreen forest in Thailand to be about

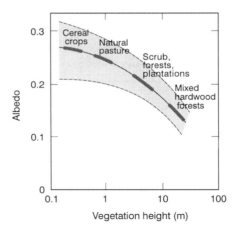

Figure 3.6 Relationship between mean albedo (α) and vegetation height (redrawn from Stanhill, 1970; note that the x axis (vegetation height) is a logarithmic scale).

0.14, with significant diurnal variation, from about 0.17 (early morning/evening) to about 0.11 at midday. There were also seasonal differences, and the albedo in a grassy clearing was generally higher (about 15%) than that of the forest. Shuttleworth et al. (1984) obtained a value of 0.14 for tropical forest in the Amazon, with some dependence on solar angle (low angles, and higher albedo).

For many purposes, φ_n can be conveniently estimated from empirical relationships of the form

$$\varphi_n = b_n \varphi_s \pm a_n. \qquad (3.7)$$

This can be established from measurements of φ_n over a forest and φ_s need not be measured nearby. Jarvis et al. (1976) gave values of b_n for coniferous forests that range from 0.71 to 0.91; a value of 0.8 will seldom be much in error although Moore's (1976) study is again worth consideration. He gave $\varphi_n = 0.67\varphi_s - 45$ (± 10) W m^{-2} for the winter months and $\varphi_n = 0.85\varphi_s - 55$ (± 18) W m^{-2} for the summer. Federer (1968) found $b_n = 0.83$ for a hardwood forest, whereas Rauner (1976) indicated that 85% of φ_s was absorbed by a dense aspen stand (i.e., $b_n \approx 0.85$) and Kalma and Fuchs (1976) gave 0.8 for a citrus orchard. Pinker et al. (1980b) found $b_n = 0.88$ for the tropical forest in Thailand, and Shuttleworth et al. (1984) obtained $b_n = 0.85$ in the Amazon.

The parameter a_n gives an estimate of the average value of φ_L and therefore depends on local climate as well as forest structure. The value of a_n for coniferous forests varies widely from -6 to -126 W m^{-2}; the median value is about -60 W m^{-2}. Federer (1968) gave $a_n = -89$ Wm^{-2}

for a deciduous forest, whereas the overall average of the monthly values given by Pinker et al. (1980b) was -54 W m^{-2}, which compares with -35 W m^{-2} given by Shuttleworth et al. (1984).

Taken overall, the values of the coefficients in Eq. (3.7) are sufficiently conservative to allow useful estimates of φ_n to be made for any closed canopy forest in any area. Working with daily values, we should note that a_n must be multiplied by time; the period between sunrise and sunset would be the appropriate interval (Δt) for daylight hours, when $\varphi_s > 0$. For example, if $\varphi_s = 15$ MJ m^{-2} day^{-1} and $\Delta t = 11.5$ hr, then if $b_n = 0.85$ and $a_n = -55$ W m^{-2}, φ_n is $(0.85 \times 15 \times 10^6 - 55 \times 11.5 \times 3600)$, i.e., ≈ 10.5 MJ m^{-2} day^{-1}.

Estimates of φ_n are an important first step toward the calculation of rates of water loss, by evaporation and transpiration, from forest canopies. The conservation equation describing the partitioning of φ_n into latent heat (evaporation or transpiration—λE, where λ is the latent heat of vaporization of water), sensible heat (H) and heat stored within the stand (G), is

$$\varphi_n + G = \lambda E + H, \qquad (3.8)$$

so that given values for φ_n, solving Eq. (3.8) for λE would provide estimates of the rates of water use by forests. This is considered under Section III. The heat storage term G, as written here, includes heat stored in the soil, in the vegetation, and in the air within the stand. It can be partitioned into these components, and detailed micrometeorolgical studies have to take them into account. Focussing on processes at the stand level, and intervals of days or longer, we can afford to ignore G, which tends toward zero on a daily basis.

Interest in albedo and its influence on φ_n [Eq. (3.6)] has increased in recent years as our understanding of the linkages between the atmosphere and terrestrial ecosystems increase. Recent models that couple land surface and atmospheric processes have shown how large-scale changes in land surface cover can affect regional energy and climate. Bonan et al. (1993) suggested that large-scale replacement of boreal forests by low-stature vegetation would increase the albedo of the region. Based on global circulation models, they suggested that the increased albedo would cause surrounding oceans to stay frozen longer, which in turn may shorten the growing season of surrounding terrestrial ecosystems. Degradation of vegetation in the Sahara and Sahelian regions of Africa has reportedly increased the regional albedo of this zone, causing decreased rainfall by pushing the intertropical convergence zone further to the south (Charney, 1975; Charney et al., 1975). Removal of tropical forests may increase the global albedo, which would cause a cooler and drier climate (Potter et al., 1975; Sagan et al., 1979; Henderson-Sellers and Gornitz, 1984). Gash and Shuttleworth (1991) predicted that large-scale deforestation of tropical forests in the Amazon would cause local increase in temperature and 25% decrease in rainfall.

B. Interception of Visible Radiation

The flux density of energy in the visible waveband, received at a point in a forest canopy, consists of beam and diffuse radiation that penetrates through gaps in the canopy. It is supplemented by radiation that is reflected from leaves and the soil and transmitted through leaves. Complete treatments of this situation are complex, and there is a large body of literature on the subject. A selection of key papers and reviews would include the early treatments by Anderson (1964, 1966) and Cowan (1968) and the work of Norman (e.g., 1980, 1982), Oker-Blom (1986), and Baldocchi (see the review by Baldocchi and Collineau, 1994). These mathematical descriptions and models can describe the radiation regimes in plant canopies with considerable accuracy if the assumptions incorporated in them are met and if there is adequate information about canopy architecture such as whether foliage is clumped or randomly distributed and leaf angle distribution. The simplest case assumes foliage is randomly distributed in the canopy space and leaf angle distribution is spherical, i.e., there is always an equal leaf area normal to the beam of radiation, whatever its direction.

Detailed models of PAR interception are essential for simulation of canopy photosynthesis over short periods, such as half an hour, and for the simulation of the diurnal course of photosynthesis. Such treatments are important because, if successful, they provide us with the assurance that we understand the processes involved in the growth of plant communities sufficiently well to allow us to simplify with confidence, knowing the implications and consequences of the simplifications. The mathematical complexity of any treatment increases rapidly when we depart from simple cases. This book is, in general, concerned with stand-level ecophysiology, where detailed treatment of PAR interception is unlikely to be justified. We will, therefore, present only the (very widely used) Beer's law equation for calculating radiation interception by plant canopies, with some discussion on its limitations.

Beer's law strictly applies to the absorption of radiation by simple systems such as uniformly turbid media (air, liquids, etc.), but it can be derived from the analysis of the path of beams through foliage (see Norman, 1979; Monteith and Unsworth, 1990). The equation is

$$\varphi_s(z) = \varphi_s(0) \exp(-kL^*(z)), \tag{3.9}$$

which says that radiation decreases exponentially through the canopy, and that the average flux at any level is a function of the leaf area index above that point. The value of k varies somewhat with species and season (sun angle), but for most forest types a value of $k = -0.5$ will not lead to serious errors (see Jarvis and Leverenz, 1983). Strictly, the equation only holds if the foliage in the canopy is randomly distributed and leaf angle distribution is spherical. If foliage is nonrandomly distributed, or clumped, Beer's law will underestimate radiation penetration (Ross, 1981; Lang et al., 1985). Clumped foliage, common in conifers, allows more light to penetrate the canopy than if the foliage was distributed randomly (Nilson, 1977; Oker-Blom and Kellomäki, 1983). In addition, the penumbra effect provides a more homogeneous distribution, which is relatively much larger in relation to needles than broad leaves, of radiation in the lower canopy (Norman and Jarvis, 1975). Nonhomogeneous foliage is most often observed in conifers and can occur at several scales. An extreme example of clumping is the canopy of boreal black spruce stands, where clumping occurs at the shoot, branch, whorl, and canopy scales (Chen et al., 1996). Chen et al. provide theory and methodology that accounts for clumping when L* is estimated using optical instruments that measure canopy gap fractions from radition transmitted through a canopy. Failure to account for nonhomogenous distribution of foliage in canopies is likely to cause inaccurate estimates of canopy photosynthesis, particularly when calculations are made over short periods (Norman and Jarvis, 1975; Kucharik et al., 1996). Despite these reservations, if the objective is to estimate the radiation absorbed over periods of days, weeks, or seasons, Eq. (3.9) gives very useful results. Integration over time, and averaging direct and diffuse radiation, tend to eliminate the errors that might arise if the equation was used to estimate absorption over short periods for canopies that do not meet the rigorous theoretical requirements.

It is worth exploring some of the implications of Eq. (3.9) in relation to canopy photosynthesis. We will use an empirical equation for leaf photosynthesis [Eq. (5.12); i.e., Eq. (4.8) from Landsberg, 1986, p. 80][3] to evaluate it. The calculations are made in terms of photon flux density (φ_p).

Assume we are concerned with a canopy with $L^* = 5$ and that above-canopy $\varphi_p = 2000$ μmol m^{-2} sec^{-1}. For z below the bottom of the live crown $\varphi_p(z) = 164$ μmol m^{-2} sec^{-1}. Because Eq. (3.9) is strongly non-

---

[3] The equation for photsynthesis is $A_{net} = (\alpha_p \varphi_p A_g)/(\alpha_p \varphi_p + A_g) - R_d$, where α_p is the quantum yield efficiency, A_g is maximum (gross, light-saturated) photosynthetic rate, and R_d is dark respiration. We have used $\alpha_p = 0.03$ mol mol^{-1}, $A_g = 10$ μmol m^{-2} sec^{-1}, and $R_d = 0.2$ μmol m^{-2} sec^{-1}. For most physiological studies, this equation has been superseded by biochemically based descriptions of photosynthesis (see Chapter 5), but it remains useful for many applications.

linear, averaging the within-canopy flux and inserting the result into the equation for leaf photosynthesis, to calculate canopy photosynthesis, is likely to lead to significant errors. In this case, φ_p(average) = (2000 + 164)/2 = 918 μmol m^{-2} sec^{-1}, which would give A_{net} = 7.14 and a canopy photosynthesis rate (A_c = 5 A_{net}) of 35.7 μmol m^{-2} sec^{-1}. A better approximation is obtained by dividing the canopy into layers, each of thickness 1 L^*, calculating φ_p for the center of each layer (i.e., for L^* = 0.5, 1.5, . . . 4.5), inserting those values into the photosynthesis equation and summing the values. This gives $\varphi_{p \cdot i}$ = 1558, 945, 573, 347, and 210, and $A_{net \cdot i}$ = 8.1, 7.2, 6.1, 4.9, and 3.7, i.e., A_c = 30 μmol m^{-2} sec^{-1} — 19% lower than the first estimate. (The greater the number of layers, the better the estimate). A value close to this can also be obtained by calculating φ_p for the middle (in terms of leaf area) of the canopy, i.e., L^* = 2.5, and inserting that value into the photosynthesis equation, which gives A_{net} = 6.1 μmol m^{-2} sec^{-1} hence—taking this as representative of the canopy—ΣA_{net} = 30.5 μmol m^{-2} sec^{-1}.

The Beer's law equation as presented here assumes that the horizontal distribution of radiation flux density at any level in the canopy is the same. The equation provides no information about the relative proportions of sunlit and shaded leaves, although simple observation of any sunlit canopy reveals that there are very large differences, at any given level, between the intensities of leaf illumination. Figure 3.7 illustrates the different patterns of radiation absorption and penetration that may occur in forest canopies. Bright sunflecks can penetrate to the floor of forest canopies that have L^* values of 8 or 9, whereas some leaves in clumps near the top of canopies may be completely shaded. Equation (3.9) also takes no account of the fact that solar radiation may be either direct or diffuse. Beam radiation is directional, the angle of incidence on the canopy being a function of earth–sun geometry (latitude, longitude, time of day, and time of year). The probability of a beam being intercepted by elements of a canopy, or of penetrating it, depends on the leaf area in the path of the beam and the average orientation of the leaves. Diffuse radiation comes from all parts of the sky. The ratio of diffuse to total solar radiation ranges from about 0.1, under clear sky conditions, to 1.0 when the sky is overcast. The ratio will vary with location, season, and atmospheric pollution. Diffuse radiation penetrates canopies more effectively than direct radiation. A good illustration of this is provided by Hollinger et al. (1994), who found that net CO_2 uptake per unit of absorbed radiation in an old-growth *Nothofagus* forest was 50% higher when radiation was predominantly diffuse as opposed to predominantly direct.

Of the models that deal with the problem of sunlit and shaded foliage, that of Norman (1980, 1982) is one of the better known and has been described in "user friendly" form by Landsberg (1986). McMurtrie's BIO-

MASS model (see McMurtrie et al., 1990: see also Chapter 9) uses a relatively simple formulation based on Norman (1980) in which the forest is represented by a randomly spaced array of crowns and the foliage is divided into three horizontal layers of equal depth. The ratio of direct and diffuse radiation, for any day, must be specified. Sunlit L^* (for which φ_p is the same as at the canopy top) is given by

$$L^*_b = (1 - \exp(-kL^*/\cos Z))\cos Z/k, \qquad (3.10)$$

where Z is solar zenith angle. This can be numerically integrated over the day. Rigorous and accurate forms are relatively complex, but to a reason-

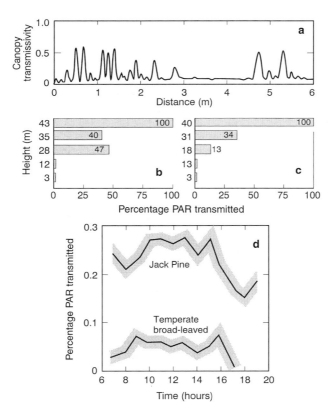

Figure 3.7 PAR transmisson through forest canopies. (a) Data from Jarvis et al. (1976), showing transmission along a transect below $L^* = 4.2$ in a Sitka spruce canopy. The 6-m transect was traversed in 3.3 sec. (b and c) Data from Torquebiau (1988) showing vertical daily average gradients of PAR in a layered tropical rainforest (b) and a nonlayered forest (c). (d) Data from Baldocchi and Vogel (1996) showing diurnal patterns of PAR interception under a boreal jack pine forest ($L^* \approx 0.7$; biomass area index ≈ 2.5) and a temperate broad-leaved forest ($L^* \approx 4.0$).

able approximation the flux density of diffuse radiation can be estimated from Eq. (3.9) applied to the diffuse component of $\varphi_s(0)$, so the radiation received by sunlit leaves is

$$\varphi_{p \cdot \text{sunlit}} = \varphi_{p \cdot \text{direct}} + \varphi_{p \cdot \text{diffuse}}, \tag{3.11}$$

and the radiation received by the shaded leaves ($L^*_s = L^* - L^*_b$, or $L^*_s = L^*(1 - L^*_b/L^*)$) is given by the $\varphi_{p \cdot \text{diffuse}}$ term. Wang *et al.* (1992) have provided a series of comparisons of canopy photosynthesis between results obtained using BIOMASS and the far more complex and complete description of canopy interception provided by the model MAESTRO (Wang and Jarvis, 1990a,b; see Chapter 9).

As we noted earlier, it is important to have detailed and accurate models to ensure that we understand the implications and consequences of simplifications. However, Eq. (3.9), used with the best available estimates of φ_s to provide the input term for a simple model such as the ε model (see Chapter 9), will generally suffice for practicing forest scientists who may be interested in growth and yield estimates for forests for which the only information about canopy architecture is height, stand density, and (perhaps) L^*.

III. Heat and Mass Transport

A. Within-Canopy Microclimate

The microclimate within stands depends on energy absorption by the foliage, branches, and soil, evaporation of water from these elements to the air in the canopy, and transfer processes from within the canopy to the reference level. Typical profiles of within-canopy conditions are shown in Fig. 3.8.

Profiles such as those sketched in Fig. 3.8 will be strongly influenced by canopy architecture as well as by current environmental conditions. In a dense canopy with high leaf area density in the upper layers, most of the incident radiant energy will be absorbed in those layers, which will also absorb most of the momentum from the wind, causing low wind speeds and ineffective exchange of scalars (heat, water vapor, and CO_2) between leaf surfaces and the air. The result will be higher temperatures in the upper than in the lower part of the canopy, and high humidity if the stomata are open and transpiration is taking place. Conditions within canopies both determine and are determined by the leaf energy balance, i.e., transpiration and leaf to air transfer processes are intimately coupled to within-canopy microclimate. Temperature and humidity in the lower part of the canopy will also depend on whether the soil is wet or dry and the rate of soil evaporation. A detailed treatment of leaf energy balance

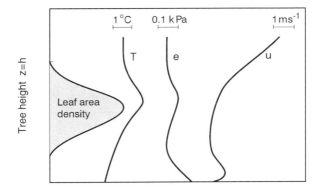

Figure 3.8 Characteristic mean day time vertical profiles of temperature ($T°$, C), vapor pressure (e, kPa), and wind speed (u, m sec^{-1}) in a forest. Note the increase in temperature in the region of greatest foliage density, where most of the energy is absorbed. The increase in wind speed near the bottom of the canopy is characteristic of forests with little understory and would appear to cause countergradient fluxes (see Denmead and Bradley, 1985). Figure is based on Fig. 13 of Jarvis et al. (1976).

has been provided by Leuning (1989), and Leuning et al. (1995) have presented a coupled set of relationships between stomatal conductance, CO_2 assimilation, and the leaf energy balance using a simplified description of radiation by sunlit and shaded leaves.

We will not be dealing in any detail with transfer processes within canopies. We simply note that, although there have been many attempts to describe them by one-dimensional flux–gradient relationships, these do not work in canopies. It is now recognized that gusts sweeping through canopies are the dominant means of scalar transfer (see Fig. 3.9). The velocity of the gusts tends to be about twice the velocity of the mean wind (M. Raupach; personal communication). Raupach (1989) provides an outline of modern theory of within-canopy transfers, which is still being developed.

B. Transpiration from Stands: The Combination Equation

Calculation of the rates of evaporation (of free water) and transpiration from stands is based on the principles of energy balance and mass transfer. Equation (3.8) indicates that the radiant energy absorbed by a canopy is partitioned into latent heat (λE), sensible heat (H), or heat stored in the biomass and the soil (G). Because heat gains during the day tend to be balanced by losses at night, we can ignore G if we are concerned with periods of days or longer.

Transport from the canopy to the overlying air can be treated as a one-dimensional process, analogous to Ficks law of diffusion, which states

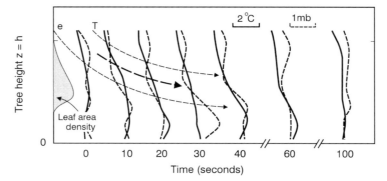

Figure 3.9 Vertical profiles of temperature (solid lines) and humidity (dotted lines) in a forest during and after the passage of a gust. The dashed arrows are the contours of constant temperature and vapor pressure. The arrow depicts the penetration of the gust (redrawn from Denmead and Bradley, 1985).

that the flux of an entity is given by the product of its concentration gradient and a diffusion coefficient. The sensible and latent heat flux densities (W m^{-2}) can, therefore, be expressed as the product of the gradients of temperature (T) and specific humidity (q) and a diffusion coefficient. The gradients are between some level in the canopy—denoted by the subscript "o"—and a reference level at height z in the boundary layer above the canopy. Specific humidity has the dimensions (kg vapor kg^{-1} air). It is directly related to vapor pressure—see, for example, Monteith and Unsworth (1990)—and is a convenient parameter to use in mass flux calculations.

If the air was absolutely still, the rate of scalar diffusion would be determined by the molecular diffusivities of the scalars—a property of the air and their molecular weights. However, most of the time the air over forests and other natural surfaces is not still, but in turbulent motion, so the diffusion coefficient depends on turbulence. Turbulence arises because, when wind blows over a surface, the molecules of air in contact with any solid (soil, leaves, and branches in the case of a forest canopy) are stationary (aerodynamicists call this the "no-slip" condition). The wind therefore exerts "drag" on the ground and plants, i.e., there is a flux of momentum to the surface (momentum has units of force: N m^{-2} or kg m^{-2} sec^{-1}). Some distance above the surface, the wind speed will have some relatively large value; therefore, between the surface and that level there must be a region of wind shear where, in any layer, air at higher levels is moving faster than the air in the layer just below [see Eq. (3.17)–(3.20)]. At all but the lowest speeds this situation is unstable and generates turbulence—eddies and gusts of air at all scales—that efficiently mixes the atmosphere. The strength of the momentum flux is related to the intensity of the turbulence produced by the wind shear. Because heat, water vapor, and CO_2 are entrained in the air stream, the diffusion coefficients that determine the rate at which they are transferred along concentration gradients depend on turbulence, which depends on the wind speed and the roughness of the surface.

We express the diffusion coefficients for heat and water vapor as an aerodynamic conductance g_a, which encapsulates the processes of molecular and turbulent diffusion of heat and vapor through the air from the leaves and from the soil surface to the reference height z. The conductances for heat, water vapor, and momentum are not strictly the same, but the assumption of similarity generally introduces only minor error. The equation for sensible heat flux is

$$H = \rho c_p (T_o - T(z)) g_a, \qquad (3.12)$$

where ρ = air density ≈ 1.2 kg m^{-3} at 20°C and c_p is the specific heat of air at constant pressure $\approx 1.01 \times 10^3$ J kg^{-1} K^{-1}), and that for latent heat flux is

$$\lambda E = \lambda \rho (q_o - q(z)) g_a, \qquad (3.13)$$

where λ is the latent heat of vaporization of water ($\approx 2.44 \times 10^6$ J kg^{-1}). The rate of water loss from a dry canopy is determined by the transpiration rates of the leaves in the canopy. Water vapor moves from the substomatal cavities, where the air is saturated at leaf temperature T_1, to the leaf surface through the stomatal apertures. This process (transpiration) can be described by an equation, similar to Eq. (3.12), for the flux of vapor:

$$E = \rho(q_{sat}(T_1) - q_{o \cdot leaf}) g_s, \qquad (3.14)$$

where g_s is the stomatal conductance and $q_{o \cdot leaf}$ is the specific humidity at the leaf surface at leaf temperature T_1. Stomatal conductance is a physiological parameter, under the "control" of the plant, responding to light, atmospheric CO_2 concentrations, atmospheric humidity, and soil water conditions. (We discuss g_s and the factors affecting it in Chapters 4 and 5, in which the influence of environmental factors and soil conditions are considered in some detail.)

If we now aggregate all the leaves in the canopy and add the vapor flux from other plant surfaces and the soil, we can write another equation analogous to Eq. (3.12) for the whole canopy, with the humidity gradient given by the difference between saturated q at T_o, the reference canopy temperature, and specific humidity (q_o) in the canopy:

$$E = \rho(q_{sat}(T_o) - q_o) g_c. \qquad (3.15)$$

3. Canopy Architecture and Microclimate

The canopy conductance g_c is no longer a purely physiological quantity like g_s but it includes fluxes from the soil and is also modified by turbulent mixing and variation of g_s within the canopy.

With some assumptions, such as that leaf and air temperatures are the same, and some approximations relating to the linearity of humidity gradients, Eqs. (3.8), (3.12), and (3.15) can be combined, using a set of now standard procedures (see Monteith and Unsworth, 1990), to yield the Penman–Monteith (P–M), or combination equation:

$$\lambda E = \frac{\varepsilon \varphi_n + \lambda \rho D_q g_a}{\varepsilon + 1 + g_a/g_c}, \tag{3.16}$$

where $D_q = q(z) - q_{\text{sat}}(T(z))$, i.e., the specific saturation deficit at the reference level z, and $\varepsilon = (\lambda/c_p)(dq_{\text{sat}}/dT)$, i.e., the dimensionless rate of change of saturated specific humidity with temperature ($\varepsilon = 2.2$ at 20°C).

The P–M equation, as we shall henceforth refer to it, neatly encapsulates the two environmental drivers of evaporation—the net radiant energy "supply" φ_n and the supply of dry air D_q—with the two essential controls: g_a, a measure of the mixing power of the atmosphere, and g_c, the (primarily) physiological control exerted by plants. An important point to note is that if canopies are wet $g_c \to \infty$ and the equation provides the rate of water loss from a wet canopy.

It is possible to obtain estimates of the aerodynamic conductance g_a by employing momentum as a tracer. This involves measuring the mean horizontal wind speed (u) above a forest, which is usually much simpler than measuring the fluxes of heat and water vapor. The measurements must be made high enough to avoid the high-frequency turbulence generated by elements of the canopy. For extensive and horizontally homogeneous canopies, the mean wind speed above a height of about 2 h_c is well described by the logarithmic wind profile:

$$u(z) = \frac{u_*}{k} \ln\left(\frac{z - d}{z_0}\right), \tag{3.17}$$

where d is known as the zero plane displacement. It is equal to the mean level of momentum absorption in the canopy and forms a new, elevated origin for the wind profile. The aerodynamic roughness length, z_0, is a measure of the momentum-absorbing capacity of the canopy, and k is a constant with a value of 0.4, called von Karman's constant. The key quantity obtained from Eq. (3.17) is u_*, the friction velocity, which provides a measure of the flux of momentum (τ) to the canopy:

$$\rho u_*^2 = \tau, \tag{3.18}$$

where τ is constant for a considerable height above horizontally uniform canopies. We can now write, by analogy with Eqs. (3.12) and (3.13),

$$\tau = \rho u_*^2 = \rho(u(z) - u_0)g_a = \rho u(z) g_{aM}, \qquad (3.19)$$

where $u_o = u(d+z_o) = 0$. Combining Eqs. (3.17) and (3.19) leads to a simple expression for g_{aM}, the aerodynamic conductance for momentum:

$$g_{aM} = ku_* \left[\ln \frac{(z-d)}{z_0} \right]^{-1}. \qquad (3.20)$$

Equation (3.17) and the consequent expression for g_{aM} are altered when there are strong temperature gradients in the atmosphere. If the ground or canopy is hot (heated by the sun), the air in contact with the surface is warmed, expands, and rises to be replaced by cooler air from above, generating turbulent mixing in the process. If the canopy or ground is cooler than the air (e.g., if it has been cooled by radiation at night), then the air at lower levels is denser and extra turbulence must be generated by wind shear to stir it up. Hence, bouyancy can enhance or decrease turbulent mixing, depending on whether the conditions are unstable (daytime) or stable (nighttime). There are well-understood equations for adjusting the diffusivities to take account of bouyancy effects (see Kaimal and Finnigan, 1994); however, the adjustment factors are different for heat, water vapor, and momentum exchange so that the assumption of similarity of conductances no longer holds if bouyancy is important. This is unlikely to be of concern except to those undertaking detailed short-term micrometeorological measurements.

Values of d have generally been found by iterative adjustments of the value used to linearize Eq. (3.17) and the use of regression to minimize least-squares deviations. This method is fraught with difficulty and uncertainty, and there are better procedures (Thom, 1975; Jackson, 1981). In most cases, a value of $d \approx 0.75\ h_c$ will give consistent results. This accords with the values reviewed by Landsberg (1986, p. 57), who also indicated a value of $z_0 \approx 0.1\ h_c$ as a useful general approximation. Jarvis et al. (1976) reviewed values of z_o and d reported from studies on coniferous forests: There was considerable variation but the ratios z_o/h_c and d/h_c were 0.08 ± 0.05 and 0.8 ± 0.09, respectively. Shuttleworth (1989) reported values of z_o/h_c and d/h_c of 0.06 and 0.86 for an Amazonian tropical forest. In general, we can expect the upward displacement of the wind profile to be more pronounced when canopies are dense than when they are sparse: In the limit of an extremely dense canopy (unrealistic in relation to radiation interception and canopy photosynthesis), d/h_c values would tend toward h_c, whereas z_o would be expected to become progresively smaller, as it does, for example, for smooth land surface covered by short grass.

Kelliher et al. (1995), evaluating the maximum conductances for global vegetation types, made a distinction between the bulk surface conductance (which they called G_s), which is essentially an aggregated measure of stomatal control, and the bulk canopy conductance (G_c), which reflects evaporation from the soil as well as from the canopy. They showed that maximum values of G_s (G_{smax}) significantly exceed G_{cmax} when L^* is less than about 3. For $L^* > 3$, $G_{smax} \approx G_{cmax}$, and G_{smax} is linearly related to maximum leaf stomatal conductance, g_{smax}, as obtained from measurements on leaves. The relationship between G_{smax} and g_{smax} was remarkably consistent, with a slope of 3, i.e., $G_{smax}/g_{smax} = 3$; therefore, in our notation, $g_{cmax} = 3\, g_{smax}$. Körner (1994) published a valuable distillation of data in the form of average values of g_{smax} for over 20 vegetation types and 200 species. Estimates of g_{smax} can be obtained from that paper. Kelliher et al. (1995) found the average value of g_{cmax} [our notation; Eq. (3.16)] to be about 18 mm sec^{-1} (0.018 m sec^{-1}) for woody vegetation. (The data they examined included coniferous forests, temperate deciduous forests, and tropical rainforests.)

The study by Kelliher et al. (1995) provides the basis for estimating maximum transpiration rates for any forest not short of water under high radiation conditions. Inserting the previous values of g_a and g_{cmax} into Eq. (3.16), with representative values of the other variables ($\varphi_n = 500$ W m^{-2}, $D_q = 6.10^{-3}$ kg kg^{-1}), gives $E = 1.3 \times 10^{-4}$ kg m^{-2} sec$^{-1} \approx 0.5$ mm hr^{-1}. If we assume sinusoidal variation in E through a 14-hr day, with 0.5 mm hr^{-1} as the maximum rate, total daily transpiration is 4.5 mm. Obviously, there will be some variation for different forests and environmental conditions.

Values of g_c are commonly estimated from stomatal conductance and L^* as $g_c = \Sigma g_s L^*$, which corresponds exactly with $g_{cmax} \approx 3\, g_{smax}$ at $L^* = 3$. However, Raupach and Finnigan (1988) demonstrated that the approximation based on L^* is not an accurate measure of g_c because—as we pointed out earlier—g_c includes an aerodynamic component and is therefore not a purely physiological parameter. The physiological conductance also does not reflect contributions to canopy evaporation from sources such as the soil and free (intercepted) water. Values of g_c for stands can be obtained from experimentally determined values of E and the solution of Eq. (3.26) for g_c, i.e.,

$$\frac{1}{g_c} = \frac{1}{\lambda E}\left(\frac{\varepsilon \varphi_n}{g_a} + \lambda \rho D\right) - \frac{\varepsilon + 1}{g_a}. \quad (3.21)$$

For example, assume that measurements over a period of weeks indicated that, for a particular stand, $E = 4$ mm day^{-1}. We will take $g_a = 0.2$ m sec^{-1}, average $\varphi_n = 18$ MJ m^{-2} day^{-1}, i.e., 18×10^6 J m^{-2} and average air temperature = 28°C, with relative humidity 65%, giving $D = 8.23 \times 10^{-3}$ kg kg^{-1}. Because 4 mm day^{-1} is equivalent to 4 kg m^{-2}, $\lambda E = 4 \times 2.44 \times 10^6$ J m^{-2} and $g_c = 0.028$ m sec^{-1} in this example.

It is clear from the foregoing discussion that good estimates of the maximum transpiration rates for virtually all forest types, when $L^* \geq 3$, can be obtained from the P–M equation with values of g_a of about 0.1–0.2 m sec^{-1} and $g_c \approx 0.020$ m sec^{-1}. The potential transpiration rate, i.e, the rate driven by atmospheric factors interacting with canopy surfaces but without the constraint of stomata, can be estimated by assuming that surfaces in the canopy are wet so that stomata exert no influence; i.e., $g_c \to \infty$. This reduces the P–M equation to

$$\lambda E = \frac{\varepsilon \varphi_n + \rho \lambda D_q g_a}{\varepsilon + 1}, \qquad (3.22)$$

which is a useful form for evaluating water balances on a long-term or wide-scale basis. Where $g_c < g_{cmax}$, for example, because of water shortage in the root zone or because the potential transpiration rate "demands" water faster than the soil–root–tree conducting system can supply it, then actual stand transpiration rates will be less than potential rates. These situations may occur on a daily (see Fig. 3.9) or longer-term basis. They are discussed in some detail in Chapter 4, Section I,D.

C. Transpiration from Stands: Bowen Ratios

Equations (3.12) and (3.13) provide us with the means of estimating the fluxes of sensible and latent heat, and Eq. (3.16) combines the energy and mass-transfer processes that drive evaporation. The ratio of these fluxes, widely used to estimate the rate of evaporation from surfaces for which φ_n is known, is called the Bowen ratio (β). Ignoring G, the flux of heat in and out of storage,

$$\beta = \frac{H}{\lambda E}, \qquad (3.23)$$

from which, substituting in Eq. (3.8),

$$\lambda E = \frac{\varphi_n}{1 + \beta}, \qquad (3.24)$$

i.e., given values for β and φ_n we can solve for λE. In principle, values for β are relatively easily obtained experimentally by using Eq. (3.23) in finite difference form ($\beta = \gamma \Delta T / \Delta e$, where γ is the psychrometric constant ≈ 0.066 kPa °C), and measuring temperature and vapor pressure gradients across some height difference (Δ). In practice, even when Δ is quite large, the gradients are often very small and difficult to measure with sufficient accuracy. Restrictions to this method are imposed by the requirement that both sets of measurements should be made high enough above the canopy for the assumptions about turbulent exchange to hold. It is also necessary that bouyancy does not cause significant differences in the eddy diffusivities, and that the measurements should be made within the layer of air above the stand that reflects the properties of the stand, i.e., within the boundary layer of the stand. Air flowing across the stand from elsewhere will carry the properties of the surfaces across which it has passed; therefore, measurements made in that airstream will not provide information about the processes taking place in the stand of interest. These requirements may be difficult to meet in forest research, except in the case of large plantations in flat areas, but β is a valuable and informative parameter, which has been widely used and measured in research on forest micrometeorolgy and water use. Jarvis *et al.* (1976) review many of the early (1960s–early 1970s) Bowen ratio measurements over forests.

The Bowen ratio can also be derived from Eq. (3.16), for which we use the procedure described by Thom (1975) (see also Kaimal and Finnigan, 1994). Rewriting the equation with resistances ($r = 1/\text{conductance}$), for algebraic convenience, gives

$$\lambda E = \frac{\varepsilon \varphi_n + \lambda \rho D/r_a}{\varepsilon + 1 + r_c/r_a}. \tag{3.25}$$

Introducing a "quasi resistance"—sometimes called a climatological resistance—

$$r_i = \frac{\lambda \rho D}{\varphi_n}, \tag{3.26}$$

and noting (as observed earlier) that the resistances to heat and water vapor transfer can be taken as approximately equal, then from Eqs. (3.16) and (3.23),

$$\frac{\lambda E}{\varphi_n} = \frac{1}{1+\beta} = \frac{\varepsilon + r_i/r_a}{\varepsilon + 1 + r_c/r_a}, \tag{3.27}$$

hence,

$$\beta = \frac{\varepsilon + 1 + r_c/r_a}{\varepsilon + r_i/r_a} - 1. \tag{3.28}$$

This provides a valuable diagnostic tool from which, given r_c, β can be estimated or the implications of values of β, in terms of r_c, can be evaluated. Alternatively, the effects of changes in r_c on β can be assessed.

There have been enough studies of the energy balance, and energy partitioning over forests, to provide a good indication of the values of β

Table 3.1 Observed Values of the Bowen Ratio ($\beta = H/\lambda E$) for a Range of Forest Ecosystems

Forest ecosystem and location	β	Notes
Tropical		
Broad-leaved evergreen, Thailand (1)[a]	6.38	Dry season
Broad-leaved evergreen, Thailand (1)	0.45	Wet season
Broad-leaved evergreen, Amazon (2)	0.43	Integrated daily values (6)
Temperate		
Loblolly pine, Southern United States (3)	0.2–2	Daily values over 4 days, spring
Scots pine, England (4)	1.38–3.06	Midday values, fine days
Deciduous mixed, Tennessee, United States (5)	0.6	Representative value, summer
Deciduous mixed, Tennessee, United States (6)	0.47	Representative value, summer
Boreal		
Jack pine, Canada (6)	6	Representative value, summer

[a]Sources: 1, Pinker, *et al.* (1980a); 2, Shuttleworth *et al.* (1984); 3, Murphy *et al.* (1981); 4, Stewart and Thom (1973); 5, Verma *et al.* (1986); 6, Baldocchi and Vogel (1996).

that can be expected under various conditions. A selection is provided in Table 3.1, and Fig. 3.10 indicates the way r_c varies.

The values of β in Table 3.1 illustrate diurnal variation as well as variation caused by conditions and species. The data analyzed by Stewart and

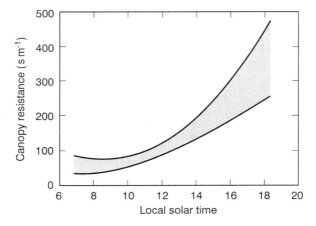

Figure 3.10 Typical diurnal patterns of forest canopy resistance (r_c) based on data of (interalia) Stewart and Thom (1973), Lindroth (1985), and Verma *et al.* (1986).

Thom (1973) showed considerable diurnal variation. The patterns were not very consistent but the general trend was for β to increase through the day and then fall toward evening; on one occasion, it reached a value of 10, indicating that more than 90% of available energy was going to sensible heat. The values from Thailand can presumably be taken as representative of tropical monsoon forests in the dry and wet season. Pinker et al. (1980b) did not say whether canopies were ever wet when they made their measurements, but it is probably safe to assume they were not. Their wet season value is consistent with that of Shuttleworth et al. (1984) for the Amazon forest. The values given by Verma et al. (1986) showed a tendency opposite to that of the Scots pine data (Stewart and Thom, 1973): β tended to fall from relatively high values early in the day to low values in the afternoon and evening. However, their "high" values were only about 0.8–1.0.

Perhaps the most interesting contrast is provided by the data of Baldocchi and Vogel (1996). They found that only about 23% of available energy (φ_n) was converted to latent heat (transpiration) by the boreal jack pine forest, compared to 66% by the broad-leaved deciduous forest in Tennessee. Some of the reason for this difference lies in the fact that the jack pine had a leaf area index of only about 0.7, compared to about 5 for the broad-leaved deciduous forest. Increasing vapor pressure deficits caused increasing transpiration rates in the mixed broad-leaved forest, where stomata are apparently not sensitive to humidity, but caused decreasing transpiration rates in the jack pine, where higher D presumably caused stomatal closure. These results are consistent with the theory put forward by Jarvis and McNaughton (1986), who explored the coupling between vegetation and the environment using the P–M equation as a basis. They developed a coupling factor (Ω), which gives a measure of the extent to which transpiration is coupled to (driven by) available energy or decoupled from it. The temperate broad-leaved deciduous forest is quite strongly coupled to φ_n, whereas the jack pine is more sensitive to variations in ambient humidity; hence, it is less strongly coupled to φ_n.

The implications of canopy resistance (r_c) values can be explored using the previous information about β and Eq. (3.28). However, it is useful to consider some data so diurnal results presented by Stewart and Thom (1973), Lindroth (1985, a Scots pine forest in Sweden), Shuttleworth et al. (1984), and Verma et al. (1986) are summarized in Fig. 3.10. There is remarkable consistency between the values obtained for different forest types—the values used to construct Fig. 3.10 were not greatly different from one another and were not normalized—and in the tendency for r_c to reach a minimum in the late morning and then rise steadily through the day. The data in Fig. 3.10 are consistent with the findings of Kelliher et al. (1993), who found maximum surface conductance (g_c) values for forests to be generally in the range 20–30 mm sec^{-1}, equivalent to the minimum resistances of about 50–80 m sec^{-1} (20–12.5 mm sec^{-1}) in Fig. 3.10. The afternoon rise in r_c is, presumably, generally attributable to increasing vapor pressure deficits (see the discussions in Chapters 4 and 5).

D. CO_2 Transfer

This section is concerned mainly with heat and water vapor transfer, but we should note that the same processes are responsible for the transfer of CO_2 to and from forest canopies.

Equations similar to Eqs. (3.12) and (3.13) can be written to describe the movement of CO_2 from the air to the canopy during the day, in response to its uptake by leaves for photosynthesis, and from the canopy to the air at night, when there is no photosynthesis but respiration continues (see Chapter 5) so that CO_2 concentrations in canopies rise. Daytime exchange is given by

$$A_c = (c_{a \cdot o} - c_i) g_{s \cdot c} = (c_a(z) - c_{a \cdot 0}) g_a, \qquad (3.29)$$

where A_c denotes canopy assimilation rate, $c_{a \cdot o}$ is the concentration at the reference level in the canopy, and $c_a(z)$ is the concentration at height z above the canopy. Stomatal conductance is written here as g_{sc} to indicate that it is not numerically the same for CO_2 and water vapor [see Chapter 5, Eq. (5.7)]. Nevertheless, stomatal conductance controls the flux of CO_2 into leaves, and water vapor out of them, providing the link between photosynthesis and transpiration. The assumption of similarity holds for turbulent transfer in the atmosphere so that the value of g_a would be (approximately) the same for heat, water vapor, and momentum. During the night, the right-hand side of the equation holds, but the gradients will be reversed: CO_2 is transferred away from the canopy, although wind speeds are commonly low at night, and turbulence heavily damped; therefore, CO_2 concentrations in canopies may rise steeply (see Grace et al., 1995).

Equation (3.29) underlies CO_2 flux measurements, but in itself it is of limited value for explaining physiological processes. Photosynthesis rates may be constrained by the "supply function" (see Chapter 5, Section I), but the process is essentially driven by visible radiation, not by the transfer of CO_2 through the air. Respiration (both autotrophic and heterotrophic) is also independent of transfer processes and radiation. The linkage between photosynthesis and radiation (photon flux) is described in detail in Chapter 5.

IV. Effects of Topography on Microclimate

Underlying most of the discussion so far has been the assumption that the land areas we are considering are, at least approximately, level and uniform. In practice, many forests and forested areas are in uneven terrain and are not uniform in their characteristics. In these situations, rigorous theoretical treatments may not be very useful, although they always provide a valuable guide.

Considering radiation, it is always possible to calculate the radiant flux on a surface of known slope and azimuth from knowledge of φ_s, the sun's position—which depends on latitude, on time of day, and date—and the ratio of direct to diffuse radiation. The equations can be easily programmed into a computer and the diurnal course of incident energy calculated. However, for forestry applications such as estimation of yield (see Chapter 9) using a simple radiation interception model (Eq. 3.9), or for the estimation of net radiation—in both cases over periods that may range from days to a season—there may be no need to run detailed equations on a day by day basis. It will usually suffice to calculate correction factors based on the best available information about the slope(s) of interest. Unless the slopes are very steep, there is no need to correct for diffuse radiation, which tends to come preferentially from high in the sky (near the sun's zenith angle; Monteith and Unsworth, 1990). For direct radiation, a correction for any given month (for example) would be based on the ratio (m_φ) of radiation received by the slope to that received on a horizontal surface, with the calculations being run for a single day in the middle of the period. The radiation received by the slope is then

$$\varphi_{s \cdot s} = \varphi_{s \cdot \text{direct}} m_\varphi + \varphi_{\text{diffuse}}. \quad (3.30)$$

Relations between φ_s and φ_n (Eq. 3.7) can be assumed to hold.

In the southern hemisphere, north-facing slopes will receive more radiation than horizontal surfaces or south-facing slopes; obviously, the converse applies in the northern hemisphere. In the case of complex land units (hills and valleys, catchments, etc.), the calculations could be done for all slope segments if a digital terrain map was available. This would involve decisions about criteria for homogeneous slopes ("what constitutes a uniform slope?") and is another example of a scaling problem.

The effects of slope on radiation balance is very important for forest ecology. Tajchman (1984) carried out an exercise of the type outlined previously for a valley in the Appalachian region in the United States. He found net radiation varied between 1.66 and 3.21 GJ m^{-2} year^{-1}, the differences being almost entirely attributable to slope azimuth. This caused changes in a water balance (aridity) index, which ranged from 0.47 to 0.88. This approach merits further investigation: There are computer packages available that, given (digitized) contour maps or digital terrain maps, plus the information necessary to calculate sun angles, can produce maps of relative slope irradiance. There is therefore little excuse for assuming that all slopes receive the same radiation load.

IV. Effects of Topography on Microclimate

The wind profile equations presented earlier provide us with understanding about transfer processes from natural surfaces, but they cannot be used to calculate exchange coefficients and scalar fluxes from uneven terrain. There are two sets of "conditions of unevenness" that we need to consider. One is flow over changing roughness, e.g., from a forest to a grassland, or vice versa, and flow in gaps; the other is flow over hills. We will not deal with these questions in detail—Kaimal and Finnigan (1994) provide an excellent, up-to-date theoretical treatment—but will outline a few of the main points relating to them.

Considering the simplest case of change in surface roughness, when air flows across one surface type and on to another there will be an abrupt change in the shear stress as the wind profile adjusts to the new underlying surface. The boundary layer of the new surface—the layer in which the profile represents the properties of that surface—adjusts over some distance that depends on the roughness length characteristic of the new surface. The problems of boundary layer adjustment are of concern in dealing with problems of advection and, of course, with wind profile measurement.

Of more concern to those concerned with forestry is the question of flow over hills. Clearly, in any topographically complicated situation wind flow will be extremely chaotic and completely unpredictable. All we know with any certainty is encapsulated in Fig. 3.11, which shows the following essential features:

- the flow decelerates slightly near the base of the hill before accelerating to the hilltop.
- the wind reaches its maximum velocity at the hilltop then decelerates behind the hill. If the hill is steep enough downwind, a separation bubble forms and a wake region develops behind the hill with a marked

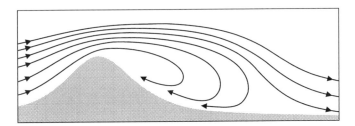

Figure 3.11 Wind flow over a hill. Note that the streamlines are compressed across the top of the hill, indicating acceleration, and there is separation of streamlines and reversed flow in the lee of the hill. The extent to which this occurs depends on hill shape, surface cover, and wind direction, but it indicates why there may be regions of high turbulence in the lee of barriers (diagram redrawn from Kaimal and Finnigan, 1994, Fig. 5.3).

velocity deficit extending for many hill heights downwind (Kaimal and Finnigan, 1994).

The principles can be applied, in a very general sense, to get some idea of how prevailing winds will affect forests on hillsides.

Air flow in gaps or clearings may be of concern in forestry or forest ecology. The problem can be defined in terms of the size of the gap and the height and density of the forest surrounding it. Gusts, or eddies, penetrating the stand upwind of a gap will be slowed by momentum absorption by the vegetation, whereas air flowing over the top of the canopy will be forced down into the gap or clearing, decelerating slightly downwind of the leading edge. The result will be a region of slow, probably confused flow immediately downwind of the edge of the stand. Flow above the floor of the clearing will be accelerated by the downward transfer of momentum from faster-moving higher air layers; when the flow strikes the upwind edge of the gap, it is forced up and through the canopy elements, as well as being lifted over the edge. There will be some back circulation near the ground, and a pool or eddy of slow moving air will tend to form there. Above the canopy, on the downwind side, there will be a very turbulent zone where the slower air flow coming up through the canopy mingles with the faster flow being accelerated over the edge (see Fig. 3.12).

The previous description will only be (approximately) correct if the gap or clearing is sigificantly larger than the canopy height; otherwise, the gust size that can penetrate will not reach ground level.

The last matter we will note in relation to topography and microclimate is cold air drainage. This occurs at night in cool or cold regions, where heat loss by radiation from hill or mountain tops results in cold, viscous air that drains into the valleys below and collects there—as demonstrated by mist and fog on winter mornings. Depending on the

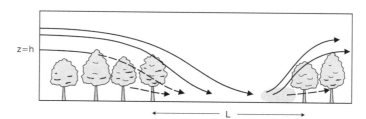

Figure 3.12 Air flow in a gap. The patterns of flow in any particular situation will depend on the size of gap, density of the surrounding forest, and wind speeds. The region of highest turbulence will tend to be the shaded area on the downwind side of the gap, particularly if the forest is dense and the linear dimension (l) of the gap is significantly larger than the forest height (h).

topography of the area, cold air collects in pools and may persist for some time, particularly in valleys sheltered from the sun until late in the day. This may result in temperature conditions in such valleys being less favorable for the growth of some plants than at higher altitudes. Under extreme conditions, cold air drainage can cause a reversal in forest vegetation zones, with more cold-tolerant species occuring at lower elevations, typically at the bottom of slopes where cold air accumulates. Cold air drainage can be disrupted by large fans that mix the air—a management technique used in some areas for high-value crops, such as fruit orchards, but not in forestry.

V. Concluding Remarks

Microclimatology provides the basic discipline and theoretical underpinning for the calculation of evapotranspiration and radiation interception by canopies. Virtually all the process-based models of forest productivity (Chapter 9) are based on radiation interception by canopies and the conversion of radiant energy into carbohydrates. Because forest growth is as much dependent on the availability of water as on temperature and soil properties, a clear understanding of the factors affecting stand, forest, and regional water balances is essential.

In more immediate terms, the consequences of clear-cutting forests, in terms of the changed conditions for seedling regeneration—as opposed to conditions within canopies—can be evaluated in terms of microclimatological processes, particularly the surface energy and water balances. The principles outlined in this chapter also provide the basis for evaluating the consequences of practices, such as thinning or clear-felling, or the destruction of an area of forest by fire in terms of energy balance and transfer processes. Clearly, the energy loads on, and energy balance of, bare or sparsely vegetated soils are vastly different than those on soils under canopies, which has massive implications for water balance and for seedling regeneration. The implications are ecological as well as managerial.

On a global scale, we are concerned with land–atmosphere interactions and the role of forests in the global carbon balance. The probable trajectory and consequences of global climate change are being explored using global circulation models, but one of the problems with such models is the poor coupling with land surfaces. Forests comprise one of the major land surface types, and any knowledge that we can acquire about forest–atmosphere interactions will be of value in improving this coupling and hence, the performance of those models. For example, the unexpectedly large heat fluxes observed from the boreal jack pine forest

(Baldocchi and Vogel, 1996) appear to cause significant differences in the development of the planetary boundary layer over those forests relative to the temperate broad-leaved forest. Baldocchi and Vogel comment that the effects of temperature, humidity, and surface wetness may need to be considered when scaling gas exchange from stands to regions.

Recommended Reading

Jarvis, P. G., James, G. B., and Landsberg, J. J. (1976). Coniferous forest. In "Vegetation and the Atmosphere" (J. L. Monteith, Ed.), Vol. 2, pp. 171–264. Academic Press, New York.

Jarvis, P. G., Monteith, J. L., Shuttleworth, W. J., and Unsworth, M. H. (Eds.) (1989). "Forests, Weather and Climate." The Royal Society, London.

Jones, H. G. (1992). "Plants and Microclimate," 2nd. ed. Cambridge Univ. Press, Cambridge, UK.

Kelliher, F. M., Leuning, R., and Schulze, E.-D. (1993). Evaporation and canopy characteristics of coniferous forests and grasslands. Oecologia **95,** 153–163.

Landsberg, J. J. (1986). "Physiological Ecology of Forest Production." Academic Press, London.

Monteith, J. L., and Unsworth, M. H. (1990). "Principles of Environmental Physics," 2nd ed. Edward Arnold, London.

Oker-Blom, P., and Kellomäki, S. (1983). Effect of grouping of foliage on the within-stand and within-crown light regime: comparison of random and grouping canopy models. Agric. Meteorol. **28,** 143–155.

Waring, R. H. (1983). Estimating forest growth and efficiency in relation to canopy leaf area. Adv. Ecol. Res. **13,** 325–354.

4

Forest Hydrology and Tree–Water Relations

Forest hydrology and tree–water relations are intimately and inextricably linked. Forest hydrology is concerned with the movement of water through forested landscapes. The water balance of stands depends on precipitation, interception, runoff, and evaporation; with the exception of precipitation all these processes are strongly influenced by tree populations, stand structure, and canopy architecture. Over a period of time, stand water balance influences the growth of trees and hence stand structure and architecture.

The principle underlying all hydrological processes is mass balance. The objective of hydrological studies or calculations must be to account for all the water coming into a system, whether it comes as rainfall, snow, overland flow in rivers, runoff from outside the system, or as belowground flow. Water will leave the system as evaporation from open water, soil, or from other wet surfaces after rain as transpiration by plants, as overland flow, or as drainage out of the system. In the case of a forest stand, drainage out of the system means drainage out of the root zone of the trees. This includes lateral subsurface flow through the soil into streams. Figure 4.1 provides a schematic representation of the main facets of stand water balance.

The question of scale is of primary importance in hydrology. We may consider any scale from the water status of plant tissue to the global water balance. At each scale there is variability. Forests are undoubtedly important in the global hydrological cycle, but their significance at that scale is difficult to estimate because of the overriding effects of the oceans. Their significance at regional levels has been illustrated recently by studies using global circulation models (GCMs) such as those carried out by Dickinson and Henderson-Sellers (1988) and Shukla et al. (1990). Both these groups simulated the consequences of converting the Amazon basin from tropical forest to degraded grassland. Dickinson and

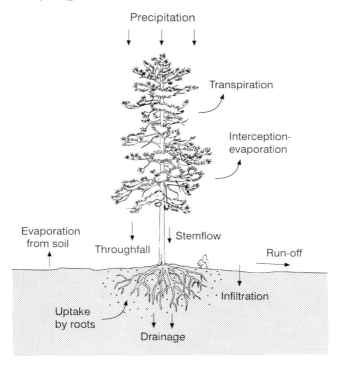

Figure 4.1 Diagrammatic representation of the main components of the hydrologic balance [see Eq. (4.1)]. The problem in forest hydrology is to quantify the various terms for different forest ecosystems and soil types. The size of the various terms, in relation to precipitation, will vary with stand density and canopy architecture. For a catchment, the runoff and drainage terms determine the catchment water yield.

Henderson-Sellers used the changes that would occur in albedo and surface roughness, canopy interception, and storage (see Section I,B) in a simulation of a 13-month period. They predicted higher air and soil temperatures, reduced evaporation (up to 50%) and precipitation, as well as lengthening the dry season after conversion. Shukla *et al.* coupled a GCM with a model of biosphere function and response and carried out the same experiment. They obtained similar results. Numerical experiments of this sort, using models, are the only way to forecast the significance of such changes in land cover. Although they may still involve many inaccuracies and arguable assumptions in the way they are formulated, these models represent a considerable step forward in our capacity to evaluate the consequences of deforestation, particularly in terms of the feedbacks to the atmosphere.

At smaller scales, the effects of the forest on the atmosphere are less important, although we have to recognize that there are always feedbacks[1] because the heat and water vapor fluxes from vegetation influence the local planetary boundary layer (see McNaughton and Jarvis, 1983, and the comment in Chapter 3 about the influence of the boreal jack pine forests, with their high heat flux) and hence the coupling between vegetation and the atmosphere. However, for most purposes, forest ecologists and managers will (reasonably) assume that local climates are imposed on, and independent of, the forests in an area, and that the influences of concern are those of the weather acting on the vegetation system.

From the point of view of the forest manager, the most important issues associated with forest hydrology are the questions of thinning and clear-cutting and the effects of the water balance of forests on the growth of trees. Forested catchments are usually associated with high water quality and stable water outflows and catchment deforestation is widely regarded as an environmentally damaging practice, likely to lead to soil erosion and lower water quality. These problems may indeed occur, but they are not axiomatic (Binkley and Brown, 1993). Tree removal will, almost invariably, lead to increased water yield; whether there are problems associated with that depends, to a great extent, on the topography of the catchment, the rainfall of the area, the type of forest removed, the way it was removed, and the subsequent management of the catchment. Bruijnzeel (1990) gives a good overview and discussion of all these points in relation to tropical forests. Here, we review briefly some of the more important findings in the literature and, to provide a basis for the understanding and interpretation of empirical results, deal with the principles and processes underlying the hydrology of forested catchments.

Under Section III of this chapter, we discuss tree–water relations and their effects on growth. It is intended to provide a cursory coverage of a vast field of research. The problem is that not much of that research has produced results of direct value to the forest scientist who needs to understand the influence of water on growth and productivity or ecosystem function. The treatment presented here provides an outline of the basic biophysical processes underlying plant–water relations, and of the rigorous approach possible when dealing with problems in the field. There is

---

[1] Feedback occurs when the effects of an input to a system (the receiving system, in this case, is the forest) result in alterations to the level of the input. Feedback may be positive, when the level of the input is enhanced, or negative, when the level is reduced; therefore, the system is damped. For example, evaporation into the planetary boundary layer, driven by the vapor pressure deficit of the air mass, tends to cause negative feedback: The water vapor content of the air mass is increased so that evaporation rate tends to be reduced.

also emphasis on the relationships between measures of plant water status and tree growth; models that describe these relationships and that can be used to calculate the effects of given events or sets of conditions are presented.

The next section presents the basic mass balance equation that describes all fluxes in forest water cycles. The various terms of the equation are first discussed in a general sense, then dealt with in more detail in subsequent sections of the chapter. Soil–water relations are important as background to all aspects of forest hydrology, although there is an emerging recognition among hydrologists that detailed, theoretically rigorous, physically based models, using equations known to work in well-defined, spatially homogeneous situations, are often of limited value in catchment hydrology (Beven, 1989). Therefore, although it is essential to understand the basic physical processes involved in soil–water movement, and to be able to describe them in rigorous, quantitative terms, the main benefit of this understanding may be to serve as a guide to the development of relatively simple, but physically soundly based, models of catchment hydrology.

I. Hydrologic Balance

The hydrologic balance is given by the equation

$$P - I - q_R - q_D - \Delta\theta - E = 0, \tag{4.1}$$

where P is precipitation (rain, snow, or fog), I is intercepted water that is evaporated from the canopy, q_R is surface runoff, q_D is drainage out of the root zone, $\Delta\theta$ is the change (Δ) in soil moisture content (θ), and E is evaporation. The evaporation term includes both transpiration and evaporation from the soil. These are not considered separately here.

Equation (4.1) holds for a single precipitation event, or series of events, although the time scales across which the different processes operate vary considerably. The shortest time scale usually applies to water intercepted by a forest canopy, which may evaporate from that canopy over periods of minutes to hours, depending on conditions and the structure of the canopy. Surface runoff may continue for hours or days. The change in soil moisture content depends on the intial moisture content and the rate of infiltration relative to the rate of drainage out of the root zone, which in turn depends on soil hydraulic properties, the depth of the soil, and root penetration. Clearly, when there is no precipitation and the canopy is dry so that $I = 0$, if q_R and q_D are zero, then $E = \Delta\theta$. Differences in the water balances of different vegetation types, e.g., grassland and forest, can be analyzed and understood in terms of Eq. (4.1), with appropriate formulation of, and parameter values for, the various terms. We deal with the terms in Eq. (4.1) in some detail in the following sections.

A. Precipitation

Precipitation in any region is usually described in terms of the mean annual amount received—millimeters of rain or meters of snow. However, precipitation is enormously variable, both spatially and temporally. The amounts received in any given month in a particular area will vary considerably, even in regions considered to have relatively stable climates, whereas in areas where the precipitation patterns are erratic—particularly low rainfall areas—the precipitation received in any particular month may vary by orders of magnitude. The seasonal distribution of precipitation can influence the composition, structure, and function of terrestrial ecosystems. For example, the lack of precipitation during the growing season in the Pacific Northwest in the United States favors evergreen conifers compared to deciduous trees (Waring and Franklin, 1979). The pronounced monsoonal climate in tropical regions (see Fig. 2.1h) favors the deciduous leaf habit that enables trees to avoid desiccation, and the equitable distribution of precipitation and generally moderate temperatures, typical of most temperate regions in the southern hemisphere, favor the evergreen habit (Axelrod, 1966).

Spatially, amounts of precipitation in particular areas within a region are affected by the patterns of storms—which may be frontal or convective—and weather systems and by topography interacting with these factors. Mountains exert dominant effects on regional rainfall by interacting with cyclonic or convective processes, forcing air upwards and causing orographic rain, and by causing rain shadows. The great north–south mountain ranges of the world (Andes, Rockies, Urals, Southern Alps) provide dramatic examples.

It is not only the amount of precipitation that affects hydrologic relationships, it is also the intensity with which the precipitation occurs. In the case of snow, there may be accumulation over winter, when the only term in Eq. (4.1) that is not effectively zero is I, because small amounts of snow are lost by evaporation (sublimation) from canopies during the winter. Large amounts may accumulate on the ground—with some held in the canopy—until temperatures rise in spring and the snow melts, resulting in a large input of water to the system over a relatively short time. During snowmelt, the rate of input (P) to the system is the rate at which the snow is converted to water. Consequently, the soil is likely to be saturated, and runoff and drainage will be high until the snow has dissipated. In some dry forests, snow provides a significant proportion of the water available to the trees during the growing season and annual productivity in these forests is positively correlated to winter snowfall (Hunt et al., 1991; Gower et al., 1992).

In the case of rainfall, there is a considerable difference, in terms of effects on forests, between a high-intensity storm, which delivers a large amount of water quickly, and the same amount of water delivered by light rain over a much longer period of time. If precipitation occurs as a series of intermittent light showers, with time between them to evaporate the water intercepted by and held on the foliage, interception losses are likely to be high relative to those from the same amount of precipitation occurring as a single heavy shower (Calder, 1990). There will also be differences in the patterns of infiltration into the soil of the rainfall that reaches the ground.

The effectiveness of rainfall in relation to the water status of forests also depends on the amount of water held in the root zone at the beginning of each precipitation event. If the soil is saturated, then all rainfall reaching the soil will be lost from the system either as runoff or drainage. Rain after a dry period is more likely to be stored in the soil and will, therefore, be more useful to the trees than the same amount of rainfall received when the soil is wet. Therefore, the probability of particular patterns of precipitation is a factor that must be taken into account when analyzing the hydrology of an area.

The rainfall at any site may be analyzed in terms of the variance about monthly mean values and the probability of the amount likely to be received in any month deviating significantly from those values. We can also assess the probability, at any season, of storms of specified size (precipitation amount) and intensity and calculate the cumulative probabilities of periods of wet or dry weather. These statistical characteristics should be recognized and may need to be taken into account in both evaluating research results and the potential productivity of forests in any particular place. We should also note that extreme events—floods and, particularly in the case of forests (or any other ecosystems), droughts—may have much greater impact on the long-term potential and performance of an ecosystem than annual variations that do not deviate too far from the mean.

B. Interception

Rain falling on a forest canopy may pass through gaps in the canopy or be intercepted by, and drip from, the foliage (both called throughfall). It may also be intercepted by branches and foliage, to be retained until lost by evaporation, or it may be channeled down the branches and trunks as stemflow (Fig. 4.1). Rainfall intercepted by and evaporated from the canopy contributes nothing to the soil moisture. Litter layers also intercept rainfall; the amount of water that can be lost to the soil in this way will obviously depend on the mass and physical characteristics of that layer. The principles of calculation are the same as those for the canopy.

There are a great many reports of interception loss studies based on direct measurements of rainfall, throughfall, and stemflow. These, not surprisingly, give a wide range of values for interception losses. For example, Feller (1981) gives interception loss values of between 10 and 20% of total rainfall for several *Eucalyptus* stands, and 21–30% for *Picea radiata;* Ford and Deans (1978) give 30% for a Sitka spruce stand, Dunin *et al.* (1985) found that interception by *E. maculata* was 13% of rainfall, and Waterloo (1994) obtained values between 13 and 19% for *P. caribaea.* The variation is not surprising because interception depends on stand structure, canopy architecture, and L^*—more dense stands will intercept and store larger amounts of water, but very high foliage densities will reduce the effectiveness of turbulent exchange between the wet foliage and the air. The influence of stand density was quantified by Myers and Talsma (1992), who found that net interception by *P. radiata* was strongly related to canopy mass. It ranged from 16% of total annual rainfall in sparse canopies to 24% in dense canopies. They also found that average interception decreased from 55% of P during small rain events (less than 5 mm) to less than 10% during events >40 mm.

To explain the observed variation and provide a basis for prediction, interception losses from forest canopies over short periods were modeled by Rutter (1975). Gash (1979) derived an analytical solution to Rutter's model that enabled it to be used over longer time periods. This version [Eq. (4.2)] has been widely adopted and applied in many studies. The equation is

$$E_{\text{int}} = [P_g'(1 - p_t - p_s) - S] + [E/R(P_g - P_g')] + S, \quad (4.2)$$

where P_g' is the amount of rain needed to saturate the canopy and P_g the total amount during the event, p_t and p_s are the fractions of rain reaching the forest floor as throughfall and stemflow, R is the mean rainfall rate during an event, and E is the rate of evaporation during rainfall events. The evaporation rate from a wet canopy, either during or after rain, can be estimated using the Penman–Monteith equation [(Eq. 3.16) with g_c set to infinity (see, for example, Stewart, 1977). The two terms in square brackets relate to the wetting up and saturation periods, and the last term (S) is the loss of water stored on the canopy after a rain event.

The logic of the model is straightforward: When rain commences the canopy begins to "wet up," although some drops will pass through unintercepted. Stemflow and drip increase as the canopy nears saturation, and there is continual evaporation of the free water on the foliage, although obviously evaporation rates are not high during rain. Therefore, to calculate accurately the interception losses from a forest requires knowledge of the canopy storage capacity (S, mm) and estimates of the mean rainfall rate (R, mm hr^{-1}), of the evaporation rate (E, mm hr^{-1}), and of throughfall and stemflow. Given these parameter values from measurements above and below the canopy, the interception loss (E_{int}) from any rain event is the sum of the losses during the period of wetting up, plus losses during the period when the canopy is saturated—if it reaches this stage—and the period of drying.

As a general value, it appears that canopies hold about 1 mm or less of water (i.e., $S \approx 1$ mm). Shuttleworth (1988) tabulated values of S from a number of studies; they range from 0.3 mm for oak in winter to 2.5 for Sitka spruce. Some specific values are the following: Amazonian rainforest, $S = 0.74$ mm (Shuttleworth, 1988); *P. caribaea*, $S = 0.8$ mm (Waterloo, 1994); and *E. maculata*, $S = 0.3$ mm (Dunin et al., 1985). Calder (1986) proposed a different formulation for the interception model based on the stochastic manner in which the individual elements of the surface of a tree are struck and wetted by individual raindrops. Calder *et al.* (1986) applied this model to a tropical rainforest and used statistical techniques to optimize the parameter values. They arrived at a canopy storage capacity of 4.5 mm—far higher than the values reported for other stands using simpler models.

Rates of water loss from canopies are strongly dependent on the effectiveness of turbulent exchange. This was studied by Teklehaimanot *et al.* (1991), who measured interception in Sitka spruce stands with trees spaced at intervals of 2, 4, 6, and 8 m. Average interception losses over a 17-week period were 29, 23, 13.8, and 8.9% for those spacings, respectively. Figure 4.2 shows the relationship between boundary layer conductance and number of trees per hectare. Teklehaimonot *et al.* noted that, as expected, there was more throughfall at the wider spacings, but they also found that the boundary layer conductance increased linearly with spacing so that interception losses per tree were greater at the wider spacings because of the more effective turbulent exchange between the wide-spaced trees and the air. The empirical relationships established by Teklehaimonot *et al.* for Sitka spruce will not hold for other species and configurations, but the principles almost certainly will.

The principles underlying canopy interception apply equally to ground vegetation and litter, but these layers can be enormously variable even within a single forest type and there are relatively few data available on interception losses from them. Doley (1981) cites two studies on tropical forest understories that indicated that litter could intercept about 1–1.5 mm water per kilogram dry litter. Waterloo (1994) measured interception losses under *P. caribaea* in Fiji to be between 9.5 and 11.5% of total rainfall and Kelliher *et al.* (1992) reported that 11% of precipitation was intercepted by litter compared to 19% by the canopy of a broadleaved evergreen forest.

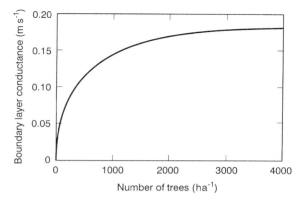

Figure 4.2 The relationship between boundary layer conductance per tree (g_a) and number of trees per hectare, established by Teklehaimanot et al. (1984). The rate of evaporation of intercepted water is strongly dependent on the value of g_a. Reprinted from *J. Hydrol.*, 123, Teklehaimanot et al., Rainfall interception and boundary layer conductance in relation to tree spacing. 261–278 (1984) with kind permission of Elevier Science-NL Sara Burgerhartstraat 25, 1055 KV Amsterdam, The Netherlands.

C. Soil Hydraulic Characteristics

Soil is a heterogeneous, porous medium that holds water at a range of water contents. The energy with which the soil holds water is measured in terms of water potential (ψ, MPa), which has units of pressure, or energy per unit volume, equivalent to a force per unit area. It is soundly based in thermodynamic theory [see Slatyer, (1967) and Passioura, (1982) for a discussion of an area where confusion has arisen, and Tyree and Jarvis (1982) for a detailed discussion in relation to plants]. Because water potential gradients indicate different energy states, the concept is fundamental to analysis of water movement through the soil plant system. The subscript 's' is used here to distinguish soil water potential (ψ_s) from foliage water potential (ψ_f).

Soil water potential (ψ_s) is the sum of gravitational, pressure, or matric and osmotic potentials. The units are negative: They represent the amount of work that must be done to move a unit quantity of pure, free water at the same temperature and elevation to the point in the soil in question. In the case of soils, the matric potential is a very important component, being a function of particle size and arrangement. As soil dries, the energy required to extract water from it increases rapidly, and water movement becomes slow and restricted. Soil wetness is usually expressed in mass or volume terms: Water content (mass water per unit mass of soil) is related to volumetric water content (θ, volume water per unit volume soil) by the soil bulk density (ρ_s) and the density of water (ρ_w). θ can be obtained from gravimetric measurements (mass water, m_w, per unit mass soil, m_s) determined by weighing and oven-drying samples:

$$\theta = \frac{(m_w/m_s)\rho_s}{\rho_w}. \qquad (4.3)$$

Note that θ, multiplied by the appropriate soil depth—for example, the depth of the effective rooting zone of trees, z_r—gives the depth equivalent of the water held in the soil. Soil porosity is the ratio of the volume of pore space per unit volume of soil. Porosity values range from about 0.3 in sands to 0.5 in clay soils. Saturated soil water content tends to be (roughly) proportional to soil porosity; however, available water content tends to reach maximum values on soils of intermediate porosity or texture (see below).

Soils are spatially highly variable in terms of the fractions of coarse and fine sand, clay, silt, and organic matter. Their properties generally change with depth and the depths at which significant changes occur will almost always vary from point to point in the landscape (see Fig.4.4). The equations given in this section, describing soil moisture characteristics, although firmly based in high-quality research and measurement, will always provide only approximate values for the way water is held by and moves through soil in the field. The equations are generally derived from studies on sieved, homogeneous soil samples. In the field, matters are complicated by cracks, channels, holes, plant roots, and peds (blocks of soil). Nevertheless, it is important that the physical relationships be described and understood.

Soil water content and potential are related by the soil moisture retention curve (see Fig. 4.3), which can be described over much of the range by the empirical relation (Gardner et al., 1970):

$$\psi_s = a_s \theta^{-n}. \qquad (4.4)$$

Equation (4.4) has been thoroughly tested by Clapp and Hornberger (1978) and by Williams et al. (1983). Figure 4.3, drawn from Williams et al., illustrates the range of water-holding properties typical of soils. The soil described by the curve in Fig. 4.3a (Soil 1) contained 21% clay, 21% silt, 16% coarse sand, and 42% fine sand. This would be a free-draining soil; because of the large amount of coarse sand the upper (saturated) limit of θ (θ_{sat}) would be about 0.35–0.4 m^3 m^{-3}, with most of the water held at relatively high potentials. By the time the water content has fallen to 0.2 m^3 m^{-3}, ψ_s is falling rapidly. The lower limit of water-holding capacity for that soil would be about $\theta = 0.10$–0.12 m^3 m^{-3}. Soil 2 (Fig. 4.3b) contained 28% clay, 38% silt, 4% coarse sand, and 30% fine

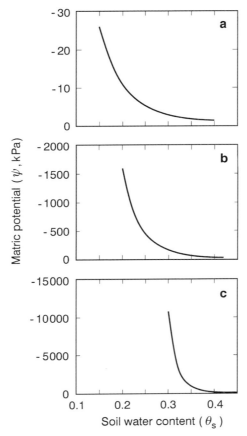

Figure 4.3. Soil water characteristic curves: matric potential as a function of volumetric water content. The curves were calculated from equations given by Williams et al. (1983); they are soils 1, 5, and 8 in their Table 5. The proportions of clay, coarse sand, and fine sand were: (1) 21, 16, and 42%; (5) 28, 4, and 30%; (8) 55, 4, and 17%.

sand. The ψ_s values are lower for equivalent water contents at the "wet end" of the range; e.g., $\psi_s \approx -400$ kPa at $\theta = 0.25$ m^3 m^{-3}, whereas in the sandier soil $\psi_s = 25$ kPa at $\theta = 0.25$ m^3 m^{-3}. The lower limit of water-holding capacity for that soil would be about $\theta = 0.20$ m^3 m^{-3}. Soil 3, with 55% clay, 24% silt, 4% coarse sand, and 17% fine sand, is completely different. θ_{sat} for that soil would be about $\theta = 0.4$–0.45 m^3 m^{-3}, but any reduction in water content causes rapid decrease in potential.

In general, plants can extract water more easily at high potentials than at low potentials. The old concept of the "permanent wilting point" was based on work that indicated that plants wilted when ψ_s fell to about

−1500 kPa. This is not necessarily true in the field, but it does provide a guide to the lower limits of plant-available water, whereas the curves in Fig. 4.3 show clearly why it is necessary to have information about the soil moisture retention curves. In this respect, the work of Williams et al. (1983) is very valuable because they provide detailed descriptions of the (relatively easily measured) soil physical properties, which enable estimates to be made of the water retention characteristics. However, we should note that empirical relationships between soil texture and soil water content will vary, depending on organic matter content. Clapp and Hornberger (1978) provide similar data for a range of soils, with less detailed description of the soil physical properties.

Water movement through soil depends on the difference in ψ_s between any two points in the soil and on the hydraulic conductivity, K_s. It can be described by Darcy's law:

$$J_s = K_s \, (d\psi_s/dx), \tag{4.5}$$

where J_s is the volume flux of water through unit cross-sectional area per unit time (flux density) in the direction of the lower potential, and x is distance. Darcy's law was originally formulated to describe flow through saturated media. Hydraulic conductivity falls rapidly as the soil wetness falls below saturation so that if Eq. (4.5) is to be used to describe flow in unsaturated soil, K_s must be made a function of the water potential, i.e., $K_s = K_s(\psi_s)$.

Over the range of interest in plant studies ($\psi_s \approx$ 3–5 to 1500–2000 kPa), K_s may be reduced by several orders of magnitude. Campbell (1974) provided the very useful equation

$$K_s = (K_s(\psi_s)/K_{sat}) = (\theta/\theta_{sat})^{2b+3}, \tag{4.6}$$

where b is an empirical coefficient. Clapp and Hornberger (1978) note that Eq. (4.6) has proved to be reasonably accurate over a range of values of b between 0.17 and 13.6. They determined values for soils ranging from sand (clay fraction = 0.03, $b \approx 4$) to loam (clay fraction = 0.19, $b \approx 5.4$) and clay (clay fraction = 0.63, $b \approx 11.4$).

From the point of view of hydrology—as opposed to studies on plant physiology—the parameter of most interest and importance is the saturated hydraulic conductivity K_{sat} because, as Eq. (4.5) shows, when θ falls below θ_{sat}, $K_s(\psi_s)$ falls very rapidly. Talsma and Hallam (1980) measured the hydraulic conductivity in four catchments, chosen for (apparent) uniformity, in the Australian Capital Territory. Their data show that K_{sat} values are very nearly log-normally distributed (Fig. 4.4). Large variability was found within each catchment, but most of the variability was present in very small areas.

I. Hydrologic Balance

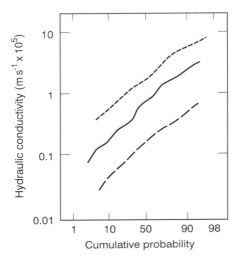

Figure 4.4 Distributions of hydraulic conductivity at different depths across a uniform catchment. The upper (dotted) line is for 0.15–0.25 m, the middle line is for 0.3–0.6 m, and the lower (dashed) line is for 0.7–1.0 m (drawn from a diagram by Talsma and Hallam, 1980).

1. Runoff, Infiltration, and Drainage Runoff from nonsaturated soils tends to be unimportant in forests because of the obstructions to movement that exist on the forest floor and also because, with their high organic matter content, high activity of microfauna, and the large root masses they contain (see Chapters 6 and 7), forest soils are likely to have high porosity and infiltration rates and can accept high precipitation rates. Enormous changes may be brought about when forests are clearcut (see Section II), but these will depend on the subsequent treatment of the site: There will be great differences between a site left with logging residues and a site at which these are burnt and the vegetation is completely changed (Fredriksen *et al.*, 1975; van Lear *et al.*, 1985). Management should take account of the fact that, in general, organic matter will increase infiltration rates and possibly the water storage capacity of soils.

The influence of cracks, surface litter, inequalities in the surface, and the modification of water input rates by drip from foliage and stemflow render any rigorous and detailed treatment of infiltration somewhat superfluous in a text concerned generally with processes as they operate at stand and forest level. We need only note here the rather obvious facts that infiltration will be faster into sand than clay soils—in the absence of soil cover the rates will be around 20 mm hr^{-1} for sands and about 1–5 mm hr^{-1} in clayey soils (Hillel, 1980), although these figures will vary enormously depending on the soil surface condition. Infiltration rates also depend on the amount of water already present in the soil: Much of the elegant (but not very practically useful) theory developed in relation to this process deals with wettting fronts and changes in K_s at those

---

Box 4.1. Soil Moisture Measurement

Improvements in soil moisture measurement techniques over the past 25–30 years have contributed greatly to understanding the effects of soil moisture on plant growth as well as to the understanding and quantification of hydrological processes. The neutron moisture meter provides the capacity to measure soil moisture content, to a high level of accuracy, at a range of depths at any point where a suitable tube ($\approx$4 cm in diameter) can be installed in the soil. The disadvantages of this instrument are that the measurements are quite slow to take—the instruments are generally too expensive to leave *in situ*, so they must be moved from tube to tube—and, even if measured daily (almost universally impractical), do not provide information about the short-term dynamics of soil moisture, although the vertical fluxes of water through the soil can be deduced from sequential measurements. The neutron sources may also be unsafe if used improperly. Neutron moisture meters will reflect the point to point variability of soil moisture; if this is very high it may make the results difficult to interpret. A technique called Time Domain Reflectometry (TDR), which allows virtually continuous observations, derived from simple probes inserted in the soil (Topp *et al.*, 1980), has been developed in recent years (Hook *et al.*, 1992). TDR is more suited to problems such as evaluation of the dynamics of soil moisture in the root zones of plants than neutron moisture meters and can be expected to supersede neutron moisture meters for most soil moisture measurement applications in the near future. The systems are readily automated; switching and data logging techniques are improving rapidly. TDR systems are now commercially available. Other techniques, such as (electrical) capacitance systems, are also becoming available. All have advantages and disadvantages. None, as yet, provide the large-scale integrated measure (over meters or tens of meters) of soil moisture that is needed to deal with the problems of soil heterogeneity and spatial variability, although falling costs and advances in TDR multiplexing technology can be expected to alleviate this problem.

fronts, the redistribution of infiltrated water, and so on (Hillel, 1980). Because of the spatial variability that will be encountered, it is not easy to measure infiltration rates in forests, but the collection of as much data as possible should be encouraged.

By drainage we mean drainage of water out of the root zone vertically, by downward flow, or horizontally across impermeable, or less permeable, layers. Because hydraulic conductivity falls very rapidly as soil moisture content falls below θ_{sat} [Eqs. (4.4) and (4.5)], drainage in unsaturated soils is slow. Under saturated conditions, drainage through holes and cracks can be an important pathway for water and solute transport (Wang et al., 1996). Drainage is difficult to measure, although estimates of water movement in the vertical can be made from sequential measurements with instruments such as neutron moisture meters or time domain reflectometry (TDR) (see Box 1). In catchment hydrology experiments, it is important to work in "sealed" catchments, where there is no vertical drainage out of the catchment. Horizontal drainage is measured using flow-gauging structures or wiers at the catchment outlet. In most water balance models, drainage is calculated by allowing the soil profile to fill with water and then "discarding," as runoff or drainage, precipitation that occurs when the profile is full.

2. Rooting Depths and Distribution The effectiveness of water uptake by trees depends on the effectiveness with which the soil is exploited by roots (defined as root length per unit volume of soil, L_v), contact between roots and soil, and the hydraulic potential gradients between roots and soil.

Tree root architecture varies between species (Lyr and Hoffmann, 1967; Karizumi, 1974, 1976; Phillips and Watson, 1994) and is affected by soil type and growing conditions. Root systems are dynamic: Trees do not simply produce a root system of a particular type and conformation and sustain it under all conditions—there is considerable evidence that the proportion of carbohydrates produced by trees that is allocated to root growth is higher in poor growing conditions than where water and nutrients are less limiting (Landsberg, 1986; Comeau and Kimmins, 1989; Gower et al., 1992; Haynes and Gower, 1995; see also Chapter 5).

From a water collection and transport point of view, the optimum system will be a considerable root mass, with high L_v, in the surface layers of the soil to harvest as much water as possible, with larger, more widely spaced roots deeper in the soil. These will absorb water more slowly, but the deeper soil layers do not dry as fast as the surface layers; therefore, deep roots are likely to play an important role in maintaining water uptake and transpiration during dry periods. Fowkes and Landsberg (1981) provided a theoretical analysis that supports these arguments. Soil textural characteristics and hard pans can greatly alter the development and vertical distribution of large roots.

A high density of fine roots ensures that the pathlength from the soil to the nearest root is always short so that even when the roots have extracted much of the water in their immediate vicinity, and $K(\psi_s)$ falls, water can still move to the roots along the potential gradient between the root surface and the soil fast enough to provide a useful contribution to the transpiration stream. Calculations with the model presented by Passioura (1982) indicate that root length density is unlikely to be a major limiting factor until L_v falls below about 2 or 3×10^4 m m^{-3}, and even then it is not likely to be important unless the soil is dry (low K_s) (see Newman, 1969). However, fine roots have high resistance to water flow and have to be relatively short; otherwise they cannot carry enough water to the main roots and the stem. Dry soil conditions will cause fine roots to stop growing or to die back (Teskey and Hinckley, 1981; Eissenstat and Yanai, 1996) and they are expensive to maintain in terms of respiration (Marshall and Waring, 1986; see Chapter 5). Maintaining a large system of fine roots and mycorrhizae increases respiration costs; consequently, seasonal changes in fine root mass are common, with maximum values occurring at times of high water and nutrient availability, with dieback when conditions become adverse (Deans, 1979; Persson, 1980a; Grier *et al.*, 1981).

Gerwitz and Page (1974) collated data on root length distribution under crops and found that, in virtually every case, L_v declined exponentially with depth (z) so that the root length density at any depth can be described by

$$L_v(z) = L_v(0) \exp(-k_L z), \qquad (4.7)$$

where $L_v(0)$ is the density in a layer just below the surface. The data used by Gerwitz and Page (1974) all came from vegetables and agricultural crops, but Gale and Grigal (1987) collected published data on the root distributions of North American tree species and found that they could all be fitted by the equation

$$\Sigma F_R = 1 - B^z, \qquad (4.8)$$

where ΣF_R is the cumulative root mass fraction from the surface to depth z (cm) and B is an estimated parameter. Values of B were in the range 0.92–0.95, which indicates that most (>95%) roots of those species are in the top half meter of the soil. Eq. (4.8) is consistent with Eq. (4.7): If we assign a length/mass conversion factor m_L to L_v, we can show that $\Sigma F_R(z) = 1 - \exp(-k_L z)$, which implies that $B = \exp(-k_L)$. This gives values of k_L of about 0.07 when z is in centimeters. Using the usual SI measure of meters, B becomes ≈ 0.001, giving $k_L \approx 7$.

The exact values of these parameters are not very important. The conversion factor m_L will vary with root diameter and probably with species. It is unlikely that the values that describe root distribution in the North American trees studied, which were mainly conifers, will be appropriate for deep-rooted species such as many of the Australian eucalypts. For example, Carbon et al. (1980) examined the distribution of roots under jarrah (*E. marginata*) in Western Australia, in soils with a sandy layer 1-m deep overlying about 2–2.5 m sandy loam and up to 20 m clay. They presented two representative profiles derived from sampling at 25 sites and showed an exponential reduction in $L_v(z)$ through the sand and sandy loam, but large vertical roots penetrated to the 20-m limit of their sampling. Carbon et al. commented on the "almost total dominance of root length by fine to very fine roots." Studies in tropical forests have shown fine roots at a 20-m depth (Nepstad et al., 1994).

There are few values of L_v for forests in the literature. Vogt (1991) collated a great many data on root mass under forests, and some of these data could be used to make estimates of root mass distribution using Eqs. (4.7) and (4.8). The data of Carbon et al. (1980) show $L_v > 1 \times 10^4$ m m^{-3} for the top 2 m of the jarrah forests, and it is safe to assume that this will be exceeded by a considerable amount in forests where most of the roots are in the top 1 m. Therefore, it will generally be safe, when analyzing the effects of soil moisture on tree growth, to assume that soil water content (and hence potential) in the top half meter will provide a good measure of the moisture content relevant to the plants (see Tan et al., 1978). In calculating the water balance of stands, the best estimates of rooting zone will be based on soil type and depth and the best available information about species rooting characteristics. Wherever possible, such estimates should be supported by measurements and observations.

D. Transpiration

Transpiration rates vary with weather conditions, leaf area and canopy architecture, and soil moisture. The process has been dealt with in some detail in Chapter 3, Section III, where the dependence of forest transpiration rates on canopy conductances is clear from Eq. (3.16). The main purpose of this section is to consider the effects of soil and atmospheric conditions that cause g_c to be less than g_{cmax} so that actual transpiration rates are less than the possible maxima (It is important to avoid confusion between maximum and potential—as defined by Eq. (3.22)—transpiration rates.) Because canopy conductance is dominated by stomatal conductance (we noted in Chapter 3 that g_c is commonly calculated as

Box 4.2 Stomatal Behavior

Although research on stomatal behavior has, for the past 20 years, been focused on the effects of light, atmospheric factors [vapor pressure deficit (strictly leaf to air)], CO_2 concentration), and leaf water status, consideration of flow pathways and resistances [see Eq. (4.9)] indicates that we should be concentrating more on these areas. If the potential rate of atmospherically driven transpiration, with some maximum value of g_s, is greater than the rate at which water can move from the soil to the roots, through the plants to the leaves, then unless stomata close, the plant will dessicate. Therefore, vapor pressure deficit (D) in particular can be regarded as a surrogate for evaporation rate: Stomata will close when the plant cannot sustain the rate of water loss driven by vapor pressure deficit. This suggests that all the empirical relationships established between g_s and D must have been influenced by the capacity of the soil–plant hydraulic conducting system. Recognition of this does not solve the problem of the actual mechanism of closure, and it also means that prediction of closure is more difficult: We cannot calculate leaf transpiration rate unless we know g_s, but g_s depends on the resistances in the flow pathways, including those between the leaves and water stored in plant tissues. This also indicates that there is unlikely to be any unique relationship between leaf water potential (ψ_f) and g_s—and indeed none has been found—because ψ_f is itself strongly influenced by flow rates and resistances to flow through the plant.

We will, unavoidably, continue to use established relationships to predict g_s, but the point discussed here highlights the need to take into account the soil moisture situation and the need for studies on stomatal behavior in relation to measurements of whole plant transpiration rates.

$g_c = \Sigma g_s \cdot L^*$), fluctuations in g_s will generally cause commensurately large fluctuations in λE. Because of its relevance to photosynthesis, stomatal conductance and the way it responds to environmental factors is discussed in Chapter 5; it is considered here with particular emphasis on its importance in relation to transpiration and the interactions between g_s and plant—or soil—water status (Box 4.2).

The three major environmental influences on stomatal conductance at the canopy level are light (φ_p), vapor pressure deficit (D), and the water status of the leaves (ψ_f). The equations given by Leuning [1995; see

Chapter 5, eqs. (5.9) and (5.10)] describe the responses to φ_p and D; they demand a unique value of stomatal conductance to satisfy the rate of photosynthesis required by the current value of φ_p and internal leaf CO_2 concentration, c_i. This provides the critical link between g_s, transpiration, and photosynthesis.

Interpretation and prediction of stomatal responses to D are complicated by the fact that the response appears to be to transpiration rate (E) rather than to D (Monteith, 1995). Monteith has shown that virtually every published set of measurements purporting to show a nonlinear relation between g_s and D can be interpreted as a linear relation between g_s and transpiration rate, with the general form

$$g_s = a - bE, \tag{4.9}$$

or, in nondimensional form

$$g_s/g_{max} = 1 - E/E_{max}, \tag{4.10}$$

where g_{max} and E_{max} are extrapolated maximum values of g_s at $E = 0$ and $1/b$, respectively. However, this relationship is not particularly helpful to those faced with a pragmatic need to predict transpiration on the scale of forest stands or relatively small land units, and it seems likely that we will continue to rely on empirical relationships between g_s and D for some time to come (see Chapter 5).

In considering the influence of foliage or soil water status on stomatal conductance, and hence transpiration, we have to note that there is strong evidence that the effects of leaf water status can be overridden, and certainly modified, by chemical signals from the roots or by changes in the conductance of the hydraulic pathways. The "traditional" explanation for the effects of leaf water potential on stomatal conductance is that there is a negative feedback between g_s and leaf water potential (see Ludlow, 1980), but this does not explain decreases in stomatal conductance while leaf water potential remains constant. Other scientists have proposed a feedforward control of leaf water potential (i.e., stomata respond directly to variables that influence leaf water potential rather than responding to leaf water potential *per se* as proposed by Ludlow). It has been suggested that a feedforward response to soil drying involves the transport of a chemical message (abscisic acid) from roots to foliage, causing reduced stomatal conductance independent of leaf water potential (Wartinger *et al.*, 1990; Zhang and Davies, 1991; Tardieu *et al.*, 1992).

Rapid reductions in stomatal conductance, independent of changes in leaf water potential, can be induced by decreased hydraulic conductance (Teskey *et al.*, 1983; Meinzer and Grantz, 1990; Sperry *et al.*, 1993) caused by severing part of the stem or root or cooling the roots. This speculation is supported by results from an experimental shading treatment of well-watered *P. radiata* trees. Whitehead *et al.* (1996) found that shading the lower canopy decreased tree canopy conductance immediately. When the cover was removed, conductance in the upper canopy decreased with a concommitant increase in the lower canopy. Bulk leaf water potential changed little while these changes occurred. These results suggest that hydraulic conductances in the xylem were altered, which may have stimulated the production of chemical messengers to regulate stomatal conductance.

Additional evidence both supporting and refuting the control of stomatal conductance by hydraulic conductance as opposed to chemical messages has been obtained from experiments using a device to apply pressure to root systems of plants in sealed chambers (e.g., Passioura and Munns, 1984). Increasing the atmospheric pressure inside the chamber increases the hydraulic and pneumatic pressure of the soils and root system equally, but outside the chamber the shoot experiences an increase in hydraulic pressure. Pressurizing the soil has been shown to have no significant effect on stomatal conductance of herbaceous plants in dry soils (Gollan *et al.*, 1986; Schurr *et al.*, 1992) but has been shown to rapidly reverse decreasing stomatal conductance of woody plants grown in drying soil (Fuchs and Livingston, 1994; Saliendra *et al.*, 1995). The apparent discrepancy may, in part, be inherent to the type of plant (herbaceous or woody) studied. Tardieu and Davies (1993) have proposed that an integrative chemical and hydraulic signal control stomatal conductance and leaf water potential. Saliendra *et al.* (1995) speculated that chemical root signals may be less important in woody than herbaceous plants because the long transport time in tall woody plants would make this mechanism ineffective for short-term stomata regulation. In summary, stomatal control by a single factor seems unlikely and is inconsistent with the multiple-control systems that operate in trees for almost all processes regulating carbon, water, and nutrient flow.

Even without this information, it is clear from our earlier discussion about root distribution that the water status of foliage must be strongly influenced by the average soil water potential in the root zone and by the resistances in the flow pathways from soil to roots through the plant to the atmosphere. Flow through plants is driven by transpiration, and if the rate at which water can move from the soil to the roots, and through the plant to the evaporating sites in the leaves, is equal to or greater than the atmospherically driven (potential) transpiration rate, then g_s can be at or near its maximum value, and the plants will not become dessicated. However, if the supply rate cannot meet the demand, then stomata must close to reduce losses from the plant to the point where they can be met by supply from the soil. If stomata do not close enough, the plants will dessicate. The mechanisms we have considered can be regarded as having evolved as reponses to this basic constraint.

Experimental results illustrating the effects of soil water potential, or content, in the root zone on transpiration and stomatal conductance have been presented by Tan et al. (1978), who measured stomatal resistance ($r_s = 1/g_s$) through the canopies of Douglas fir trees in a thinned stand over a period of about a month, in summer, when there was no rain. They showed that the effects of D were enhanced by dry soil. Within the three soil wetness ranges (moderately wet, $0 > \psi_s > -350$ kPa, to dry, $-950 > \psi_s > -1250$ kPa) at a vapor pressure deficit of 1.5 kPa, $r_s = 0.78$, 1.57, and 5.31 mm sec^{-1}, i.e., $g_s = 1.27$, 0.44 and 0.18 mm sec^{-1} so that (from $g_c = \Sigma g_s \cdot L^*$) g_c for a canopy with $L^* = 4$ would be (approximately) 5.1, 2.5, and 0.75 mm sec^{-1}. These values are very much lower than the maximum values identified by Kelliher et al. (1995), suggesting that even the "moderately wet" soil was exerting a considerable effect on stomata and hence on transpiration rates.

In contrast, Calder (1978) obtained results that indicated no dependence of g_s on soil moisture. They used a drainage lysimeter to measure the water balance, and hence transpiration rates, of Sitka spruce in

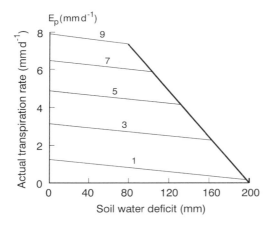

Figure 4.5 Effects of soil water deficit on transpiration rates of *Eucalyptus maculata*. The diagram shows that, at any potential transpiration rate E_p (calculated from the P–M equation with a standard value of g_c), the actual transpiration rate fell slightly with increasing soil water deficit until a critical value was reached, after which actual transpiration rates fell rapidly. The critical soil water deficit values increased as E_p decreased, indicating that, if E_p was low, the trees were able to meet atmospheric demand even when the soil was dry (diagram redrawn from Dunin et al., 1985).

Wales. An optimization procedure was used to establish the values of g_s that in the Penman–Montieth (P–M) equation, gave the closest agreement with measured values of transpiration. The results indicated that g_s was a function of atmospheric vapor pressure deficit D, but no dependence on soil moisture deficits was found, down to deficits of 200 mm, equivalent to about -600 kPa in the root zone. One of the reasons for this lack of effect was, almost certainly, the very low transpiration rates during the experimental period—down to 0.02–0.03 mm day^{-1}. In these circumstances, water can, presumably, always move through the soil fast enough to supply transpiration demand.

Dunin et al. (1985) used a large weighing lysimeter, on the south coast of New South Wales, Australia, to evaluate transpiration rates from a segment of *Eucalyptus* forest dominated by *E. maculata*. L^* on the lysimeter, and in the surrounding forest, fluctuated over the period of measurements (more than 12 months) but was generally about 3. For 102 days without rain, when $L^* > 3$, Dunin et al. plotted measured daily (actual) transpiration rates against soil moisture deficit in the lysimeter, which contained about 210 mm (depth equivalent) of water in the 1.8-m deep root zone. They also calculated transpiration using the P–M equation, with $r_c = 60$ m sec^{-1} ($g_c \approx 17$ mm sec^{-1}), calling this the potential transpiration rate. It is interesting to note that the value of g_c used by Dunin et al. to calculate "potential" transpiration rates was almost exactly the same as the average value of g_{cmax} for woody plant communities that Kelliher et al (1995) obtained 10 years later. The results from Dunin et al. are reproduced in Fig. 4.5.

Figure 4.5 indicates that, when the soil moisture deficit was less than about 140 mm—or about 60–65% of available water in this soil—the rate of water loss by the trees approximated the maximum rate: Observed rates fell about the maximum rate lines until the soil moisture deficit exceeded this "critical" amount. The rates of transpiration then fell rapidly. The range of soil water across which transpiration was apparently unaffected by soil moisture content was considerable. It would have been possible to derive information about canopy resistance values in relation to soil moisture in the root zone, similar to that of Tan et al. (1978), by solving Eq. (3.27), but after the critical point the effects of low soil water contents on transpiration rates were clear enough.

Obviously, actual transpiration rates can vary from zero to the maximum rates achievable with particular values of L^* and g_s (see Chapter 3, Section III,B). For example, Myers and Talsma (1992), in Australia, measured daily values ranging from 1 or 2 mm day^{-1} in a *P. radiata* plantation in winter to 2 or 3 mm day^{-1} in spring, with values of 6–8 mm day^{-1} in an irrigated and fertilized plantation in summer. Shuttleworth (1989) measured rates of 3 or 4 mm day^{-1}, with little day-to-day variation, from

Amazon rainforest, and McNaughton and Black (1973) measured values ranging from 1 to 4.5 mm day^{-1}, with considerable day-to-day variation, from a Douglas fir forest in British Columbia, Canada.

II. Catchment Hydrology

Because forested catchments are important sources of water for many human population centers and aquatic ecosystems, it is essential that we have some appreciation of catchment hydrology and the implications of various forest management practices on the groundwater. The presence or absence of forest cover on part or the whole of a catchment, and cover type, will have a considerable effect on the catchment water yield (Fahey and Rowe, 1992) and may also affect water quality, particularly immediately after disturbance (Binkley and Brown, 1993). There have been many studies of the effects of forest clearance on streamflow, nutrient fluxes in stream water, and sediment transport. The main findings in terms of streamflow can be summarized as follows (see Fig. 4.6):

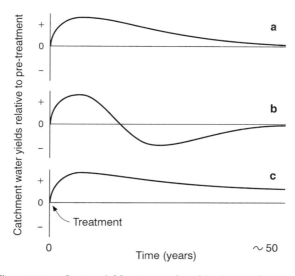

Figure 4.6 Time course of water yield patterns after thinning or clearing forested catchments. The initial increase will depend on the proportion of the tree cover removed in the treatment. Subsequent patterns of water yield depend on the rate of recovery of the vegetation and its treatment. (a) Gradual recovery of vegetation to its original state; (b) results of heavy regrowth, leading to higher leaf area than in the original stand, hence greater water demand and lower water yield (self thinning over a long period tends toward the original conditions); (c) a stand cleared and maintained as, for example, pastoral land.

- Clearing a forested catchment generally results in an initial increase in streamflow, with a gradual return to original levels as the vegetative cover reestablishes itself. The size of the increase depends on the proportion of the vegetation removed and on the potential evaporation rates, taking account of slope azimuth and angle and solar radiation on the slopes.
- In some cases, there is an initial increase in streamflow after clearance, followed by a long period when the streamflow is below the original baseline flow.
- There is a sustained increase in streamflow after clearance, particularly likely to be the case where land use is changed to pasture or agricultural crops.

We provide some background to these results and review some of the literature that describes them. As an introduction, we restate the point made in relation to Eq. (4.1); that is, that the differences in catchment water yield under different vegetation types can be understood and explained in terms of differences in the various terms of that equation and the way they are affected by the structural and functional characteristics of the vegetation. Each comparison or analysis must take into account possible differences in the amount of water intercepted and evaporated from canopies, the pattern and duration of L^* [a function of stand density and leaf habit (e.g., deciduous, evergreen)], and the depth of the rooting zone, which determines the capacity of the vegetation to access deep groundwater. The rates of transpiration from the vegetation, at any particular season, will be determined by the factors we have discussed at some length—the energy balance, aerodynamic, and canopy conductances. Runoff and drainage are strongly dependent on the condition of the soil surface and the water content of the root zone as well as precipitation patterns.

Classical concepts divide water yield from catchments into *quickflow*, the peak outflow that occurs during and shortly after storms, and *baseflow*, the more or less continuous, long-term outflow from a catchment. The reasons for these flows, and the pathways followed by the water, have been the subjects of study and debate among hydrologists for a long time. Ward (1984) has provided an excellent treatment of the situation, of which the following is a summary.

Catchments are highly variable, with differences in slope, topographic shape, and soil depths dictating the runoff response. A hypothesis that held sway for many years was that of Horton (1933). This said that if rain falls on a catchment at a rate greater than the rate at which it can be absorbed, the excess water will flow over the ground as overland flow (runoff). However, it is now recognized that in humid regions, overland flow very rarely occurs until there are areas of saturated soil in the catchment. These cannot accept more water and act as sources for quickflow. These saturated areas link with one another through above- and belowground channels. It appears that all rainfall infiltrates and is then transmitted through the soil profile, with upslope rainfall recharging the soil moisture store, providing for subsquent baseflow, and downslope rainfall and channel precipitation providing most of the quickflow. Throughflow (movement of water through the soil profile) occurs through channels and connections between saturated areas at much higher rates than would be predicted on the basis of assumptions of soil matrix flow. It also appears that, in some cases, when catchments are saturated each new increment of rainfall displaces all preceding increments, causing the oldest water to exit from the bottom end of the system.

Catchment experiments are usually of two types: they are either based on a period of precalibration before treatment or based on paired catchments. Pereira (1973) provided a useful general treatment of forest hydrology experiments, which includes a number of illustrations of the effects, in terms of erosion, of forest clearance. A famous study by Bormann and Likens (1979) shows the dramatic effects of clear-felling the forest on the hydrology of a watershed (Hubbard Brook). Summer streamflow increased by a factor of about 4, reaching a peak during the second year after cutting. Bruijnzeel (1990) has provided a compendium of the results of experiments on moist tropical forests and a collation of extant information about the effects of land use change in those areas. He summarized his findings—which included information on catchment water yields and flow patterns, nutrient fluxes, erosion, and sediment transport—in 37 points, to which he did not allocate priority or relative importance. However, point 35 is worth emphasizing: "The information summarised in this report leads to the observation that the adverse environmental conditions so often observed following "deforestation" in the humid tropics are not so much the result of "deforestation" *per se* but rather of poor land use practices after clearing the forest."

In relation to water yield alone, Bosch and Hewlett (1982) produced a review of 94 catchment experiments from all over the world. All these, with the exception of one, showed that catchment water yield increased following reductions in vegetation cover. On average, for each 1% of cover of conifer, mixed hardwood, and scrub removed, annual streamflow increased 4.2, 2.0, and 1.2 mm, respectively. (Note: there is huge variation associated with these statistical averages, which should not be used as general quantitative estimates.) The exception to the general pattern was a series of long-term observations reported by Langford (1976) who showed that, following regeneration after major fires, the water yield from *E. regnans* forest catchments in Victoria, Australia, first increased but then decreased significantly 3–5 years after the fire. The decreases were of order 200 mm per year in areas with annual rainfalls ranging from 900 to 2000 mm. The indications were that it will take up to 100 years for the catchment water yields to return to their original levels. Work by Cornish (1993), on logged *Eucalyptus* catchments, is showing the same trends—early increase in catchment water yield, then a steady decrease. The initial increases in runoff after logging were proportional to the percentage of the catchment that was logged. There are now data from Hubbard Brook (HB) that show similar patterns, reported by Hornbeck *et al.* (1993), who presented a diagram showing that water yield from HB Catchment 2 increased 350% after clear-cutting and treatment with herbicides for several years. It then declined steadily until, about 12 years after treatment, water yield fell below pretreatment levels. Twenty-five years after treatment, water yield was still below pretreatment yields.

Langford (1976) was not able to offer any satisfactory explanation for his observations, but the explanation now accepted is that the decreases in water yield observed a few years after logging or burning are caused by the large increases in leaf area associated with *Eucalyptus* regrowth. L^* in regrowth may become significantly higher than in old growth forest (see Chapter 8). This is supported by the work of Jayasuriya *et al.* (1993) who used a heat pulse technique to measure sap flow velocities in *E. regnans* on the catchments studies by Langford. They showed that differences in the transpiration rates of the trees could be entirely accounted for by differences in sapwood area [and hence leaf area; see Chapter 3, Eq. (3.2)]

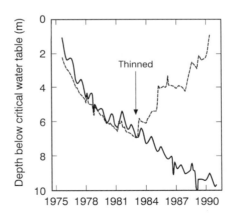

Figure 4.7 The effects of thinning on groundwater levels in two catchments in Western Australia. The data have been normalized. Reprinted from J. Hydrol. 150, G. L. Stoneman, Hydrological response to thinning a small jarrab (*Eucalyptus marginata*) 393–407 (1993) with kind permision from Elsevier Science-NL, Sara Burgerhartstraat 25, 1055 KV Amsterdam, The Netherlands.

and that streamflow differences after thinning could be accounted for by the differences in stand transpiration rates. Hornbeck *et al.* (1993) did not explain the Hubbard Brook results in terms of leaf area, but accepted that they are consistent with such an explanation.

Illustrations of the hydrological responses of forests to thinning and the effects of land use change (conversion to pasture) come from Western Australia. In a paired catchment experiment, Ruprecht *et al.* (1991) thinned a small forested catchment from 700 to 110 trees per hectare in an area with about 1200 mm rainfall per year. Groundwater levels in the thinned catchment began to rise within the first year after thinning. Deep groundwater attained a new equilibrium after about 2 years, rising by approximately 2 m in a downslope area and about 5 m upslope. Streamflow increased from approximately 6% of annual rainfall before thinning to about 20% after thinning. Ruprecht *et al.* did not interpret these results in terms of the mechanisms discussed previously, but it is likely that the reduction in crown (canopy) from 60 to 14% would have resulted in substantial reductions in interception and transpiration losses, leading to the observed rise in average groundwater levels. These would have resulted in much quicker filling of groundwater storage at the start of a rainstorm, hence more, and more rapid, streamflow. Similar results have been reported by Sharma *et al.* (1987). An experiment by Stoneman (1993), in which one of two precalibrated catchments was thinned to about 20% of the original canopy cover, provides a dramatic illustration of the effects of thinning on groundwater levels. His results are shown in Fig. 4.7

Conversely, paired catchment studies in New Zealand have shown that the conversion of native tussock grassland to *P. radiata* plantations resulted in an initial increase in water yield, followed by a gradual decline as the canopy of the pine plantation developed.

Other changes in vegetation cover, such as native forest to pine forest, which is occuring worldwide (Chapter 2), have more subtle effects on water yield. Obviously, water yield will increase dramatically immediately after the conversion because of the reduction of L^* (see Fig. 4.6a). Removal of native podocarp forest in New Zealand initially increased water yield by as much as 535 mm year^{-1}, or approximately 30% of annual precipitation (Rowe, 1983). The increased water yield commonly occurred in the form of increased quickflow in areas of high precipitation and in the form of delayed flow in areas of lower precipitation (Fahey and Rowe, 1992). However, water balance calculations have shown that within 8–10 years of conversion of native evergreen podocarp forest to pine plantation the water yields for the two forest types were similar (Davoren, 1986; Dons, 1986 in Fahey and Rowe, 1992). In the southeastern United States, conversion of broad-leaved deciduous forest to pine plantation increased water yield, presumably due to the larger L^* of the conifer forests during the growing season and additional transpiration losses by the conifer forests while the deciduous forest was leafless (Swank and Douglas, 1974).

The complexity of the processes involved in water movement into and through forested catchments makes it clear why, as noted previously, theoretically rigorous, physically based models, using equations known to work in well-defined, spatially homogeneous situations, are often of limited value in predicting water yield. The problem lies in the difficulty of specifying appropriate parameter values for highly variable systems. There are some distributed-parameter models, utilizing terrain analysis, which are providing good predictions and, hence, confidence that we are developing the capability to predict the consequences of management actions or inadvertant modification on the basis of the mechanisms involved (e.g., Vertessy et al., 1993), but for most management and decision-making purposes we have to rely on empirical relationships and use our understanding of the underlying mechanisms to interpret the results. As a general "rule of thumb," we can say that the water yield of forested catchments will be increased by clearing or thinning by an amount proportional to the percentage of the canopy removed by the operation. The actual amount of the increase will depend on the type of plant cover that fills the gaps, on soil depth and storage characteristics, and, of course, on the precipitation patterns. The persistence of any increase will depend on subsequent land use and the treatment of the vegetation.

III. Tree–Water Relations and Their Effects on Growth

A. Tree–Water Relations

In the context of catchment hydrology, trees can be treated as important factors in the water balance in terms of their influence on interception, transpiration, and, hence, runoff and drainage. From the point of view of the forest manager concerned with wood production or ecological balance and change, the important considerations are the influence of the water balance on tree growth and indeed survival. It would be useful to be able to predict the effects of water shortage on tree growth, either in the short term, through effects on stomatal conductance, or in the longer term, through effects on the production and maintenance of foliage and biomass production. It is also important to be able to predict the effects of stand thinning on water relations and growth of trees, as well as the soil water balance of sites.

There has been an enormous amount of research on these issues. Slatyer's (1967) book was a landmark in the field of plant–water relations, and there have been many books and major reviews since, some devoted exclusively to the water relations of trees (see, for example, Vol. VI (1981) in the series edited by Kozlowski). The chapter by Whitehead and Jarvis (1981) in that volume provides a comprehensive treatment of water movement through conifers and Landsberg (1986) provided a general analytical treament of water movement through trees.

Under Section I,D, we considered the effects of soil moisture content on transpiration rates and argued that the underlying factor controlling E is the rate at which water moves from the soil to the roots, and then through the plant to the leaves. It is useful to consider this process in a formal way.

Water moves from soil to roots and through plants along potential gradients, caused primarily by changes in leaf water potential as a result of transpiration. The generally accepted model of water movement through plants is based on an analog of Ohm's law, which follows from this gradient-driven flow. Using this analog, the flow of water (J, m^3 sec^{-1}) through the tree–soil system can be represented by flow through a series of hydraulic resistances and described by the equation

$$J = (\psi_s - \psi_f) / R_s + R_r + R_x + R_f, \qquad (4.11)$$

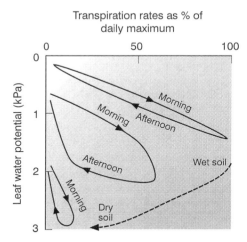

Figure 4.8 Diurnal patterns of leaf water potential showing characteristic hysteresis between morning and afternooon. The diagram shows that, at the same transpiration rate, ψ_f tends to be lower in the afternoon than in the morning, with the difference being (relatively) greater as the soil dries out. In dry soils, transpiration rates cannot be as high as in wet soil because the plants cannot absorb water fast enough to meet the atmospheric demand (diagram redrawn from Hinckley et al., 1978).

where R_s denotes resistance to water flow through the soil to the root, R_r resistance through the root to the xylem, R_x resistance to flow through the xylem, and R_f from the xylem to the evaporating surfaces in the leaves.

If Eq. (4.11) holds, there must be flow continuity and mass conservation through the system so that a given volume of water ($J\Delta t$) lost from the leaves in time interval Δt will result in the extraction of the same volume of water from the soil over that interval. The relationship between J and $\Delta\psi$ becomes nonlinear if the resistances are not constant, but there has been no evidence of this from measurements made on trees in the field, although it may appear to be the case if capacitance (water stored in tissue such as stems) is a factor (see Landsberg *et al.*, 1976; Whitehead and Jarvis, 1981; Landsberg, 1986).

If resistances are constant and Eq. (4.11) holds, it can easily be solved by rewriting it as

$$\psi_f = \psi_s - J\Sigma R_i, \quad (4.12)$$

This is essentially an equation for a tree with a single leaf (at potential ψ_f) to which all the water flows. More rigorously, it would be necessary to subtract the summed products of the partial flows and resistances, which change in a branched system. This was discussed by Richter (1973). However, Eq. (4.12) may provide a good first approximation to the average value of ψ_f. If it does, plotting $\Delta\psi$ [$= (\psi_f - \psi_s)$] against J should yield a straight line with a slope ΣR_i, giving the sum of the resistances in the flow pathway. If the soil is wet, ψ_s can be taken as approximately zero; therefore, the plot reduces to ψ_f against J (or against transpiration rate, which may be taken as an estimate of J). Equation (4.12) has, on some occasions, been adequate for this type of analysis (Landsberg *et al.*, 1975), but in most cases, particularly if the diurnal course of transpiration is plotted against (for example) hourly average values of ψ_f, such plots yield a hysteresis loop (Fig. 4.8). This appears to be a consequence of the fact that tissues such a sapwood may store significant quantities of water, i.e., exhibit capacitance (Holbrook, 1995).

Figure 4.8 shows relationships between transpiration and leaf water potential at different values of ψ_s. It can be interpreted as follows: As the transpiration rate increases during the first part of the day, water is withdrawn from the soil and from tissue storage. In nonsaturated soils, water will have moved, during the night, to rewet the regions around the roots from which it had been withdrawn the previous day. The extent of the drying out, the distance the water has to move, and the wetness of the soil and its hydraulic properties [see Eq. (4.4)–(4.6)] all influence the extent to which the soil surrounding the roots is recharged. We would expect that, before transpiration commences ($J = 0$), leaf, xylem, and soil water potential would all be approximately equal. This is the assumption underlying the commonly made measurement of predawn water potential (ψ_{pd}). If the soil is relatively dry, the intercept on the ψ axis of the downward (morning) arm of the ψ_s/J plot can be taken to give an indication of average ψ_s in the root zone. It should correspond closely to ψ_{pd}.

In general, the diurnal course of ψ_f follows the same path, falling from ψ_{pd} to minimum values (ψ_{min}) through the day—the pattern of reduction being determined via Eq. (4.11) by transpiration rate—and rising again in the evening (see Fig. 4.8). It would appear that, in the absence of comprehensive data on flow resistances through trees, it will not usually be possible to apply Eq. (4.11), but we note that ψ_{min} is a conservative quantity: in *P. radiata* trees in Australia, in treatments with widely different soil moisture, ψ_{min} varied from about -1.4 MPa in the irrigated treatment to -1.9 MPa in the dry (Myers and Talsma, 1992); in *P. sitchensis* trees in Scotland, under totally different conditions, ψ_{min} reached -1.6 MPa (Hellkvist *et al.*, 1974); and in *Quercus alba* in Missouri, ψ_{min} reached about -2.3 MPa (Hinckley *et al.*, 1978). It therefore seems likely that values of R_i [Eq. (4.12)] will also be relatively conservative, although they will vary with stand density and possibly with various environmental conditions (see Mencuccini and Grace, 1995).

Among the various problems encountered in analyses of plant–water relationships is the fact that most measurements of plant water status are instantaneous—they provide a value for the variable at the moment of measurement, which reflects the interactions of dynamic processes discussed previously. The water status of the plant at any time may influence the rates of plant growth processes taking place at that time, but it may not provide much information about the way those processes have acted in the past; therefore, analysis of plant growth in terms of measurements of plant water status made at points in time may not be very enlightening. Because the state of the plant at any time, in terms of its mass and the distribution of that mass, is the end result (integral) of rate processes over a period (strictly, from germination), what is required is a measure of water status integrated over the period of interest.

Myers (1988) made a useful contribution to the solution of this problem by introducing the water stress integral

$$S_\psi = \Sigma(\psi_{pd} - c)n, \qquad (4.13)$$

where ψ_{pd} is the mean value of predawn water potential over any interval, c is a datum value (the maximum value of ψ_{pd} measured (or calculated) during the period), and n is the number of days. Myers (1988) showed, for a stand of *P. radiata*, that needle lengths and basal area increments of trees subjected to five water and fertilizer regimes were closely related ($r^2 = 0.90$ and 0.91, respectively) to S_ψ over the growing season. This approach deserves more attention and development.

B. Effects of Water Stress on Growth

We are concerned here with stand-level effects. To describe these in quantitative terms, we need relatively simple relationships that can be used in a predictive way. These will inevitably be empirical but, if they are soundly based on the detailed mechanistic knowledge that we have of the dynamics of plant–water relationships, they should be reproducible and therefore of value in a predictive sense. To illustrate, we outline below the type of relationships that we believe could be established and should be developed. We also indicate how such relationships could be translated into information that would be of value in practical management and policy decisions.

The underlying assumption is that growth is directly related to absorbed radiant energy (see Chapter 9), and that the efficiency of energy utilization is affected by plant water status. In effect, we propose that growth (W), which may be estimated as dry mass increment per unit area per unit time or as a surrogate such as stem volume, can be expressed as

$$\frac{dW}{dt} = f(\varphi, N), g(\psi), \qquad (4.14)$$

from which

$$W(t) = f(\varphi, N) \int g(\psi) \cdot dt. \qquad (4.15)$$

Here, N denotes nutrients in general—not only nitrogen. The relationships are general and of course Eq. (4.15) is not strictly true because $f(\varphi, N)$ is not invariant with time, but they illustrate the point.

We assume, following Myers (1988), that the integral of predawn tissue water potential over time provides a parameter (S_ψ) that will influence growth patterns in easily predictable ways. The problem therefore becomes one of providing a method for calculating ψ_{pd}.

The starting point will be the assumption that ψ_{pd} will be related to soil moisture potential in the root zone. A study by Fahey and Young (1984) indicates that this assumption is correct; their results also illustrate the need for information about soil moisture characteristics—the nonlinearity of the ψ_s/θ_s relationship [Eq. (4.4)] makes it very unlikely that relationships consistent across soil types will be obtained between θ_s and ψ_{pd}. More such studies are required. Given the values of the coefficients of Eq. (4.4), the average value of ψ_s in the root zone of a stand can be estimated from values of θ_s derived from water balance calculations [using Eq. (4.1) with a daily time step]. We would expect ψ_{pd} to be linearly related to ψ_s; ψ_{pd} will be lower because of the resistances in the flow pathways [Eq. (4.11)], although how these act when $J = 0$ is difficult to say. Again, the results of Fahey and Young support this assumption.

Given that we can estimate ψ_{pd} on the basis of water balance calculations, it follows that we can obtain values of S_ψ for any period across which growth measurements have been made and use them to explain variations in growth. It can be argued that the most important mechanism by which water status influences plant growth (in terms of dry matter production) is reduction in g_s, which leads to reduction in CO_2 assimilation and, hence, dry matter production. In the short term (hours and days), this seems likely to be correct; therefore, there is a case to be made for expressing the effects of ψ_{pd} in terms of effects on stomatal conductance; Sala and Tenhunen (1994) have published the results of a study in which such relationships were investigated. However, on longer time scales, water stress affects growth through its effects on leaf area both through effects on leaf expansion and retention and through effects on carbon allocation (see Chapter 5). The evidence for this is not strong, but it is supported by data from irrigation studies that showed that increased water availability increases carbon allocation to foliage and decreases carbon allocation to fine roots (Gower et al., 1992). It therefore seems likely that a complete model of the effects of water status on growth would have at least two components: relationships expressing the (short-term) effects of ψ_{pd} on g_s and, hence, on carbon assimilation, and relationships expressing the (longer-term) effects of ψ_{pd} on growth in terms of leaf expansion and stem diameter growth.

One of the important effects of thinning is to change the water relations of stands. Reduction in the number of trees is likely to improve the water balance of the stand by reducing interception—so that more rain reaches the ground—and reducing transpiration losses from the stand as a whole. Whether stand transpiration rates remain lower than those of a similar unthinned stand will depend on the relative leaf areas of the stands as well as the transpiration rate per unit leaf area. The thinning experiment done by Stoneman (1993), cited earlier (see Fig. 4.7), provided an interesting demonstration of the effects of thinning on groundwater levels and streamflow and the results of Jayasuriya et al. (1993) are also relevant.

A very thorough and a useful study of the effects of thining on a stand of oak (*Quercus petraea*) was carried out by Bréda et al. (1995). They found that, across two seasons following thinning, ψ_{pd} was generally higher in the thinned stand, although midday values of ψ_f were about the same. Sap flux in trees in the thinned stand was considerably higher than in the controls, which implied that the hydraulic path conductances were higher. Calculations indicated that these conductances were higher (statistically significantly) in the thinned trees in the second season, but

Bréda *et al.* attributed this to decreasing conductances in the control trees. The cause of the differences may have been the improved soil water conditions in the thinned stand. Stand transpiration was lower in the thinned stand in the first year after thinning, but it was the same in the thinned and control stands in the second year, although there was no significant change in leaf area. A thorough theoretical treatment of the processes involved in these relationships has been given by Whitehead and Jarvis (1981).

Expanding from the stand level to regional estimates of the effects of water on forest growth, the only feasible approach is through water balance calculations. The most important information needed is soil type, rainfall, and enough weather data to permit the calculation of transpiration. The procedure would then be to solve the hydrologic equation [Eq. (4.1)] as accurately as possible for each stand or forest block and calculate the periods for which soil moisture content can be expected to be limiting to growth. The reduction in dry mass production as a result of periods of drought would have to be established empirically by measurements and records maintained over long periods; ideally, these would be consistent with more detailed information and models of the type described in the previous paragraph. Applying such calculations to long-term weather records provides the basis for estimates of the probability of drought periods of specified duration and intensity.

IV. Concluding Remarks

We noted in the introduction to this chapter that, from the point of view of the forest manager, there are two major issues associated with forest hydrology and tree water relations. These are the effects of forest manipulation (logging, burning, etc.) on catchment water yield and quality (see Chapter 7) and the effects of stand water balance and soil water content on forest growth. The analysis of both issues requires the ability to calculate stand water balance.

Much of the research done in hydrology has been empirical, unaccompanied by the measurements needed to identify and quantify the processes contributing to the results obtained. Classical catchment hydrology and streamflow measurements are of this type, which perhaps explains why Langford's (1976) *E. regnans* results were initially regarded as something of an anomaly. The reason for the results seems, in retrospect, relatively obvious, but it is only in recent years that the importance of leaf area index and the reasons for the three categories of catchment response to clearing (see Fig. 4.5) have been clearly recognized and appreciated. The increase in measurements of water flow up trees (e.g., Jayasuriya *et al.*, 1993) and in understanding of the role of stomatal con-

IV. Concluding Remarks

ductance and aerodynamic exchange mechanisms has led to increasing recognition of the need to measure these variables as well as soil water content and streamflow. We note Beven's (1989) argument, mentioned earlier, that rigorous, physically based models are often of limited value in catchment hydrology; however, we also note that, although Beven's argument was general, he appeared to focus primarily on the soil—transpiration rates and vegetation were not mentioned. We would argue, from the evidence that we have reviewed and the information presented elsewhere in this book, that it should be possible to estimate quite accurately the effects of tree clearance on the water yield of catchments using relatively simple models. Managers should also be able to make qualitative estimates of the effects of clearing from the information presented here.

Vertessy *et al.* (1993) have demonstrated that physically based models can provide accurate simulations of the runoff and water yields of complex catchments. Theirs is esentially a "bottom up" approach: The system is described in detail using equations that require many parameter values and attempting to account for all the components of the system for each small subunit of the catchment. This is excellent from a research point of view, but such a model is not useful to managers. However, it can serve the very valuable purpose of testing and evaluating much simpler models to determine how much complex model(s) can be simplified without losing the capacity to produce useful results. At the time of writing, Vertessy *et al.* had not made much progress on evaluating the extent to which their model (TOPOG) could be simplified and used in this way, but it is one of their research objectives. The alternative approach is to develop a simple model from first principles—perhaps based on the areas in a catchment where L^* is within particular limits, some simple, general descriptions of soils, and simple topography. This would be tested against models such as TOPOG.

In any environment where the potential (atmospherically driven) evaporation [Eq. (3.22)] exceeds precipitation, trees will tend to use all the water available to them, which will be total precipitation less losses caused by interception, runoff during high-intensity rain events, and evaporation from the soil surface or from litter layers. If rainfall is highly seasonal, such as in some tropical or subtropical areas that may support deciduous forests (see Chapter 2), then the losses are likely to be greater because soil storage may be filled during the rainy season, and losses by drainage and runoff may be significant. In areas where the precipitation exceeds potential evaporation, annual water use by trees will be equal to cumulative transpiration, which will depend on atmospheric conditions and their interaction with leaf area, stomatal conductance, and canopy architecture. In the first instance, (water-limiting) tree growth will depend on water availability: When soil water in the root zone is limiting to the extent that water cannot move to the roots fast enough to meet the atmospheric demand, modified by stomatal responses to light, stomata will close so that transpiration rates do not exceed supply rates. This must result in restriction of growth rates. In the limit, growth will cease. From the point of view of production forestry, the objective must be to be able to identify the periods when this will happen.

Recommended Reading

Borman, F. H., and Likens, G. F. (1979). "Patterns and Progress in a Forested Ecosystem." Springer-Verlag, New York.

Hornbeck, J. W., Adams, M. B., Corbett, E. S., Verry, E. S., and Lynch, J. A. (1993). Long-term impacts of forest treatments on water yield: A summary for north-eastern USA. *Hydrol.*, **150**, 323–344.

McNaughton, K. G., and Jarvis, P. G. (1986). Predicting effects of vegetation changes on transpiration and evaporation. *In* "Water Deficits and Plant Growth" (T. T. Kozlowski, Ed.), pp. 1–47. Academic Press, New York.

Pallardy, S. G., Cermack, J., Ewers, F. W., Kaufman, M. R., Parker, W. C., and Sperry, J. S. (1994). Water transport dynamics in trees and stands. *In* "Resource Physiology of Conifers: Acquisition, Allocation and Utilization." (W. R. Smith and T. M. Hickley, eds), pp. 301–389. Academic Press, San Diego.

Pereira, H. C. (1973). "Land Use and Water Resources in Temperate and Tropical Climates." Cambridge Univ. Press, Cambridge, UK.

Tardieu, F., and Davies, W. J. (1993). Integration of hydraulic and chemical signaling in the control of stomatal conductance and water status of droughted plants. *Plant Cell Environ.* **16**, 341–349.

Verstessy, R. A., Hatton, T. J., O'Shaughnessy, P. J., and Jayasuriya, M.D.A. (1993). Predicting water yield from a mountain ash forest catchment using a terrain analysis-based catchment model. *J. Hydrol.* **150**, 65–70.

Ward, R. C. (1984). On the response to precipitation of headwater streams in humid areas. *J. Hydrol.* **74**, 171–189.

5

Carbon Balance of Forests

This chapter is, in a sense, the central chapter of the book. Forest growth is a matter of carbon sequestration and distribution: The standing biomass at any time reflects net primary production integrated over the life of the plants. All the processes we discuss in detail in other chapters—radiation interception, nutrient uptake, and water relations—affect forest carbon balance. The carbon balance of forests is important in the global carbon balance; Landsberg *et al.* (1995) point out that the forested areas of the world account for 80–90% of plant and 30–40% of soil carbon (see data in Schlesinger, 1991). The patterns of carbon uptake by forests, and the release of CO_2 as a result of land clearance and forest destruction, are significant factors in the global carbon balance, although the net effect of changes in the uptake of CO_2 by forests is unlikely to be significant. We comment on this in more detail at the end of this chapter.

The carbon balance of a forest is the net result of CO_2 uptake for the fundamental biological process of photosynthesis and CO_2 emission as a product of autotrophic respiration, which is often divided into growth and maintenance respiration. Over any time interval, the difference between leaf photosynthesis, net photorespiration, and the carbon lost by autotrophic respiration is called **net primary production** (NPP). The carbohydrates formed as a result of photosynthesis are allocated to the component parts of plants [e.g., foliage, branches, stems (sapwood and bark), and roots]. Carbon allocation is poorly understood but of central importance. The proportion of the carbon fixed by trees that goes to their various component parts determines the growth pattern of the trees, their potential for future growth, and their ability to tolerate environmental stresses. Our ability to predict forest productivity and the consequences of natural disturbances and management practices that affect forest growth will remain limited until we understand the factors influencing carbon allocation and can predict its effects on growth patterns.

Net ecosystem production (NEP), the net flux of CO_2 to or from a forest ecosystem, is the end result of carbon fixation by photosynthesis (gross primary production) and losses by autotrophic and heterotrophic respiration. Heterotrophic respiration is essentially the process of oxidation of organic matter on the forest floor and in the soil, known as decomposition (Chapter 6). The dominant inputs of organic matter to the forest floor are above-ground (leaf, branch, stem, and reproductive structures) and below-ground (mycorrhizae and fine and coarse root turnover) detritus.

The processes outlined previously are depicted diagrammatically in Fig. 5.1. Although we are not, for the most part, concerned with measurement methods and procedures in this book, it is worth briefly outlining some of the methods used to measure the major variables discussed in this chapter. This outline is provided in Box 5.2 (p. 142).

In the following sections, we provide background information on photosynthesis at the leaf level and review the current state of knowledge

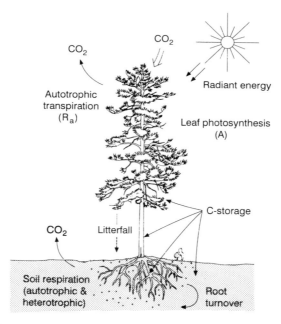

Figure 5.1 Schematic diagram illustrating the major components of a forest carbon budget. CO_2 is absorbed from the atmosphere and fixed by photosynthesis (**A**), resulting in carbon storage in leaves, stems, and roots. Autotrophic respiration (R_a) returns CO_2 to the atmosphere. Litterfall and root turnover provide carbon to the soil, where heterotrophic respiration also returns CO_2 to the atmosphere. Net primary production (NPP) = GPP-R_a.

> **Box 5.1 Measurements of Stomatal Conductance**
>
> Stomatal conductance is normally inferred from measurements of the flux of water vapor from leaves/needles, either in gas-exchange systems in the laboratory or using porometers in the field. Laboratory gas-exchange equipment has not changed, in terms of its principle of operation, in more than 20 years, but the control—and to a lesser extent the measurement—systems have improved considerably. Porometers and field gas-exchange systems, on the other hand, have been tremendously developed and improved to the point that it is now possible to buy instruments with several different chamber designs (for laminar leaves and needles), with control of temperature, radiation, and the properties of the air stream that is passed over the leaves, and computer-recorded output, in terms of g_s or r_s. Control of the air stream includes CO_2 concentration. These systems make it possible to collect a great deal of high-quality data in a relatively short time and develop stomatal conductance response surfaces for a variety of environmental parameters while keeping other parameters constant.

about stomatal conductance (g_s). Stomata provide the means whereby plants expose a wet surface to the air to capture CO_2, and a mechanism for controlling the inevitable water loss that results from this. Because of their key role, there has been a vast amount of research on stomata and the way they function and respond to environmental factors, but much of the current knowledge can be encapsulated in relatively simple mathematical models (Jarvis, 1976; Collatz et al., 1991; Leuning, 1995). Knowledge about photosynthesis and g_s, combined with models of radiation interception by canopies, can be used to calculate canopy photosynthesis (see Section III). We review the information currently available about photosynthesis and respiration and the factors affecting carbon allocation, NPP, and NEP. We deal with autotrophic respiration in Section III and with net primary production and carbon allocation in Section IV. The role of forests in the global carbon budget is considered briefly at the end of the chapter. Heterotrophic respiration, a major determinant of net ecosystem production; is discussed in detail in Chapter 6.

For the forester or forest manager concerned with wood production, "productivity" means getting the maximum possible amount of biomass into the stem; therefore carbon allocation is important and litterfall, included in the ecologists' estimate of net primary production, is of little interest except in relation to its importance as a source of organic matter for mineralization. For ecologists concerned with ecosystem energetics, net ecosystem production provides a measure of the net change in carbon content in the whole ecosystem (vegetation, detritus, and soil) and is the variable of primary concern. A small fraction of the annual carbon budget of trees is allocated to secondary compounds to reduce herbivory. Despite the fact that they comprise a relatively small proportion of the total mass of carbon, these compounds have significant effects on carbon and nutrient cycling processes in the soil (Chapters 6 and 7).

I. Leaf Photosynthesis

Photosynthesis comprises light and dark reactions that involve the removal of electrons from water—resulting in the release of O_2—and donation of these electrons to CO_2, leading to reduced carbon compounds (CH_2O) with a gain in free energy. The process takes place in the chloroplasts. The overall process may be written

$$H_2O + CO_2 + \text{energy} \rightarrow O_2 + (CH_2O).$$

The primary photochemical processes take place when light energy is absorbed by the photosynthetic pigments, which raises the energy level of the light-harvesting chlorophyll molecules to an excited state. A specialized chlorophyll molecule donates electrons to electron carriers. The electrons flow down the transport chain and their energy is used to generate adenosine triphosphate (ATP) and nicotinamide adenine dinucleotide phosphate ($NADPH_2$). The photophysical and photochemical light reactions proceed at a rate that depends only on the wavelength and light intensity. These reactions are not affected by temperature or CO_2 concentration. In contrast, electron transport is strongly dependent on temperature because it occurs through the chemical reactions of molecules bound to the chloroplast membranes.

The dark reactions use the energy (ATP) and reducing power ($NADPH_2$) produced by the light reactions to reduce CO_2 to carbohydrate (CH_2O). The acceptor of CO_2 is ribulose-1,5-bisphosphate (RuBP), the reaction being catalyzed by the enzyme RuBP carboxylase-oxygenase (Rubisco). The first carbon reduction product in most trees is a 3-carbon (C_3) compound, 3-phosphoglyceric acid, reduced by ATP and $NADPH_2$, which further metabolizes to form sugars. The RuBP is regenerated in the Calvin–Benson cycle. [There is a large and important group of plants—including some trees—in which the first carbon reduction product is a C_4 compound. However, these will not be considered in this book; readers interested in this alternative biochemical pathway of CO_2 assimilation are referred to Salisbury and Ross (1992).] RuBP carboxylase constitutes a major fraction of leaf protein, but because it has a relatively low affinity for CO_2 and is competitively inhibited by oxygen, it has been implicated as a factor that limits the rate of photosynthesis.

The relationship between photosynthesis (equated with CO_2 assimilation rate A) and the intercellular concentrations of CO_2 (c_i) takes the form of an asymptotic curve (Fig. 5.2) that has been designated the "demand function," whereas the line connecting the ambient CO_2 concentration c_a to A is the "supply function." The slope of the supply function line is $-g_s$ [Eq. (5.8)] and the downward projection from the point of intersection of the demand and supply functions to the c_i axis gives the value of c_i. The regeneration of RuBP appears to be dependent on the partial pressure of CO_2 at the carboxylation sites ($p(CO_2)$). If $p(CO_2)$ is low, CO_2 assimilation is not limited by the amount of the enzyme (Rubisco). As $p(CO_2)$ increases, electron transport reactions, and therefore the capacity to regenerate RuBP, become limiting. The linear portion of the curve has been designated RuBP saturated (Farquhar and Sharkey, 1982); in this part of the curve there is ample RuBP and Rubisco and any increase in c_i results in activation of more enzyme, which increases the rate at which CO_2 is fixed. However, if the rate of RuBP carboxylation is increased sufficiently, the capacity to regenerate the substrate becomes

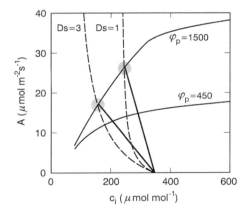

Figure 5.2 Hypothetical A–c_i curves showing the photosynthetic demand function [Eq. (5.2)] for two photon flux density values (φ_p = 1500 and 450 µmol m^{-2} sec^{-1}). Supply is constrained by stomatal conductance. The dotted curves are supply constraint functions, determined by vapor pressure deficit at the leaf surface (D); their points of intersection with the demand functions give the equilibrium assimilation rates, where demand for CO_2 and supply are in balance. Vertical projection from the intersection points gives the equilibrium value of c_i, and the slope of the dark lines connecting these points to the ambient CO_2 concentration on the c_i axis gives $-g_s$. (see Leuning, 1990, 1995; We are grateful to Dr. Ray Leuning for the diagram).

limiting, and any further increase in c_i does not lead to concomitant increase in A. This suggests that there is some optimum value of g_s that will lead to maximum photosynthesis for a particular leaf condition (e.g., nitrogen status).

In recent years, the Farquhar and von Caemmerer (1982) model has become the most widely used as a basis for the analysis of photosynthesis by C_3 plants (see, for example, Leuning, 1990, 1995; McMurtrie et al., 1992a; McMurtrie and Wang, 1993; Wullschleger, 1993; Wang and Polglase, 1995). This model gives the net rate of leaf photosynthesis as

$$A = V_c(1 - \Gamma/c_i) - R_d \tag{5.1}$$

(μmol m^{-2} sec^{-1}). Γ is the CO_2 compensation point in the absence of day respiration, c_i is the intercellular concentration of CO_2, and R_d is the rate of day respiration. V_c is given by

$$V_c = \min(W_j, W_c, W_p), \tag{5.2}$$

where W_j, W_c, and W_p are the rates of carboxylation limited by RuBP regeneration, by Rubisco activity, and by triose phosphate utilization, respectively. Limitation by triose phosphate utilization occurs when the utilization of triose phosphates for the production of starch and sucrose does not keep pace with the rate of production of triose phosphates in the Calvin cycle. It appears, from Wullschleger's (1993) data, that this may occur in only a few plants, none of which are trees. Most treatments ignore W_p and we will do so here: There are many uncertainties in the calculation of leaf and canopy photosynthesis that can lead to errors far greater than those caused by neglecting this term.

The RuBP regeneration-limited rate of carboxylation is

$$Wj = J/(4 + 8\Gamma/c_i), \tag{5.3}$$

where J is the potential electron transport rate (μmol m^{-2} sec^{-1}), calculated from a nonrectangular hyperbola given by

$$\Theta J^2 - (\alpha_p \varphi_p + J_{max})J + \alpha_p \varphi_p J_{max} = 0, \tag{5.4}$$

where Θ is a shape coefficient, which takes values between 0 and 1, and α_p is the quantum requirement for electron transport; Eq. (5.4) tends to a rectangular hyperbola, as Θ tends to zero, and to a Blackman curve— with a clear transition point at which $J = J_{max}$—when $\Theta = 1$. The rate of electron transport is calculated as the smaller and positive solution of Eq. (5.4).

The Rubisco-limited rate is

$$W_c = V_{cmax} c_i / (c_i + K_c(1 + O_i/K_o)), \tag{5.5}$$

where K_c and K_o are Michaelis coefficients for CO_2 and O_2, respectively, and O_i is intercellular O_2 concentration. Wullschleger (1993) analyzed

A/c_i curves in the literature and derived V_{cmax} and J_{max} values for 109 C_3 plant species. He found that the two parameters are strongly (linearly) related, indicating that C_3 species preserve a close relationship between the carboxylation and electron transport processes. The values for broad-leaved forest trees were statistically significantly higher than for conifers (see Table 5.1). The parameters α_p and Γ are specific to the Rubisco enzyme and can be taken as the same for all C_3 species. Values are given in Table 5.1, together with reference values of the Michaelis constants, J_{max}, and V_{cmax}. Γ and R_d vary with temperature: Leuning (1990), McMurtrie et al. (1992a), McMurtrie and Wang (1993), and Wang and Polglase (1995) give functional relationships for the temperature dependencies.

To use Eq. (5.1), (5.3), and (5.5), we need values for c_i. The rate of CO_2 supply to leaves can be described as a process of diffusion across the boundary layer and stomatal resistances. It can be written

$$A = g_s(c_a - c_i)/P, \qquad (5.6)$$

where c_a is the partial pressure of ambient CO_2 concentration (currently about 350 μbar) and P is atmospheric pressure. If r_a and r_s are the boundary layer and stomatal resistances for water vapor, respectively, then

$$g_s = 1/(1.37 r_a + 1.6 r_s). \qquad (5.7)$$

(The molecular diffusivities for water vapor and CO_2 in air are different; therefore, the r_a and r_s values used to calculate g_s for CO_2 must be corrected by the ratios of the diffusivities, i.e., by 1.6. The ratio for boundary layer resistances is 1.37 because of the influence of turbulence.) If we as-

Table 5.1 Average Values (± 1 Standard Error, SE) for the Parameters of the Leaf Photosynthesis Equations[a]

Parameter	Mean ± SE	Unit of Measure
J_{max}: broad-leaved temperate	104 ± 64	μmol mol^{-1}
J_{max}: evergreen conifers	40 ± 32	μmol mol^{-1}
V_{cmax}: broad-leaved temperate	47 ± 33	μmol mol^{-1}
V_{cmax}: evergreen conifers	25 ± 12	μmol mol^{-1}
K_c	300	μmol mol^{-1}
K_o	250	mmol mol^{-1}
Γ	42	μmol mol^{-1}
α	0.385	μmol mol^{-1}

[a] The J_{max} and V_{cmax} values, and estimates of their standard errors, are from Wullschleger (1993). The values of the Michaelis constants, Γ, O_i, and α are representative of the values used in a number of publications (see text).

sume that $r_a \gg r_s$ (see Chapter 3), then CO_2 concentration at the leaf surface (c_s) can be equated to c_a, and Eqs. (5.6) and (5.7) simplify to

$$A = g_s(c_a - c_i), \qquad (5.8)$$

There have been many models of stomatal conductance and the factors affecting it; some comment is provided in the following section. The current most complete, and apparently accurate, model is the modified version, developed by Leuning (1990; 1995), of the Ball et al. (1987) equation. Eliminating a constant, and a correction for Γ, which accounts for behavior at low CO_2, it can be written

$$g_s = a_1 A / (1 + D_s/D_o) c_s, \qquad (5.9)$$

where D_l is the vapor pressure deficit at the leaf surface and a_1 and D_o are empirical parameters for which Leuning gives a range of values for *E. grandis* from 20 to 43 for a_1 when $D_o = 350$ Pa. Substituting for g_s from Eq. (5.8) leads to

$$c_i/c_s = 1 - (1 + D_s/D_o)/a_1, \qquad (5.10)$$

which indicates that the ratio of internal to ambient (leaf surface) CO_2 concentrations varies with D_s. The conservative nature of this ratio is consistent with the idea that stomata respond to the environment in such a way that c_i is maintained more or less constant (Wong et al., 1979). By the same argument used to equate c_s and c_a, we can take $D_s \approx D$, where D is the atmospheric vapor pressure deficit in the region of the leaf. It follows that, given values for D_o and a_1—and we note Leuning's warning that there is considerable variation in these—we can estimate c_i/c_s ($\approx c_i/c_a$), and hence c_i, for insertion into the photosynthesis equations. We should also note here that this model makes no allowance for the effects of leaf water status and soil water content on stomatal conductance. These are discussed in Chapter 4.

A. Stomatal Conductance

Although the model outlined previously [Eqs. (5.9) and (5.10)] provides an (apparently) sound mechanistic description of variations in stomatal conductance in relation to photon flux density, vapor pressure deficit, and ambient CO_2 concentrations, it has the disadvantage that an estimate of A is required before estimates of g_s can be calculated. It is therefore worth briefly reviewing some more empirical models of stomatal conductance.

Jarvis (1976) presented a model to describe the responses of stomata to environmental variables and applied it to temperate conifers. He used results from controlled environment studies to choose the empirical function that best described the response of stomata to each variable and combined them in a multiplicative model of the form

$$g_s = f_1(D) \cdot f_2(\varphi_p) \cdot f_3(\psi_f) \cdot f_4(T). \tag{5.11}$$

The general forms of the functions used by Jarvis (1976) have been found to be suitable for use with a number of plants other than temperate conifers (see, for example, Whitehead et al., 1981), although the values of the coefficients may vary. The model has been widely used. A similar model was developed by Thorpe et al. (1980). This omits the foliage water potential term (ψ_f)—because water stress does not become a factor until ψ_f falls quite low (Landsberg et al., 1976; Beadle et al., 1978)—and the effects of varying CO_2 concentrations, which need not be included in empirical models for plants well coupled to the environment. Thorpe et al. expressed their model as a single equation:

$$g_s = g_{ref}(1 - aD)/(1 + b/\varphi_p), \tag{5.12}$$

where a and b are empirical "constants" and g_{ref} is a reference conductance. The parameter values may be determined from measurements of stomatal response to φ_p at low values of D (e.g., from 0.5 to 1.0 kPa) and responses to D when φ_p is not a limiting variable. In both cases, the analysis uses values of g_s normalized to the highest observed value, i.e., that value is taken as unity. This greatly reduces the scatter in data. The reference value (g_{ref}) is then the value of g_s expected when both D and φ_p are nonlimiting, i.e., it is the maximum value of g_s. Körner et al. (1979) list maximum leaf conductance values for 294 species. For woody species, the values range from 1 to 5 mm sec^{-1}. If we take $a = 0.3$ kPa^{-1} and $b = 70$ μmol m^{-2} sec^{-1}, then $g_s = 0$ when $D = 3$ kPa and $g_s = 0.5\, g_{ref}$ when $\varphi_p = 70$ μmol m^{-2} sec^{-1}. Schulze et al. (1994) summarized stomatal conductance values for major vegetation biomes and demonstrated several useful scaling algorithms, and their theoretical background, for stomatal conductance and water vapor and carbon dioxide fluxes at the canopy level. They reported a strong positive linear correlation between stomatal conductance and leaf nitrogen concentration for 15 major vegetation cover types in the world, although they concluded that the relationship within a vegetation type was relatively conservative. Reich et al. (1992) reported a negative exponential relationship between g_s and leaf longevity.

II. Canopy Photosynthesis

Equations (5.1)–(5.10) provide a coupled model of assimilation and stomatal conductance. To apply them to the calculation of canopy photosynthesis requires calculation of photon flux density (φ_p) at any level in the canopy and values of the vapor pressure deficit (D) for the air in the canopy. The procedure for calculating φ_p in the canopy has been described in Chapter 3. More accurate results will be obtained for canopy photosynthesis if the canopy is divided into layers rather than being treated as a single layer with an average value of φ_p (see Chapter 3).

Photon flux density in the middle of any layer in a canopy is applied to the leaf area in that layer and Eq. (5.4) solved for J. For calculations over relatively short periods, such as days, it is important that the radiation absorbed by the canopy be separated into diffuse and direct components [see Eqs. (3.10) and (3.11)], because shaded leaves receive diffuse radiation only, whereas sunlit leaves receive diffuse plus direct radiation. Without this separation, canopy assimilation is likely to be overestimated because it will be assumed that all the leaves, in any layer of the canopy, receive the average φ_p for that layer. For calculations over longer periods, errors resulting from treating radiation simply as a total flux are likely to balance out and be negligible. (See the comments in Chapter 3 relating to radiation absorption by canopies and the relative effectiveness of direct and diffuse radiation.)

Equation (5.10) is used, with the average value of D ($\approx D_1$) for the period of interest, to calculate c_i/c_s from which, assuming $c_s \approx c_a$, we obtain c_i. Inserting this in Eqs. (5.3) and (5.5) with appropriate values of J and the other parameters gives values for W_j and W_c; the smaller of these is the rate-limiting step and is used in Eq. (5.1) to calculate the leaf photosynthesis rate. If required, this can be used in Eq. (5.8), with c_s and c_i, to solve for g_s. Canopy photosynthesis is the sum of the rates in each layer multiplied by the leaf area in those layers. Alternatively, other models, such as Eqs. (5.11) or (5.12), can be used to estimate values for g_s, in which case the value is inserted in Eq. (5.8) and used to calculate c_i. The procedure is then as outlined.

The approach described here has the advantage that it can be scaled up in space and time. For large-scale calculations, canopies would be treated as single layered. If calculations are to be made over long periods, such as months or seasons, greater accuracy will be achieved if daily average values of shortwave incoming radiation (φ_s) are used as the basis for estimating φ_p, with average daily values of vapor pressure deficit (D). Because we are dealing with nonlinear processes, long-period averages are more likely to lead to significant error than summation of results derived from a series of short periods.

Leuning et al. (1996) have used a somewhat more complex version of the procedures outlined here to calculate canopy photosynthesis. Leaf photosynthesis is related to leaf nitrogen (N) concentration (e.g., Reich et al., 1992; Gower et al., 1993a) and leaf N has been shown to be distributed vertically through canopies in a manner that suggests a positive correlation to average φ_p at any level (Hollinger, 1989). Leuning *et al.* calculated the distribution of N through a canopy and adjusted the values of the parameter describing Rubisco activity ($V_{c\cdot max}$) through the canopy (see Fig. 5.4). They also inserted the values of g_s, obtained as described here, into the Penman–Monteith equation, which leads to a solution for leaf temperature and, hence, better values for the temperature corrections to the photosynthesis parameters. Their model analysis allowed them to examine the relative importance of the various parameters of the model to obtain insight into the behavior of real plant canopies. Leuning *et al.* compared their results with observations in a wheat crop. Analyses of that complexity are, perhaps, not yet justified in forestry, but may be valuable as a basis for evaluating the parameter values appropriate for use in simple models that will be applied over large areas and relatively long periods and for evaluating the impact and implications of the various ecophysiological factors that affect photosynthesis (see Section II,C).

For many purposes, it is convenient to use empirical equations to model canopy photosynthesis rather than the mechanistic equations of the Farquhar and von Caemmerer (1982) model. The two most commonly used are the rectangular and nonrectangular hyperbolae. Thornley and Johnston (1990) have provided a detailed mathematical treatment of the properties of the nonrectangular hyperbola and its use in modeling photosynthesis. For gross photosynthesis, it is Eq. (5.4), with CO_2 assimilation (**A**) substituted for electron transfer *J*. The shape factor Θ can be expressed in terms of the diffusion (r_s) and carboxylation (r_x) resistances:

$$\Theta = r_s/(r_x + r_s), \tag{5.13}$$

and the rectangular hyperbola, with R_d included to give net assimilation rate, is

$$A = \alpha_p \varphi_p A_{max}/(\alpha_p \varphi_p + A_{max}) - R_d. \tag{5.14}$$

α_p is given by the initial slope of the assimilation/photon flux response curve and the photon-saturated assimilation rate is given by

$$A_{max} = (c_i - \Gamma)/r_x. \tag{5.15}$$

Landsberg (1986) tabulated values for α_p and A_{max} for various tree species (see also Holbrook and Lund, 1995). They are moderately variable, depending on the method of measurement as well as on whether the leaves were acclimated to sun or shade conditions. Typical values of α_p are 0.03–0.05 mol mol^{-1} ($\approx$ 1.4–2.1 g mol^{-1}) and for A_{max} 10–15 μmol m^{-2} sec^{-1} ($\approx$ 4.5–6.8 × 10^{-2} mg m^{-2} sec^{-1}).

A. Relations between Canopy Photosynthesis and Productivity

Several of the models discussed in Chapter 9 consist essentially of procedures for calculating canopy photosynthesis and respiration, and hence stand productivity, utilizing knowledge about the processes discussed in this chapter. Figure 5.3 is a plot of gross primary productivity—calculated using the model BIOMASS (McMurtrie et al., 1990), which is essentially a canopy photosynthesis model—against absorbed photosynthetically active radiation (φ_{abs}) after the φ_{abs} values had been corrected to account for environmental conditions that reduce photosynthesis and hence the radiation utilization efficiency. The figure is presented here to illustrate the very strong positive relationship between canopy photosynthesis and the radiant energy absorbed by the canopy.

Despite the relationship illustrated in Fig. 5.3, it is unclear whether there are simple relationships between canopy photosynthesis and productivity; as we noted in the introductory section of this chapter, canopy photosynthesis is only one factor in the equation. To calculate forest productivity (as NPP), we need estimates of autotrophic respiration losses as well as total net photosynthesis over the period of interest. We deal with

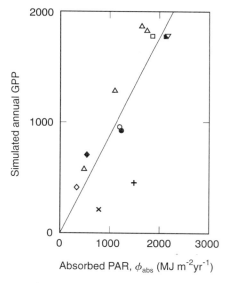

Figure 5.3 Canopy gross primary production as a function of absorbed photosynthetically active radiation (φ_{abs}) corrected for temperature and vapor pressure effects. The figure (redrawn from McMurtrie et al., © 1994, OIKOS) illustrates the close relationship between canopy photosynthesis and absorbed radiation. The symbols relate to pine stands in different locations (Australia, New Zealand, USA, and Sweden) and different water and nutritional treatments.

autotrophic respiration under Section III and with NPP and carbohydrate allocation under Section IV.

B. Ecophysiological Aspects of Leaf Photosynthesis

The factors affecting leaf photosynthesis are leaf age, water relations—primarily through their effect on stomatal conductance—acclimation to sun or shade conditions, and nutritional status. A_{max} almost universally declines with leaf age within a tree (Troeng and Linder, 1982; Teskey *et al.*, 1984). This decline is related to increased shading by new leaf cohorts (Schoettle and Smith, 1991; Schoettle and Fahey, 1994) and retranslocation of nutrients from aging needles (Son and Gower, 1991).

Figure 5.4 illustrates the commonly observed light response of leaf photosynthesis for various nitrogen concentrations. The curves are typical hyperbolic photon flux/CO_2 assimilation response curves and simple mathematical theory links leaf photosynthesis, nitrogen concentration, and radiation regime to predict vertical distribution of nitrogen in the canopy needed to maximize canopy photosynthesis (Sellers *et al.*, 1992; Leuning *et al.*, 1995). Many stand- to global-level models use the relationship between leaf photosynthesis and nitrogen concentration to scale carbon assimilation from the leaf to canopy level (e.g., Running and Gower, 1991; Aber and Federer, 1992; Schulze *et al.*, 1994).

Despite the relationships shown in Fig. 5.4, it appears that, in the case of conifers, the contribution of increased leaf photosynthesis (on a leaf area or weight basis) to the growth responses to fertilization is small; the reasons for the responses may lie in other processes. Fertilization has

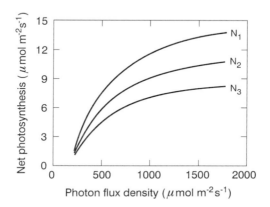

Figure 5.4 Hypothetical photosynthesis–photon flux density (φ_p) response curves for leaves of broad-leaved species with three different nitrogen concentrations: $N_1 > N_2 > N_3$. The values of N_1-N_3 would be expected to be in the range $1-3$ g N m^{-2} (leaf area) [The diagram is consistent with data of Hollinger (1989), Abrams and Mosteller (1995), and Field (1983)].

been reported to have little or no stimulatory effect on photosynthesis of *P. radiata* (Sheriff *et al.*, 1986), *Pseudotsuga menziesii* (Van den Driessche, 1973) and *P. contorta* (Reid *et al.*, 1983); however, other scientists have found that fertilization increased *A* for *P. sylvestris* (Smolander and Oker-Blom, 1990), *Pseudotsuga menziesii* (Brix and Ebell, 1969; Brix, 1981), and *P. radiata* (Thompson and Wheeler, 1992). In a study involving four conifer species, there were no consistent differences in *A* of cut branches, measured under optimal environmental conditions, between control and fertilized trees, although fertilization increased needle N concentration by 40–60% (Gower *et al.*, 1996a). Similarly, Teskey *et al.* (1995) found no differences in *A*, measured under controlled environmental conditions in shoots, of a *P. elliottii* stand where fertilization had increased foliage mass and above-ground NPP and resulted in a relative decrease in carbon allocation to fine roots. It seems likely that, in these experiments, fertilization resulted in luxury consumption of N, which was stored in the foliage in the form of amino acids (Yoder *et al.*, 1994) instead of Rubisco.

Numerous studies have demonstrated that leaf nitrogen concentration, specific leaf area, and leaf longevity are linked (see Lambers and Poorter, 1992; Reich *et al.*, 1995, and papers cited therein), and a general model connecting leaf longevity, structure, and carbon balance is beginning to emerge. Long-lived leaves have lower specific leaf area than short-lived leaves (see review by Reich *et al.*, 1995; Fig. 5.5) because of

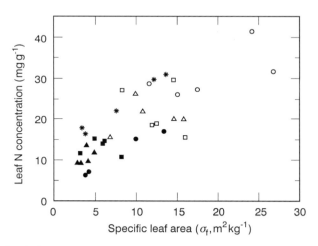

Figure 5.5 Leaf nitrogen concentration as a function of specific leaf area in a range of ecosystems. The symbols denote temperate needle-leaved deciduous (△), temperate needle-leaved evergreen (▲), temperate broad-leaved deciduous (□), temperate broad-leaved evergreen (■), tropical deciduous (○), tropical evergreen (●); The asterisks denote five tree species grown in adjacent plantations (Gower *et al.*, 1993b).

the greater concentration of structural carbon-based constituents and thicker cuticle. This results in a positive relationship between leaf nitrogen concentration and specific leaf area. Survey studies have indicated that leaf N concentration is inversely related to leaf lifespan across a broad range of plant groups in contrasting (Field and Mooney, 1986; Reich et al., 1992) and in similar environmental conditions (Gower et al., 1993a). Greater photosynthetic rates measured in deciduous conifers (*Larix* spp.) than in co-occurring evergreen conifers are consistent with these results (Kloeppel et al., 1995). This is useful for understanding the relationship between leaf photosynthesis (on a weight basis) and the lifespan of leaves on plants in the field, whereas the causal relationship between leaf N concentration, photosynthesis, and leaf longevity suggests that it may be possible to derive general empirical relationships to estimate maximum net photosynthetic rates based on leaf N and leaf lifespan. In general, plants growing in harsh climates or on nutrient-poor soils tend to have longer-lived leaves. The notable exception is the circumpolar importance of *Larix*, a deciduous conifer, in the northern hemisphere (Gower and Richards, 1990).

III. Autotrophic Respiration

Autotrophic respiration (R_a) involves the oxidation of organic substances to CO_2 and water, with the production of ATP and reducing power (NADPH):

$$O_2 + CH_2O \rightarrow CO_2 + H_2O. \tag{5.16}$$

Total autotrophic respiration is commonly divided into three components: maintenance ($R_{a \cdot m}$), growth ($R_{a \cdot g}$) and ion uptake ($R_{a \cdot u}$). $R_{a \cdot m}$ is the cost of protein synthesis and replacement, membrane repair, and maintenance of ion gradients (Penning de Vries, 1975); it is the component of R_a that is most sensitive to environmental change (Ryan, 1991). The most important environmental factor affecting maintenance respiration is temperature because protein synthesis rates increase exponentially with increasing temperature. The temperature dependence of R_a is commonly expressed in terms of the Q_{10}, or the change in rate with a 10°C rise in temperature:

$$R_a = R_{a \cdot ref} Q_{10}^{[(T-T_{ref})/10]}, \tag{5.17}$$

where $R_{a \cdot ref}$ is the respiration rate at some reference temperature. Q_{10} values for the tissues of trees vary less than agricultural crops and commonly range from 2.0 to 2.3.

Because the relationship between respiration and temperature is nonlinear [Eq. (5.17)], ignoring daily and seasonal variations in temperature of the various biomass components will result in underestimates—in some environments the bias can be sizable. For example, a temperature difference of only 10°C between the maximum and minimum daily temperature will cause a 12% underestimate of daily respiration (Ågren and Axelsson, 1980). To avoid this bias, a sine function can be used to correct daily estimates, as described by Ryan (1991):

$$R_d = \tau_r R_o \exp(\beta_r T_d) I_o(\beta_r A_1), \qquad (5.18)$$

where R_d is daily total respiration, R_o is respiration at 0°C, β_r is $\ln(Q_{10}^{[(T-T_{ref})/10]})$, T_d is average daily temperature, A_1 is daily temperature amplitude [$(T_{max} - T_{min})/2$], and τ_r scales R_o to a daily flux. For a $x < 2$,

$$I_o(x) = 1 + 0.25x^2 + 0.016x^4 + 0.0004x^6. \qquad (5.19)$$

R_o can be calculated by dividing a respiration rate at a given temperature by $\exp(\beta_r)$. A similar approach can be used to account for seasonal variation in annual respiration estimates:

$$R_t = \tau_r R_o \exp(\beta_r T_a) I_o(\beta_r A_1) I_o(\beta_r A_2), \qquad (5.20)$$

where R_t is the total annual respiration, T_a is the average annual temperature, A_2 is the annual temperature amplitude, and τ_r scales R_o to an annual rate.

Because most organic N in plants is in the form of protein, protein synthesis and replacement, membrane repair, and the maintenance of ion gradients account for more than 60% of $R_{a \cdot m}$ (Penning de Vries, 1975). Maintenance respiration is therefore strongly influenced by tissue N concentrations (Amthor, 1989; Ryan et al., 1994). $R_{a \cdot m}$ varies among tissues, probably because of differences in protein type and amount (Ryan, 1991), so that changes in carbohydrate allocation may alter the maintenance respiration rate of trees.

Growth respiration ($R_{a \cdot g}$, also referred to as construction respiration) includes the carbon cost of synthesizing new tissue from glucose and minerals. Several approaches have been used to estimate growth costs, ranging from theoretical analysis of anabolic biochemical pathways (Penning de Vries et al., 1974; Chung and Barnes, 1977) to elemental analysis (McDermitt and Loomis, 1981) and heat of combustion (Williams et al., 1987). Ryan (1991) suggested that in the absence of data a reasonable approximation to growth costs can be made by assuming that $R_{a \cdot g}$ consumes 25% of the carbon allocated annually to each biomass component.

Ion uptake respiration is associated with the carbon cost of moving ions into the roots and across membranes. Few estimates of $R_{a \cdot u}$ exist. Veen (1980) estimated that it constitutes between 18 and 60% of total root respiration for maize, whereas van der Werf et al. (1988) reported slightly lower proportions for sedge. Veen (1980) recommended that, in the absence of experimental data, it is reasonable to assume 1.02 mol C mol^{-1} N for calculations of $R_{a \cdot u}$.

Measurements of the respiration of forest trees are usually made with chambers attached to stems or roots (see Box 5.2). One of the most comprehensive studies was probably that of Linder and Troeng (1981), who made continuous measurements of stem and coarse root respiration of a 20-year-old Scots pine stand in Sweden from January to November. Hourly respiration rates were exponentially related to temperature with a Q_{10} close to 2. There was pronounced variation in the rate of respiration at a given temperature at different times of the season. In recent years, Ryan has made measurements on a range of tree species, from which a number of useful relationships have emerged (Ryan, 1991; Ryan and Waring, 1992, Ryan et al., 1995). There is large temporal and spatial variation in chamber measurements, but it appears that stem respiration is highly correlated to live sapwood volume (Fig. 5.6) so that using allometric relationships to estimate stem sapwood volume from stem diameter, with published or measured Q_{10} values and temperature data, it is possible to calculate annual estimates of autotrophic respiration of woody biomass. Ryan et al. (1995) reported that the fraction of net canopy photosynthesis allocated to stemwood increased linearly with mean annual temperature for four temperate conifer stands.

Views on the importance of respiration as a fraction of net canopy photosynthesis at the stand level have changed as more respiration data

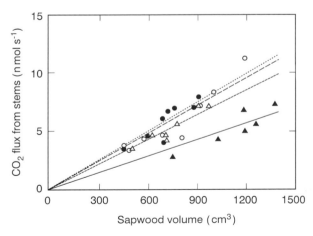

Figure 5.6 Relationship between stem maintenance respiration rate and sapwood volume for *Pinus resinosa* (△), *Pinus ponderosa* (○), *Tsuga heterophylla* (▲), and *Pinus elliottii* (●) in Wisconsin, Montana, Oregon, and Florida, respectively (adapted from Ryan et al., 1995).

Box 5.2 The Measurement of Photosynthesis, Respiration NPP, and NEP

Our knowledge of photosynthesis and its responses to environmental factors, at the leaf level, is good, and the literature is voluminous (see, for example, Lawlor, 1987). Photosynthesis is, and has been for many years, measured by gas exchange in the laboratory, usually on the leaves of seedlings, in chambers in which the (attached) leaves are enclosed and subjected to carefully controlled environmental conditions. There are now commercially available portable infrared carbon dioxide analyzers with a variety of chambers that can be used in the field, some of which provide environmental control. Autotrophic respiration can also be measured using portable gas analyzers and chambers that may be modified to fit stems, branches, and coarse roots. There have been fewer measurements of respiration than of photosynthesis, but there is an accumulating body of data on rates of CO_2 output from this source (see Section III). Measurements of total soil surface CO_2 flux is discussed in Chapter 6 (see Box 6.2).

NPP can be estimated from sequential sampling of standing biomass and litterfall. There have been many studies using these methods in forests, but because of the difficulty of estimating belowground biomass and fine root turnover, most biomass data are for above-ground biomass and NPP components only. However, there are large amounts of data on root mass, NPP, and turnover for temperate forests (see the review by Vogt *et al.*, 1986b), largely based on sequential soil coring or so-called "root ingrowth" methods, which consist of taking soil cores and filling the holes with root-free soil. These are extracted later, at intervals, and the rate of root growth into the cores is assessed. Root turnover has also been estimated by the "nitrogen turnover" method (Aber *et al.*, 1985; Nadelhoffer *et al.*, 1985). A method of studying fine root turnover that has recently been developed to the point of practicality utilizes a miniaturized video camera and automated system for scanning the roots along the sides of transparent tubes inserted in the soil (Hendrick and Pregitzer, 1992). The photographs are scanned and the results processed for root length and surface area estimates by computer. The technique provides a good basis for extending results obtained by the much more laborious techniques of manual extraction, washing, weighing and counting.

> The best estimates of NEP come from the modern, high-technology methods of measuring the net flux of CO_2 to (or from) extensive areas of forest by the micrometeorological technique known as eddy correlation. This involves one- or preferably three-dimensional, fast-response anemometers to measure the upward and downward (wind) eddies, accurate gas-exchange equipment, and considerable on-line computing capability (see Fitzjarrald and Moore, 1995). The net flux of CO_2 crossing the plane of the instruments is calculated as the mean covariance between fluctuations in vertical velocity (w) and the density of CO_2 (c): $F_c = \rho_a(\overline{w'c'})$, where the prime denotes deviations from the mean and the overbar signifies a time average. (See Kaimal and Finnigan, (1994), for a complete treatment, including discussion of instruments)
>
> The outstanding examples of work of this type are the studies made by Wofsy et al. (1993), over a temperate mixed hardwood forest, by Hollinger et al. (1994), over temperate broad-leaf evergreen forest in New Zealand; by Grace et al. (1995a,b), over Amazonian forest; and by Fan et al. (1995) and Baldocchi and Vogel (1996), over boreal forest (see Fig. 5.2). In 1994, there were a large number of measurements over boreal forests as part of the BOREAS project. The results were not published at the time of writing but will be published in a special issue of the *Journal of Geophysical Research*.

have become available. An earlier generation of ecologists assumed that, because respiration increases exponentially in relation to temperature while photosynthesis has an asymptotic relationship to temperature, a greater fraction of canopy photosynthesis would be allocated to R_a for tropical than temperate or boreal forests (Whittaker and Woodwell, 1975). However, carbon budgets for several temperate coniferous forests in contrasting climates indicate that the ratio of R_a/(net canopy photosynthesis) is surprisingly stable, averaging around 0.4 to 0.5 (Fig. 5.7; see also Gifford, 1994). These data imply that forests are highly conservative. The absence of marked trends in tree respiration in relation to climate may be explained by adjustments in carbon allocation to various biomass components (Ryan et al., 1994); also, plant respiration rates appear to acclimate to changes in temperature (Strain et al., 1976; Drew and Ledig, 1981). One exception to the general pattern is a *P. banksiana* stand near a prairie–boreal forest ecotone; in this stand, R_a/(canopy photosynthesis) ratio is 0.70. There is a view that R_a/(canopy photosynthesis) ratios may serve as indicators of stress; if this is the case, the observed ratio of

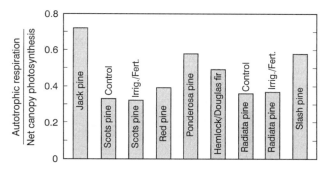

Figure 5.7 Fraction of annual net canopy photosynthesis allocated to autotrophic respiration (R_a). Values ranged from 0.3 to 0.72 and averaged 0.52. It is interesting to note that the highest value (0.72) was reported for a boreal *Pinus banksiana* stand located near a boreal forest–prairie ecotone. It is possible that this may serve as an indicator of stress related to climate change (data are from Gower *et al.*, 1996b; Ryan *et al.*, 1996; Baldocchi *et al.*, 1996; Linder and Axelsson, 1982).

0.7 for the boreal forest may have some significance for the carbon budgets of future boreal forests.

IV. Net Primary Production

NPP is defined as the difference between gross primary production (GPP) and autotrophic respiration R_a. It is therefore the amount of carbon accumulated by a tree over a specified time interval—usually 1 year. In theory, NPP can be calculated as the difference between gross primary production and R_a, but it is more commonly calculated from the relationship

$$\text{NPP} = \Delta W + w_{\text{det}} + w_{\text{herb}}, \tag{5.21}$$

where ΔW is biomass increment, w_{det} is detritus production, and w_{herb} denotes loss of tissue mass due to herbivory. Biomass increment is calculated from repeated measures or estimates of the biomass of a stand. Two approaches are commonly used to estimate annual biomass increment: Diameter growth can be followed for several years for all trees in a plot, or radial growth over the period of interest can be determined retrospectively by collecting radial increment cores from the trees in a plot or area of interest. Above-ground detritus is commonly separated into fine litter (leaves, twigs, reproductive tissue, etc.) and large litter (large branches and stems). Fine litter is measured using traps of varying size (e.g., 0.25–1.0 m^2), whereas large litter is usually determined from permanent tree mortality plots.

Although there are fewer data for large than fine litter, several patterns are evident: Leaf litter comprises 70–80% of the annual total fine litter (Bray and Gorham, 1964) but appears to average around 60% for *Eucalyptus* forests (Herbohn and Congdon, 1993). Large detritus comprises 10–15% of total detritus (Vogt *et al.*, 1986b; Raich and Nadelhoffer, 1989), although it can comprise the greater part of total aboveground detritus in old-growth forests (Chapter 8). Herbivory is often ignored in forest production estimates because it is assumed to be a relatively minor factor (<10–15% of NPP) in normal noninfected stands (Schowalter *et al.*, 1986). Herbivory can be measured by either measuring the amount of tissue removed or by measuring insect frass mass in litter traps and assuming a utilization efficiency (see Box 5.2)

A. Carbon Allocation within Trees

1. General Relationships NPP is of ecological interest because it provides a measure, as the name implies, of primary productivity. From the point of view of the growth patterns of trees, we are concerned with the way the carbon is allocated to the component parts of the trees, which affects their growth patterns and capacity to compete with other plants in their environment, as well as NEP, through the amount of detritus and fine roots produced. The factors determining carbon allocation are not well understood—study of these processes remains one of the major challenges facing environmental physiology and plant ecology. In this section, we discuss carbon allocation in general terms; the following section focuses on the matter of allocation to roots and root turnover, which are of particular importance. Because of the difficulty of making measurements of root systems, there are fewer empirical data, relating to roots, that can be used to evaluate models.

Various mechanisms have been proposed for carbon allocation and models developed to describe it (see later), but empirical allometric relationships are widely used to estimate the mass of the various tree components. The usual allometric equation—which holds across a remarkable range of species and conditions—is given in Chapter 3 [Eq. (3.2): $w_i = c_i d_b^n$, where w_i denotes the mass of any component i, c_i is a constant relating to that component, and the exponent n of diameter at breast height, d_b is also likely to vary, as discussed below.] Reference should also be made to Fig. 3.3, which illustrates the form of the relationship and variation of the type that can occur.

Allometric ratios are determined by the patterns of carbon allocation within trees, which are affected by a number of environmental factors. Soil fertility strongly affects foliage growth, especially that of new foliage, and hence the relationship between d_b and foliage mass. Because the concentration of nitrogen is greater in foliage than in many other tree tissues, nitrogen fertilization tends to increase the allocation of carbon to foliage. In conifers, this results in greater needle length, more needles per shoot, and the production of more new shoots by fertilized than by control trees (Brix and Mitchell, 1983; Grier *et al.*, 1985; Gower *et al.*, 1993a). It follows that increased nutrient availability tends to increase the ratio of new to total foliage mass. Water availability has a similar effect on the allometric relations for foliage (Brix and Mitchell, 1983; Gower *et al.*, 1992; Raison *et al.*, 1992). Gower *et al.* (1992) demonstrated that irrigated Douglas fir trees allocated more carbon to new foliage production than did similar diameter control trees. Grier and Running (1977) and Gholz (1982) showed that foliage mass or area was positively correlated to water availability for western (United States) conifer forests; a similar relationship exists for mixed and deciduous forests in Wisconsin (Fassnacht, 1996). For similar-diameter trees, shade-tolerant species support a greater foliage mass than shade-intolerant species (Grier and Logan, 1977; Chapman and Gower, 1991) and species with greater leaf longevity support greater foliage mass than species with shorter leaf lifespans (Schulze *et al.*, 1977a; Gower *et al.*, 1993a). However, for similar-diameter trees, leaf area is often not as different as foliage mass for tree species with different leaf lifespans because long-lived foliage has a lower specific leaf area (foliage area per unit dry weight) than short-lived foliage (Schulze *et al.*, 1977a; Reich *et al.*, 1992, 1995; Gower *et al.*, 1993a).

Allometric equations can be used in tree or stand growth models to derive partitioning coefficients η_i (allocation is the process; carbon is partitioned between the different parts of the trees) as follows. If the rate of dry mass production is dW/dt, then the rates of growth in leaf (w_f), root (w_r), and stem (w_s) mass are

$$dw_f/dt = dW/dt \cdot \eta_f - \gamma_f w_f \qquad (5.22a)$$

$$dw_r/dt = dW/dt \cdot \eta_r - \gamma_r w_r \qquad (5.22b)$$

$$dw_s/dt = dW/dt \cdot \eta_s - R_a. \qquad (5.22c)$$

The second term on the right-hand side of Eqs. (5.22a) and (5.22b) denotes litterfall and fine root turnover, respectively. The partitioning coefficients can be obtained by differentiating Eq. (3.2) to give $dw_i/d_b = c_i n_i d_b^{n-1}$, where d_b is a surrogate for total tree mass. Normalizing the values obtained (they must sum to unity) gives values for the partitioning coefficients. [Eqs. (5.22a) (522b) and (522c) were originally written in that form by McMurtrie and Wolf (1983). They have since been used in numerous models; see, for example, McMurtrie (1985) and McMurtrie and Landsberg (1992) for an application to analysis of experimental data.]

There have been many studies of tree root mass, but few allometric relationships have been established for coarse root biomass; where they have been, the equations are remarkably similar among tree species (Santantonio *et al.*, 1977; Haynes and Gower, 1995). The relationships for smaller diameter roots vary among species and are strongly affected by environmental conditions (see below).

Approaches to models of carbon allocation have been based on the so-called balanced growth approach (Davidson, 1969), which assumes that

$$w_s/w_r = \sigma_r c_{nut}/\sigma_s c_c, \tag{5.23}$$

where w_s and w_r are shoot (top growth in general) and root mass, respectively, σ_r is the specific activity of the roots (the rate of nutrient uptake per unit root mass), and σ_s is the rate of carbon assimilation per unit shoot mass; c_{nut} and c_c are the nutrient and carbon composition of new material. This model implies, among other things, that root/shoot ratios will decrease if the root specific activity increases, e.g., with improved nutrition. Several forest ecosystem process models use this procedure to allocate carbon to foliage and roots (Weinstein *et al.*, 1991; Running and Gower, 1991).

A more mechanistic treatment of carbon and nitrogen allocation within plants was developed by Thornley (1972a,b). Thornley initially derived partitioning coefficients [analogous to η_i in Eq. (5.22)] using the concept of specific root and shoot activities, and later extended the model to a transport-resistance network in which fluxes of carbon and nitrogen are driven across resistances by substrate concentration gradients. The concentrations and resistances can be adjusted so that the model is indepedent of the environment (Reynolds and Thornley, 1982) or modified to enable C/N ratios of root and shoot mass to maximize relative plant growth rate (Johnson and Thornley, 1987) and incorporate the feedback effect of leaf nitrogen concentration on shoot photosynthesis (Hilbert *et al.*, 1991; Gleeson, 1993). In all cases, a decrease in nitrogen supply will result in an equivalent decrease in nitrogen in roots and shoots, but the product of carbon and N in the shoots will decrease more than the product of carbon and nitrogen in the roots so that a decrease in nitrogen availability produces a relative increase in carbon allocation to roots. This is consistent with observations in forests (see Section IV,A,2).

Ågren (1983) used the concept of nitrogen productivity—the amount of biomass produced per unit nitrogen taken up by a stand—to model carbon allocation. Ågren and Ingestad (1987) proposed that the balance between shoots and roots in plants is based on an equilibrium between an internal growth sink (specific activity) for carbon substrate, the strength of which is defined by the total N content, and the photosynthetic capacity of the plant. They quantified these relationships as

$$f_{sh} = \frac{W_n N_p}{\sigma_c / f_c}, \qquad (5.24)$$

where f_{sh} is the shoot fraction of the plant, W_n is the biomass produced per unit N (nitrogen productivity), N_p is plant N concentration, σ_c is the specific activity for carbon, and f_c is the fraction of the plant dry matter that is carbon. This approach to carbon allocation predicts different qualitative responses of the root/shoot ratio for changes in supply of different nutrients, depending on the effect of the limiting nutrient on photosynthesis relative to its effects on the internal growth sink. The major difference between the procedure used by Thornley and co-workers and that of Ågren and Ingestad (1987) is that the Thornley models assume sink strength is controlled by carbon and N substrate concentration, whereas Ågren and Ingestad define sink strength based on the total plant N content.

2. Roots and Root Activity Roots are classified into two major categories: coarse and fine roots. Coarse roots are generally considered to be greater than 5–10 mm in diameter and are largely responsible for support. Fine roots are commonly considered to include roots less than 2–5 mm in diameter and are largely responsible for nutrient and water uptake. Mycorrhizae—fungi that live in symbiotic relationships with plant roots—appear to be essential for almost all plants (Law, 1985; Koide and Schreiner, 1992), largely because they lead to increased nutrient uptake by the host plant (Allen, 1991). Two major types of mycorrhizae occur: endomycorrhizae and ectomycorrhizae. Both forms occur with trees. Ectomycorrhizae are more important in many of the most commercially important tree genera such as *Pinus, Eucalyptus, Picea,* and *Abies.*

Fine root mass is positively correlated to environmental conditions favorable for photosynthesis (Vogt *et al.*, 1986b), although local site differences in nutrient availability cause variation in this relationship. Within a similar climate, fine root mass is inversely related to the most limiting nutrient. This inverse relationship between fine root mass and nutrient availability has been widely observed in boreal (Linder and Axelsson, 1982), temperate (Keyes and Grier, 1981; Vogt *et al.*, 1987; Comeau and Kimmins, 1989; Nadelhoffer *et al.*, 1985), and tropical forests (Gower, 1987; Gower and Vitousek, 1989). Nitrogen is the most common nutrient influencing fine root mass in temperate and boreal forests, whereas phosphorus, and perhaps calcium and magnesium availability, affect fine root mass in tropical forests, except primary successional forests, which are N limited.

The effects of nutrient availability on mycorrhizae parallel those for fine roots: The colonization frequency and degree of mycorrhizal colonization is higher for plants well supplied with nutrients, especially phosphorus, than for plants growing in infertile soils (Menge et al., 1977; Menge and Grand, 1978; Braunberger et al., 1991). It is important to note that the effect of nutrient availability at a microsite on fine root biomass and mycorrhizae will differ from that of soil fertility at the stand level. Soil nutrient availability is notoriously heterogeneous and microsites of high nutrient availability are important to the nutrient status of plants. As a result, plants support a disproportionately higher density of fine roots and mycorrhizae at these localized areas of high nutrient availability (St. John et al., 1983; Jackson and Caldwell, 1993; Duke et al., 1994).

Progress toward understanding below-ground processes, such as fine root production and turnover, has been slow because of the sampling problems besetting all current techniques used to estimate below-ground carbon fluxes (Nadelhoffer and Raich, 1992). As a result, there is less agreement (than in the case of fine root biomass) on how environmental factors affect fine root production. Keyes and Grier (1981) first reported that fine root production was greater for infertile than fertile Douglas fir (*Pseudtsuga menziesii*) stands. Their results are supported by numerous comparative (Comeau and Kimmins, 1989; Kurz, 1989) and experimental studies (Vogt et al., 1990; Gower et al., 1992). However, Nadelhoffer et al. (1985) and Aber et al. (1985), using essentially the same data, reported that fine root net primary production was positively correlated to nitrogen mineralization and above-ground NPP. The studies by Nadelhoffer et al. and Aber et al. differ from the other studies mentioned because they used the N budget technique to calculate fine root NPP, and species composition was not the same among the different forests they studied. This is important because below-ground carbon allocation patterns may differ among species (Vogt et al., 1986b; Walters et al., 1993). Several different techniques have recently been used to estimate fine root production and total root carbon allocation for temperate conifer forests in contrasting climates, and it has been found that fertilization decreases the allocation of carbon to fine roots on a relative basis for all forests and on an absolute basis for red pine (*P. resinosa*) (Haynes and Gower, 1995; Gower et al., 1996b).

Useful insights into the influence of nutrition on carbon allocation to fine roots are obtained by expressing the allocation of carbon to foliage growth as a fraction of that allocated to fine root NPP. Collating data of Linder and Axelsson (1982), Gower et al. (1996b) and Ryan et al. (1996)

from experiments with fertilizer and control treatments gave the following values:

	P. radiata	P. sylvestris	P. resinosa	P. ponderosa	P. elliottii
	(Foliage: fine root NPP)				
Control	0.4	0.8	0.6	1.4	0.4
Fertilized	1.0	4.8	0.8	2.3	1.6

It is clear that, on a relative basis, fertilization increases allocation of carbon to foliage and decreases allocation of carbon to fine roots.

Water availability also influences below-ground carbon allocation patterns. Annual carbon allocation to fine roots is greater in xeric than in mesic conifer forests (Santantonio and Hermann, 1985; Comeau and Kimmins, 1989). Gower et al. (1992) reported fine root production in an experiment with Douglas fir that included an irrigated treatment and an irrigated treatment with wood chips added to produce high water availability but low N availability. Fine root production was lower in the irrigated treatments than in the control treatment, suggesting that water availability directly controls carbon allocation to fine roots and mycorrhizae.

Few attempts have been made to quantify the C allocated to mycorrhizae or to determine whether the benefits provided by mycorrhizae exceed the C costs to the plant (Koide and Elliott, 1989). Vogt et al. (1982) reported that 150 g C m^{-2} year^{-1}, or 15% of total net primary production, was allocated to production of mycorrhizae in an *Abies amabilis* stand, and Bevege et al. (1975) found that 15 times more ^{14}C-labeled assimilate was allocated to mycorrhizal-infected than to nonmycorrhizal roots. The respiration costs of mycorrhizae are largely unknown.

Rygiewicz and Andersen (1994) used a root mycocosm and ^{14}C labeling to measure C fluxes for intact (i.e., symbiotic relationship between root and fungus was maintained) mycorrhizal and nonmycorrhizal roots of ponderosa pine seedlings. They found that the total below-ground respiration rate was 2.1 times greater for mycorrhizal than for nonmycorrhizal seedlings, confirming results from previous studies that small amounts of fungi have a substantial effect on carbon allocation in plants (Miller et al., 1989; Dosskey et al., 1990) and strongly affect the residence time of carbon below ground. Growth rates are often 5- to 10-fold greater for inoculated as opposed to nonmycorrhizal conifer seedlings (Lamb and Richards, 1971).

There is general agreement that on a global basis fine root NPP is positively correlated with the length of favorable growing season. Vogt et al. (1986b) reported that fine root turnover was positively correlated to mean climatic ratio (precipitation/mean annual temperature) for both temperate and boreal deciduous and evergreen forests. Raich and Nadelhoffer (1989) have proposed a carbon balance approach to quantify the total amount of carbon allocated to roots and mycorrhizal turnover and respiration. By assuming the soil carbon content is in steady state (i.e., C inputs = C losses), they estimated that total allocation of C below-ground should roughly equal annual soil surface C flux minus above-ground detritus C content. Using this method, they reported that total below-ground carbon allocation increases from boreal to tropical forests and is positively correlated to annual aboveground detritus. Although the assumption of steady state should be carefully considered for each forest, the approach does provide an upper limit on the amount of carbon allocated to fine roots (Fig. 5.8).

B. Storage

Stored carbon is used to construct new tissue and repair damaged tissues. When trees experience mild water stress or nutrient deficits, or a decrease in temperature, cell expansion is more adversely affected than photosynthesis, thus producing excess carbohydrates that the tree stores. The vigor of seedlings or trees is related to the amount of stored carbo-

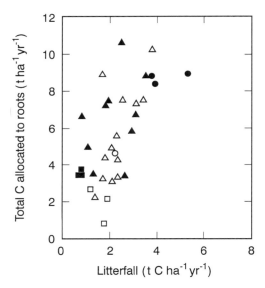

Figure 5.8 The relationship between the amount of carbon allocated to roots—defined as root + mycorrhizal turnover and respiration—and litterfall for forests at a number of locations (data from Raich and Nadelhoffer, 1989; Gower et al., 1995, 1996b). Symbols denote temperate evergreen (▲), tropical evergreen (●), temperate deciduous (△), tropical deciduous (○), boreal deciduous (□) and boreal evergreen forests (■).

hydrates so that trees with depleted carbohydrate reserves are more disposed to die (Sprugel, 1976; Puttonen, 1986). In contrast, vigorous trees store sufficient carbohydrates to survive environmental stresses such as insect attack, pathogen infestations, or drought (Waring, 1987).

Starch is the main storage form for carbohydrates in conifers; other forms include quinic and shikimic acids, hemicellulose and, of lesser importance, sucrose, cyclitols, and monosaccharides. The importance of the storage forms is influenced by both tissue type and species. Interested readers are referred to the review by Kozlowski (1992) for more detailed information on other storage forms of carbohydrates.

Carbohydrate storage patterns differ between conifer and deciduous tree species in that conifers store larger amounts of carbohydates in foliage and tend to store less carbohydrates in wood than deciduous species and accumulate them later in the season. Seasonal patterns of starch storage in conifers appear to be similar for all tissues, with the maximum occurring shortly before new growth begins in the spring (see, for example, Cranswick *et al.*, 1987; Gholz and Cropper, 1991). The relative importance of the various tissues in terms of whole tree carbon storage varies with age because biomass components change with age. Needles and fine roots are the primary storage organs for conifer seedlings (Glerum, 1980), but coarse roots and stems are the primary carbohydrate storage sites for mature trees (Cranswick *et al.*, 1987; Gholz and Cropper, 1991). The amount of carbohydrates in conifer needles varies with seasonal duration of shoot growth; maximum starch concentrations occur in pines with the shortest growing season (Kozlowski, 1992). A substantial amount of carbohydrate is stored in fine roots: Ericsson and Persson (1980) reported maximum starch concentrations of 30% by weight for fine roots of *P. sylvestris*.

The physiological importance of carbohydrate storage in trees is well recognized, but our inadequate understanding of mechanisms controlling carbohydrate storage and utilization has largely prevented modelers from including this component of the forest carbon budget in forest ecosystem process models (but see Weinstein *et al.*, 1991). This deficiency needs to be rectified if physiologically based models are to be used to simulate multiple stresses in any realistic manner.

C. Reproduction

Few data are available on the proportion of C allocated to reproduction by coniferous forests. Linder and Troeng (1981) estimated that a 14-year-old *P. sylvestris* tree allocated 6% of its canopy photosynthate to cones and Fielding (1960) estimated that the amount of C that *P. radiata* allocated to cones was about 16% of stem production. Cremer (1992) found that the mass of matured male and female cones on *P. radiata* was equivalent to 10% of stem growth and 5% of total aboveground growth. Many trees produce more flowers than they can sustain to maturity, consequently, when adverse environmental conditions restrict carbon assimilation, flowers are aborted (Lloyd, 1980).

The time span during which strobili are strong carbohydrate sinks for conifers ranges from one season (i.e., *Abies, Larix, Picea,* and *Pseudotsuga*) to more than one for *Pinus*. The C required to construct reproductive structures is commonly derived from stored carbohydrates during the early spring when net canopy photosynthesis is very low; however, later in the spring most of the carbohydrates are obtained from current photosynthate produced largely by 1-year-old needles (Kozlowski, 1992). Allocation of carbon to reproductive structures has been thought to decrease allocation of carbon to stem production (Fielding, 1960; Linder and Rook, 1984); however, it appears this may not be true for *P. contorta* (Dick *et al.*, 1991).

D. Secondary Defense Compounds

Herbivores generally consume from 5 to 15% of the foliage in temperate forests during normal years but can consume all the foliage in years of major insect outbreaks (Schowalter *et al.*, 1986). Low levels of herbivory can actually increase tree growth, but moderate to high levels reduce photosynthetic surface area, deplete carbohydrate storage reserves, and increase the susceptibility of trees to other environmental stresses. As a result, trees allocate a portion of their annual carbon gain to construct and maintain defense compounds and physical barriers to deter herbivory. Clancy *et al.* (1995) provide a thorough discussion on herbivory and secondary defense compounds

In general, the concentration of defense compounds appears to be inversely related to the leaf lifespan (Coley, 1988). Also, mild water stress increases the partitioning of carbon to defense compounds, as well as to storage, because cell division is adversely affected by water stress before photosynthesis. For example, Horner (1987) reported greater lignin concentrations in control than in irrigated Douglas fir trees during a drought year but little difference in lignin concentration between the two treatments for a wetter year. Futhermore, increased nitrogen availability decreased the amount of carbon partitioned to lignin, whereas decreased nitrogen availability tended to increase lignin concentration in new foliage. The allocation of carbon to volatile trace gases (e.g., terpenes, isoprenes, etc.) may comprise a significant component of the leaf carbon budget, but we have a poor understanding of the environmental and metabolic controls over their production, despite the important influence they have on the chemistry of the atmosphere (Brasseur and Chatfield, 1991).

Defense compounds are also important because they indirectly affect carbon assimilation and allocation patterns. Greater partitioning of carbon to defense compounds reduces stored carbon, constitutive defenses such as lignin reduce the net photosynthetic rate, resulting in lower net carbon gain by the canopy, and defense compounds such as polyphenolics retard decomposition (see Chapter 6). Foliage lignin/N ratios exert a strong control on N mineralization rates in forests, which in turn influences net primary production and C allocation patterns.

V. Growth Efficiency

The ratio of wood production to leaf area index, called growth efficiency, is a commonly used index to quantify and understand how abiotic and biotic factors affect stand vigor. The index has a physiological basis because wood production has lower priority for carbohydrates than most other tissues (Waring and Pitman, 1985). In general, the growth effciency of a stand decreases with increasing L^*, although maximum wood production tends to occur at intermediate values (usually about $L^* \approx 3$). The decline in growth efficiency is largely caused by self-shading, but it may also be the result of water and nutrient shortages.

Growth efficiency data have been used in designing forest management practices to increase tree vigor and thereby reduce the susceptibility of trees to pest and pathogen attack (Mitchell *et al.*, 1983). In the first year of an experiment in which they manipulated the vigor of trees by fertilization and thinning, and deliberately attracted pine bark beetles with synthetic phermones, Waring and Pitman (1985) found that insects attacked trees in all plots, presumably due to their low vigor. By the end of the second year, however, the growth efficiency of surviving trees increased as a result of greater nitrogen availability and improved canopy illumination as a result of thinning or mortality caused by bark beetle attack. The surviving trees had greater vigor, enabling them to survive bark beetle attacks of greater intensity than prior to treatment.

Pathogens can have a similar effect on the productivity and vigor of stands. Perhaps the best documented phenomenon in this respect is the natural progression during stand development from thrifty, vigorous growing stands to low-vigor stands that eventually succumb to one or more stresses. This can produce wave-like patterns on mountain slopes, sometimes referred to as wave mortality. Wave mortality commonly occurs in subalpine forests and has been reported in Japan, (Kohyama and Fujita, 1981), the northeastern United States (Sprugel, 1976), and Oregon (McCauley and Cook, 1980). In all cases, above-ground net primary productivity and growth efficiency decrease with distance from the front of the wave, often culminating in the total collapse of the stand, followed by stand reinitiation (Sprugel, 1984; Matson and Boone, 1984). In an experiment on this phenomenon, Matson and Waring (1984) innoculated mountain hemlock (*Tsuga mertensiana*) seedlings with a root rot fungi and grew them under different nutrient and light regimes. They found that the seedlings could resist the pathogen if they were supplied with suitable light and nutrients, but infection was lethal for the seedlings under suboptimal conditions. In both the bark beetle and root rot examples, the exact cause(s) of the decline in productivity and growth efficiency is complex, probably resulting from endogenous and exogenous factors and their interaction on the productivity and carbon allocation of trees. One of the more important challenges facing ecologists and foresters in the future is to gain a better understanding of the effects of multiple stresses on species composition, structure, and function of forest ecosystems.

VI. Net Ecosystem Production

NEP is defined as the net flux of CO_2 to or from a forest ecosystem—the end result of carbon fixation by photosynthesis and losses by autotrophic and heterotrophic respiration. Therefore,

$$\text{NEP} = \text{GPP} - R_a - R_h. \tag{5.25}$$

Because NEP reflects the annual change in C stored in the ecosystem (vegetation + detritus + mineral soil), it indicates whether the ecosystem is a carbon "sink" or "source" in relation to the atmosphere. NEP can be calculated on the basis of chamber measurements of carbon flux for all components of the ecosystem and extrapolation of those measurements. However, such an approach leads to many problems of sampling and scaling. Recent improvements in equipment and software have now made it possible to make reliable eddy-covariance measurements over long periods (see, for example, Wofsy *et al.,* 1993). These measurements integrate local variation in soil or vegetation properties over areas of several square kilometers, provide upper-limit values to the fluxes of CO_2, heat, and water vapor, and provide a sound basis for extrapolation to large areas.

Measurements made over long periods also provide considerable information about temporal changes in the system under study. The combination of flux and foliage, stem, and soil chamber measurements is particularly valuable: The eddy-covariance measurements provide the NEP value, whereas the chamber measurements provide estimates of the components and can be used to ascribe causes to observed temporal patterns. This is illustrated in Fig. 5.9, which shows typical diurnal courses of

NEP measured by eddy correlation over three very different forest systems: boreal forest in northern Canada, tropical rainforest in the Amazon, and a mixed deciduous forest in Tennessee. If closure can be achieved between such measurements and chamber measurements, without a large error term, we can be confident that most of the component values obtained are correct.

Few measurements of NEP, on an annual basis, currently exist, although this is likely to change very rapidly over the next few years because the theory and technology for measuring net CO_2 exchange for terrestrial systems have been reasonably well developed (Fitzjarrald and

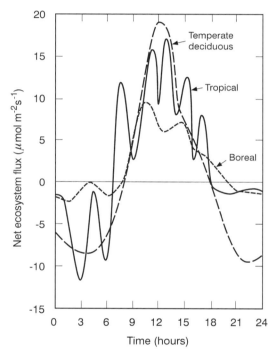

Figure 5.9 Examples of net ecosystem (CO_2) fluxes from temperate deciduous (TD), boreal (B), and tropical forests (T). Leaf area index was higher in TD than in B. Daytime radiation values reached about 1800 μmol m^{-2} sec^{-1} over TD and about 12–1500 μmol m^{-2} sec^{-1} over B. Both nighttime respiration and fluxes were lower over B than TD: respiration values reached 14 μmol m^{-2} sec^{-1} over TD; they fluctuated less, but were lower (1 or 2 μmol m^{-2} sec^{-1}) over B. Daytime fluxes (into the canopies) were considerably higher over TD: maxima reached 28 μmol m^{-2} sec^{-1} compared to maximum values of about 11 μmol m^{-2} sec^{-1} over B. Radiation and CO_2 fluxes over T were generally about the same as those over TD, although they were heavily influenced by cloud. Nighttime fluxes were characterized by storage of CO_2 in the canopy (the TD and B data for this diagram were taken from Baldocchi and Vogel, 1996; the T data are from Grace *et al.*, 1995).

Moore, 1995). Wofsy et al. (1993) measured average annual NEP of 3.7 t C ha^{-1} year^{-1} for the Harvard forest in the northeast United States (this is a mature mixed deciduous forest). [This estimate has since been revised down to 2.2 t C ha^{-1} year $^{-1}$ (Goulden et al., 1996).] Fan et al. (1995) measured NEP of 0.8 t C ha^{-1} for a 53-day period (end of June to mid-August) for a black spruce boreal forest near Schefferville, Quebec. Grace et al. (1995) measured NEP for 55 days and then used a process-based ecosystem model to derive an annual estimate of 0.9 t ha^{-1} for a mature tropical forest in Rondonia, Brazil. The stage of stand development will also strongly affect NEP—immediately following disturbance when no vegetation is present NEP will be negative (i.e., a net flux of CO_2 from the forest to the atmosphere). NEP can also be estimated from coupled canopy photosynthesis and soil decomposition models such as that used by Wang and Polglase (1995) to examine the effects of climate change on NEP of tundra, boreal forest, and tropical rainforests. Wang and Polglase found that the inclusion of interannual temperature variation in the analysis had significant effects, causing NEP to change from positive (carbon storage) to negative (carbon source). They also showed that increases in NPP resulting from elevated atmospheric CO_2 concentrations were more than offset by increases in heterotrophic respiration, suggesting that the potential for terrestrial carbon storage will decrease in response to future global warming. A theoretical treatment of how NEP is likely to change during succession is provided in Chapter 8.

VII. Forests in the Global Carbon Budget

We commented in Chapter 1 on the matter of rising atmospheric concentrations of CO_2 and other "greenhouse" gases (e.g., methane, NO_x, and halocarbons) and the implications of these changes for climate and plant growth. The gases are called greenhouse gases because they alter the absorption and transmission properties of the atmosphere to radiant energy. Radiation in the longwave bands (~8–10 µm), in which energy is reradiated from the earth, is particularly affected; absorption is increased by the greenhouse gases so that a higher proportion of the shortwave energy (solar radiation) absorbed by the earth is retained, tending to lead to higher temperatures. There is a very strong consensus among scientists that global mean surface temperatures are rising (IPCC, 1995); since the late 19th century, the global mean surface temperature has increased by between 0.3 and 0.6°C—only a small fraction of the rise can be attributed to urbanization. The warming is not expected to be globally uniform. The greatest increases in surface temperature are predicted to occur on the continents between 40° and 70° N, whereas temperature changes near the equator will be small. This pattern is not expected to change in the future. Current uncertainities that affect our ability to detect and predict future climate change include the response of clouds to warming and land use change, the role of oceans, including sea ice dynamics, and uncertainties of the magnitude of natural climate variability. The situation is complicated by increased evaporation and aerosols, which can be expected to lead to increased cloudiness and the reflection of a greater proportion of solar radiation away from the earth. A case in point is the short-term cooling observed in 1992 and 1993 that was caused, in part, by increased aerosols in the upper atmosphere resulting from the mid-1991 eruption of Mt. Pinatubo in the Philippines (McCormick, 1995).

Of more direct relevance to forestry are the probable effects of increasing atmospheric CO_2 concentrations on tree growth. There are no long-term, ecosystem-level studies to draw on, so all conclusions must be based on the results of studies on seedlings and saplings grown in pots or open-top chambers. The growth rates of seedlings and saplings in doubled CO_2 concentrations have generally increased by 20–40% (Melillo et al., 1993). The availability of water and nutrients strongly controls the magnitude of the growth response; the relative response of growth to elevated CO_2 is greater when water is moderately limiting because elevated CO_2 decreases stomatal conductance and transpiration, thereby enhancing water use efficiency (Wong et al., 1979; Gifford, 1979; Morrison and Gifford, 1984). Nitrogen limitation decreases the growth response to elevated CO_2. It has been proposed, on the basis of the known fact that increased CO_2 leads to increased plant growth—at least in the short term (see review by Gunderson and Wullschleger, 1994), that planting large areas of forest, which would sequestrate and store CO_2, could offset rising atmospheric CO_2 concentrations. However, feasibility studies (e.g., Vitousek, 1984; Freedman et al., 1989) have shown that there is not enough land available for this to significantly offset anthropogenic CO_2 inputs.

VIII. Concluding Remarks

The carbon balance of forests, as of any plant community, is the end result of CO_2 absorption through stomata and fixation by photosynthesis, of losses by photorespiration in the leaves and autotrophic respiration in leaves, branches, stems, and roots, and of litterfall and heterotrophic respiration on the forest floor and in the soil. From the point of view of the ecologist, forest productivity is a matter of total biomass production and the flows of carbon within the trees and through various parts of the system as a whole. From the point of view of the forest manager concerned with wood production, the end point of the process is the amount of carbon that can be stored in stems, although even where wood production is the main objective of forest management, it remains essential that the managers are aware of, and understand, the larger picture. It is the flows and interactions and resulting balances, over the life cycle of a stand, that determine the productivity of the forest, the transfer of organic matter and nutrients to the soil, and whether that production is sustainable in the long term.

From the point of view of science, and our knowledge of the processes involved, it is probably reasonable to suggest that our understanding of processes at the leaf level—particularly photosynthesis—far outstrips our understanding of respiration and carbon allocation. A great deal of physiological research has, over many years, been focused on photosynthesis for a number of obvious reasons, including the fact that photosynthesis is (justifiably) regarded as the fundamental biological process. However, it is only the first process in the chain that leads to standing biomass. Our inadequate knowledge of carbon allocation is now recognized as a major problem and is increasingly being addressed, at least in research on models. It will also be clear, from this chapter and Chapter 4, that despite the vast amount of research on stomatal behavior and responses, we cannot yet be confident that the interactions between g_s and photosynthesis and g_s and water relations are fully understood and can be described quantitatively in most circumstances.

Research in forestry (as opposed to physiology and ecology) has not, traditionally, focused on physiological processes, presumably because it has been difficult for foresters to see the relevance of such research to the business of managing large, long-lived systems such as forest stands. This—if it was the reason—is not as valid now as it was, because the development of models (see Chapter 9) that incorporate our understanding of the processes that govern tree growth has reached the point where those models can be used to explore the implications of various scenarios and the consequences of specifed actions. However, the models, as we note in Chapter 9, are not yet sufficiently "user-friendly" for use by foresters concerned with tree growth; this must remain an urgent objective. Also, knowledge about the physiological processes that govern the carbon balance of forests is essential as a basis for interpreting forestry experiments and explaining the results obtained from them.

One of the exciting developments in research, in recent years, is the improvement of technology for measuring gaseous fluxes above ecosystems and the ability to maintain those measurements over long periods. Enormous amounts of information will be obtained from long-term measurements, such as those of Wofsy *et al.* (1993), which give us the upper limit of ecosystem productivity; it then becomes a matter of "balancing the budget" in terms of the various components. Long-term NEP measurements will allow analysis of the contribution of the components to the system and of the consequences of altering the size or responses of those components.

Recommended Reading

Amthor, J. S. (1989). "Respiration and Crop Productivity." Springer-Verlag, New York.
Cannell, M. G. R. (1989). Physiological basis of wood production: A review. *Scand. J. For. Res.* **4,** 459–490.
Cannell, M. G. R., and Dewar, R. C. (1994). Carbon allocation in trees: A review of concepts for modelling. *Adv. Ecol. Res.* **25,** 59–104.
Gower, S. T., Vogt, K. A., and Grier, C. C. (1992). Carbon dynamics of Rocky Mountain Douglas-fir: Influence of water and nutrient availability. *Ecol. Monogr.* **62,** 43–65.
Holbrook, N. M., and Lund, C. (1995). Photosynthesis in forest canopies. *In* "Forest Canopies" (M. D. Lowman and N. M. Nadkarni, Eds.). Academic Press, New York.
Ryan, M. G. (1991). Effects of climate change on plant respiration. *Ecol. Appl.* **1,** 157–167.
Wang, Y.-P., McMurtrie, R. E., and Landsberg, J. J. (1992). Modelling canopy photosynthetic productivity. *In* "Crop Photosynthesis: Spatial and Temporal Determinants" (N. R. Baker and H. Thomas, Eds.), pp. 43–67. Elsevier, Amsterdam.
Wofsy, S. C., Goulden, M. L., Munger, J. W., Fan, S.-M., Bakwin, P. S., Daube, B. C., Bassow, S. L., and Bazzaz, F. A. (1993). Net exchange of CO_2 in a mid-latitude forest. *Science* **260,** 1314–1317.

6

Soil Organic Matter and Decomposition

Soil organic matter consists of plant and animal residue in various stages of decay and resynthesis. Inputs come from dead plant and animal tissues, often referred to as detritus, that accumulate at the soil surface. Except in old-growth forests, the organic matter in the surface layers of mineral soils generally comprises most of the total organic matter in forest ecosystems. All but the most recalcitrant fractions constantly undergo chemical breakdown by soil microorganisms that decompose and resynthesize the material to form complex carbon-based compounds. Typically, the organic matter concentration of the surface of forest soils ranges from 0.5 to 5% by weight, but it can approach 100% in the organic soils common to poorly drained forests.

Although organic matter comprises a small fraction of the soil by weight, it has profound effects on the chemical, physical, and hydrological properties of forest soils and plays a critical role in forest nutrition. The annual nutrient requirements for most forests are met by nutrients released by organic matter decomposition, and the ability of soils to retain nutrients is strongly influenced by the cation exchange capacity, which is substantially greater for humus—the name given to the colloidal, carbon (C)-based polymers that are more resistant to decay than the original tissue—than for clay minerals. Key physical characteristics of soil, such as structure, aggregation, and bulk density—which affect the soil water holding characteristics (see Chapter 4)—as well as gas and water transport in the soil, and root growth, are affected by soil organic matter. Organic material on the forest floor increases water infiltration and minimizes overland flow that causes erosion. Lastly, soil organic matter is a major energy source for soil macro- and microinvertebrates and is an important component of the global carbon cycle.

Given the numerous important processes in which soil organic matter is involved, it is essential to understand the soil organic matter (or soil carbon) cycle (Fig. 6.1). For this, it is necessary to understand the factors

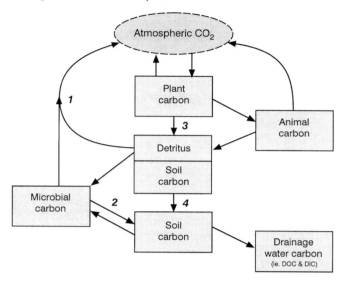

Figure 6.1 Schematic diagram of the soil C cycle. Major sources of C input to forest soils include the turnover of above- and below-ground plant tissues and to a lesser extent the death of animals. Plant and animal tissue is decomposed via physical and chemical processes (3) and degraded and resynthesized to form complex carbon polymers known as fulvic and humic acids and humin—collectively known as humus (4). Soil microbes derive their energy by decomposing the organic matter and add to it upon their death (2). Carbon losses from soil occur primarily as a gaseous loss when soil organic matter is converted to carbon dioxide (CO_2), water, and energy—this process is refered to as soil surface CO_2 flux (1). During pedogenesis, dissolved organic compounds move downward in the soil profile but are arrested in the lower soil hoizons.

that control carbon inputs (detritus production), transformations (decomposition and formation of various humus fractions), and losses (soil surface CO_2 flux and leaching). The water and nutrient cycles have a profound effect on the soil carbon cycle because they influence the type and quality of litter input, the residence time of the litter, and the accumulation or loss rates of carbon in the soil. The net difference between these carbon inputs and losses determines whether soil organic matter is accumulating or being lost from forest soils.

In this chapter, we review the major sources and rates of surface detrital input into forests, the factors controlling decomposition, resynthesis of detritus, soil CO_2 flux, the importance of carbon leaching, and soil organic matter content of world forests. We also discuss the impacts of forest management practices on the various components of the soil carbon cycle and the potential impact that climate change may have on soil carbon cycles.

I. Soil Carbon Content and Accumulation

The organic matter content of a soil represents the integrated (net) balance between detrital inputs—both above- and below-ground—and organic matter losses in the form of CO_2 flux from the soil, which reflects organic matter decomposition and root repiration, as well as by leaching and erosion (Fig. 6.1). Post et al. (1982) summarized 2696 soil profiles from almost every terrestrial biome and calculated that forests contain 34% of the global organic carbon soil content. Soil C contents were generally lowest for warm temperate and very dry tropical forests (7.1 and 6.1 kg m^{-2}, respectively) and highest for wet boreal and tropical forests (19.3 and 19.1 kg m^{-2}, respectively). They reported that, in general, soil organic matter increased with increasing precipitation, decreasing temperature, and decreasing evapotranspiration/precipitation ratio. Similar relationships have been observed in studies across local climatic gradients (Amundson et al., 1990), but total soil organic matter can vary significantly within a biome or even a forest ecosystem because of differences in parent material, aspect, topography, and past stand history. For example, the range in soil organic matter in a coastal Douglas fir ecosystem (Stow et al., 1996) is similar to the range reported among the major terrestrial biomes (Post et al., 1982; Schlesinger, 1991).

If detrital inputs exceed carbon losses, carbon accumulates in the soil, and the converse obviously holds. It is difficult to determine whether the carbon content of a soil is in steady state by measuring carbon losses and inputs because the input and output fluxes may be an order of magnitude greater than their differences and there is significant uncertainty associated with each of them. The major losses are reflected in the soil CO_2 flux (see Section IV), but measured fluxes include contributions from root respiration, which can comprise a varying fraction of the total.

Schlesinger (1991) summarized carbon accumulation rates measured in numerous soil chronosequence studies; using these data we calculated average rates of 8.7, 5.6, and 2.4 g C m^{-2} year^{-1} for boreal, temperate, and tropical forests, respectively (Table 6.1). These estimates of soil carbon accumulation are substantially smaller than values reported for afforestation and in some cases reforestation studies. The Rothamsted study (Jenkinson et al., 1992) provides a rigorous, long-term assessment of the potential influence of land use practices on soil carbon accumulation: Annual carbon accumulation rates ranged from 26 to 48 g C m^{-2} year^{-1} for agricultural fields converted to forests, compared to -2 g C m^{-2} year^{-1} for continuous cropping with wheat. We noted in Chapter 5 that planting trees to sequester carbon was limited by land availability and the problem of what to do with the carbon once it had been fixed in vegetation biomass. Few studies examining the feasibility of planting trees to

6. Soil Organic Matter and Decomposition

Table 6.1 Typical Carbon Accumulation Intervals (Year) and Carbon Accumulation Rates (g C m^{-2} year^{-1}) for Contrasting Forest Biomes[a]

Biome type	Accumulation interval	Rate of accumulation
Boreal (C)	150–3,500 (2415)	1–12 (9)
Temperate		
C	1000–10,000 (4260)	1–12 (6)
R, A	35–100 (75)	1–80 (36)
Tropical		
C	3500–8,600 (6060)	2
R, A	~50 (50)	50–200 (103)

[a] The data indicate the range of values from studies on primary successional soil chronosequences (C) summarized by Schlesinger (1991) and reforestation and afforestation sites (R, A). (Data from Lugo *et al.*, 1986; Zarin and Johnson, 1995; Boone *et al.*, 1988; Gholz and Fisher, 1982; Jenkinson *et al.*, 1992; Huntington, 1995; Schiffman and Johnson, 1990.) Means are in parentheses.

offset rising atmospheric CO_2 concentrations consider carbon accumulation in the soil, but empirical data from afforestation studies suggest this can be both significant and long term.

The mean residence time of soil organic matter, the inverse of decomposition rate, is commonly calculated as the ratio of soil organic matter content/organic matter inputs (or losses); this calculation assumes that the soil organic matter pool is in steady state or equilibrium. Raich and Schlesinger (1992) used average soil carbon content and annual soil surface CO_2 flux values to calculate mean carbon residence times for tropical lowland, temperate, and boreal forests of 38, 29, and 91 years, respectively. We used forest floor biomass and above-ground detritus production data reported in the literature to calculate mean residence times for the detritus in the major forest biomes of the world. We found that it increases from tropical to boreal forests and is commonly greater for evergreen than for deciduous forests (Fig. 6.2). The rates are, of course, substantially smaller than those reported for total soil carbon. The greater mean residence time of surface litter in evergreen as opposed to deciduous forests may be because the litter of evergreen trees tends to have higher concentrations of secondary compounds, such as lignins and tannins, and lower concentrations of essential nutrients required by decomposers. Vogt *et al.* (1986b) point out that excluding below-ground detritus input from calculations of the mean residence time of surface litter can cause overestimation of decompostion rates in forest ecosystems by 19–77%.

The testing of nuclear bombs in the early 1960s nearly doubled the radiocarbon concentration of atmospheric CO_2, which was absorbed by vegetation and transferred to the soil organic matter. This pulse of radiocarbon provides another way for scientists to examine the residence time of organic matter in the soil (Jenkinson, 1963; Rafter and Stout, 1970; Trumbore *et al.*, 1989). Such analyses consistently yield mean residence times from 400 to 3000 years for soil carbon (Jenkinson, 1963; Campbell *et al.*, 1967; O'Brien, 1984; Stevenson, 1982; Jenkinson *et al.*, 1992), which are one to two orders of magnitude more than those of surface litter (Fig. 6.2). This suggests that there is a small fraction of soil organic matter that is extremely recalcitrant or inert. The recalcitrant properties of humus (see below) result in large accumulations of humus in the soil profile so that the mass of humus often exceeds the carbon content of surface detritus and living biomass (Schlesinger, 1977). Jenkinson and Rayner (1977) used the radiocarbon evidence of very long mean residence times of the inert carbon fraction to rationalize the formulation of a three-pool soil organic matter model; this approach has been widely adopted by other modelers (Parton *et al.*, 1987; Comins and McMurtrie, 1993). The conceptual separation of soil organic matter into fractions of different mean residence time has lead to a better understanding of soil organic matter dynamics and better estimates of soil carbon cycling rates.

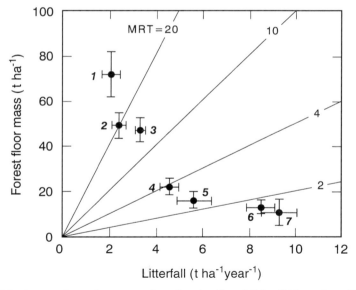

Figure 6.2 Average forest floor mass plotted against litterfall for (1) boreal needle-leaved evergreen ($n = 16$), (2) boreal broad-leaved deciduous ($n = 7$), (3) temperate needle-leaved evergreen ($n = 73$), (4) temperate broad-leaved evergreen ($n = 11$), (5) temperate broad-leaved deciduous ($n = 40$), (6) tropical broad-leaved evergreen ($n = 31$), and (7) tropical broad-leaved deciduous ($n = 2$). Assuming the forest floor is in steady state, the average mean residence time (MRT, years) can be calulated as forest floor mass/litterfall mass. The lines indicate these values. Gower and Landsberg, submitted for publication.

Table 6.2 Comparison of Cation Exchange Capacity (CEC), Surface Area, and pH Charge Dependency for Soil Organic Matter and Clay Minerals[a]

Component	Mineral type	CEC (meq 100 g^{-1} soil)	Surface area (m^2 g^{-1} soil)	pH dependence of charge
Organic matter	—	100–300	800–900	Large
Allophane	—	10–150	70–300	Large
Vermiculite	2:1	120–150	600–800	Small
Montmorillonite	2:1	80–120	600–800	Small
Kaolinite	1:1	1–10	10–20	Medium

[a] (Adapted from Bohn et al., 1979).

Townsend et al. (1995) have shown that soil organic matter turnover rates based on a three-pool soil organic matter model gave rates three times slower than those produced by a single pool model. It also appears that the labile carbon pool in the soil is more sensitive to soil warming than the large recalcitrant pool (Updengraff et al., 1995).

Even the most soluble compounds are rarely decomposed completely and the by-products often undergo further enzymatic and chemical reactions to form humus. The synthesis and composition of humus is complex and poorly understood (Stevenson, 1982; Tate, 1987), although there is general agreement regarding the probable structure. Humus is a very variable, amorphous compound that consists of highly branched units with an aromatic ring skeleton linked by O, NH, N, and S bonds. Its amorphous nature and phenolic and organic acid groups provide humus with an extremely large cation exchange capacity, surface area, and water-holding capacity relative to clay minerals (Table 6.2).

II. Sources of Soil Organic Matter

A. Detritus Production: Factors Controlling Sources and Quantity

Detritus, which derives from the death of above- and below-ground plant tissue, death of organisms, and fecal material, is the dominant source of C input to the soil under normal conditions. Insects consume less than 10% of net primary production (NPP) in forests (Schowalter et al., 1986), with only a fraction of this returning to the soil in the form of frass (insect feces). Vertebrates commonly comprise less than 1 or 2% of the total organic matter content of forests (Whittaker, 1975); their carbon input can be significant at a microscale, but averaged over a stand it is very small. Uncontrolled population growth of insects or vertebrates, however, can drastically alter the soil organic matter cycle of forest ecosystems. Pastor *et al.* (1988) provide an interesting and elegant example of the effect that moose, mediated through soil microbes, can have in increasing the soil carbon and nutrient cycling rates of boreal forests. Similar examples exist for insects (Mattson and Addy, 1975; Schowalter *et al.*, 1986; Romme *et al.*, 1986), prompting ecologists to acknowledge the importance of insects in regulating net primary production of terrestrial ecosystems (Romme *et al.*, 1986).

Most measurements of above-ground litterfall are restricted to small tissues such as foliage, twigs, and reproductive tissue, of which leaf detritus generally comprises the largest fraction (see Chapter 5). On a global scale, above-ground detritus production follows the gradient of NPP, being generally highest in tropical forests and lowest in boreal forests (Fig. 6.2; see Chapter 5). Latitudinal trends have also been reported (Van Cleve *et al.*, 1983; Schlesinger, 1977; Vogt *et al.*, 1986b). Extreme environmental conditions such as severe drought can cause abnormally high leaf litter production (Linder *et al.*, 1987; Raison *et al.*, 1992), and during mast (flowering) years the proportion of carbon allocated to reproductive tissue increases and leaf litter decreases (Burton *et al.*, 1991).

Litterfall does not appear to vary consistently among deciduous and evergreen forests in a similar climate, although the nutrient content of litter and nutrient use efficiency of the forests may be different (Chapter 7). Comparative studies in natural forests, aimed at examining the effects of species on carbon and nutrient cycling, are complicated by the fact that many variables that influence carbon cycling, such as climate, parent material, topography, and relief, are not held constant. Experimental plantations in replicated plots, commonly known as a common garden study design, can be used to examine the effect of species on forest ecosystem processes. Alban *et al.* (1982) in Minnesota and Gower and Son (1992) in Wisconsin used common garden studies to examine the effect of species on forest ecosystem processes and, in both cases, showed that leaf litter mass did not differ among species in which leaf longevity was typically very different (e.g., <1 to >6 years), indicating that climatic and edaphic conditions are more important than species effects on above-ground detritus inputs.

The significance of coarse woody detritus to the soil carbon balance of forest ecosystems, as a factor in the rate of nutrient cycling, as habitat for vertebrates and invertebrates, and as an energy source for soil microorganisms, has only recently been fully appreciated (Harmon *et al.*, 1986). The large temporal and spatial variability of coarse woody detritus makes it difficult to quantify accurately, except by using large plots and collecting data for many years. The limited data available indicate that woody litterfall comprises 17–37% of the total aboveground detritus production, the proportions being larger in tropical than in boreal forests (Vogt *et al.*, 1986b).

Although fine roots and mycorrhizae comprise a small fraction of the total forest ecosystem carbon content, because of their rapid turnover they provide a significant fraction of the total annual input of carbon and nutrients to forest soils (Vogt et al., 1986b; Vogt, 1991; Gower et al., 1995). Most estimates of below-ground detritus are likely to be underestimates because of the difficulty of quantifying ephemeral fluxes such as mycorrhizal turnover and root exudates. Vogt et al. (1986b) summarized fine root turnover rates for the major forest biomes but the averages for some ecosystems were based on only a few studies, making it difficult to draw any general conclusions about the environmental or ecological controls on below-ground detritus production. As more data have become available, several trends are emerging, many supporting those noted by Vogt et al. Above-ground detritus production increases from boreal to tropical biomes for both deciduous and evergreen forests and, for a similar biome, above-ground detritus production tends to be greater in deciduous than evergreen forests. Fine root detritus production also increases from boreal to warm temperate forests in both deciduous and evergreen forests, but it does not differ consistently between leaf habits for a similar climate. Last, the ratio of fine root:total (above-ground + fine root) detritus production is larger in evergreen than deciduous forests for a similar biome, but it is relatively constant across biomes for a similar leaf habit. Warmer soils presumably increase root respiration costs (Chapter 5) and lead to faster fine root turnover rates (Table 6.3). Marshall and Waring (1986) found that fine roots of Douglas fir contained fixed amounts of carbohydrates so that the roots of seedlings grown in warm soils died more quickly than those grown in cooler conditions. Hendrick and Pregitzer (1993) used a miniature video camera and minirhizotron tubes to monitor fine root turnover of two sugar maple stands on a north–south gradient in Michigan, from which they concluded that fine root turnover was greater in the southern than northern stand. Extreme flucuations of the soil environment, such as drought, can also cause rapid fine root turnover (Eissenstat and Yanai, 1996).

B. Detritus Production: Chemical Composition

The chemical composition of detritus will affect soil carbon dynamics. Aber and Melillo (1991) summarized information about some of the major carbon constituents of foliage, stem, and roots for a variety of boreal and temperate tree species. The common carbonaceous constituents of plant tissue include sugars, cellulose and hemicellulose, and complex phenolics such as lignins, tannins, and suberin. The energy yield of these tissues varies considerably. If the chemical bonds are difficult for microbes to break, then the net energy gain from their decomposition by microorganisms is small. Sugar molecules, such as glucose, are simple and small and therefore an excellent source of energy for microbes. Car-

Table 6.3 Above-ground and Fine Root Detritus Production and Fine Root:Total Detritus Ratio for Evergreen and Deciduous Forest in Contrasting Biomes[a]

Biome leaf habit	Above-ground (kg ha^{-1} year^{-1}) mean	range	Fine root (kg ha^{-1} year^{-1}) mean	range	Fine root:Total mean	range
Boreal						
Evergreen (8)[b]	1370	(550–3000)	1990	(600–4110)	0.58	(0.41–0.77)
Deciduous (5)	2520	(1660–3450)	2338	(500–4390)	0.41	(0.17–0.68)
Cold Temperate						
Evergreen (18)	2940	(1700–6010)	5420	(1440–15880)	0.59	(0.27–0.89)
Deciduous (11)	3980	(2840–5360)	3500	(540–6682)	0.43	(0.11–0.59)
Warm Temperate						
Evergreen (4)	3032	(250–5406)	6154	(2120–9506)	0.70	(0.60–0.95)
Deciduous (3)	4290	(3310–5300)	5730	(2190–9000)	0.54	(0.29–0.73)
Tropical						
Evergeen (4)	6200	(2430–10250)	4200	(1200–11170)	0.36	(0.18–0.52)

[a]Data summarized from the literature by Gower and Landsberg (unpublished manuscript).
[b]Sample size.

bohydrate polymers, such as starch, are slightly more difficult to break down because the longer molecules must first be cleaved into sugar units, which can then be metabolized. Sugars and starches occur in small amounts (1–7% by weight) in plant litter.

The most important constituent of plant tissue on a weight basis is cellulose. It is the primary compound of the cell wall in plants and comprises 40–80% of plant tissue (Aber and Melillo, 1991). Cellulose consists of a series of sugar units that are linked between the 1 and 4 carbon of adjacent sugar molecules. Unlike simple sugars, cellulose cannot be remobilized and retains its original form until decomposed. Numerous soil microorganisms are capable of decomposing cellulose and hemicellulose—a highly branched cellulose-like compound very similar in decomposibility to cellulose. Fungi are the primary decomposers of cellulose in humid soils, whereas bacteria are the main decomposers of cellulose in semiarid for-ests (Alexander, 1977). Various fungi, bacteria, and actinomycetes can decompose hemicellulose.

Lignins are also abundant in plant tissue. The molecules are highly variable in structure but are generally composed of aromatic rings containing primarily carbon, hydrogen, and oxygen. The exact role of lignin is not fully understood: It is highly resistant to enzymatic decay, which has led to speculation that it is the most important structural constituent of plants, with a major role in deterring herbivory (Coley et al., 1985; Horner et al., 1988). Many basidiomycetes, a type of fungi, and some aerobic bacteria can break down lignin, but the reaction is very slow because

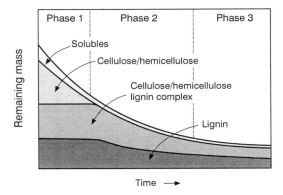

Figure 6.3 Diagram illustrating the relative composition and rate of mass loss of plant tissue during decomposition. Initially, mass loss is rapid as soluble and easily decomposable constituents are degraded. In the later stages of decomposition, mass loss is largely controlled by lignin mass loss (adapted from Coûteaux *et al.*, 1995).

of the high density of aromatic rings, numerous side chains, and complex interlinkages between lignins. It is interesting to note that lignin must be decomposed during the papermaking process—an expensive process that requires the use of extremely caustic chemicals. Paper chemists are currently trying to identify and isolate "super fungi" that can degrade lignin so that papermaking will be cheaper and require fewer envionmentally hazardous chemicals. Tannins consist of aromatic rings and are thought to deter herbivory by their low digestibility and energy return (see Horner *et al.*, 1988).

The collective effect of the various carbonaceous constituents on decomposition of organic matter is shown in Fig. 6.3. Although all the components are constantly decomposing, their relative importance differs during decomposition. Mass loss of detritus is characterized by an initial rapid weight loss of soluble compounds, followed by an intermediate weight loss of cellulose and hemicellulose and finally the slow decomposition of lignin and lignin-derivative compounds (Berg *et al.*, 1984). A negative exponential decay function is often used by ecologists to describe the weight loss of organic matter in terms of time (Olson, 1963; see Box 6.1).

III. Litter Decomposition

A. The Role of Organisms

Soon after plant and animal tissue are incorporated into the soil, microorganisms begin decomposing the material to obtain energy. A great many invertebrates and microorganisms are involved (Alexander, 1977;

Box 6.1 Estimating Decomposition Constants

Several techniques are used to calculate decomposition; the method of choice is largely determined by the type of material of interest and question to be answered. The simplest approach is the mass-balance method. The decomposition constant, k, is calculated by dividing annual litterfall or detritus production by organic matter content in the forest floor or total soil organic matter. Detritus production is expressed as a rate—mass area^{-1} time^{-1} (where the units of time are usually years) and detritus mass is expressed as mass area^{-1} so that k has units of time^{-1}:

$$k = (\text{detritus production}/\text{detritus mass})$$

A critical assumption underlying this approach is that the detritus is in steady state. Excluding fine root turnover and coarse woody debris will result in an underestimation of k (Vogt et al., 1982, 1986b).

A second approach is the use of litter decomposition bags. This method consists of placing a known mass of plant litter (m_l), usually leaves or fine roots, into nylon mesh bags, collecting the bags at intervals (Δt), and measuring weight loss ($\triangle m_l$). A simple negative exponential decay model is usually sufficient to describe the fractional weight loss over time. Written in terms of the fraction of m_l remaining after $t = \Sigma \Delta t$, (days, months, or years),

$$m_l(t) = c_l \exp(-kt).$$

k is obtained from the plot of $\ln(m_l)$ against t, from which $k = {}^-(\ln m_l - \ln c_l)/t$, where ln is the natural logarithm, c is the intercept, and if it is set to unity (the fraction remaining at the beginning of the exercise is one) then $k = {}^-(\ln m_l)/t$.

Decomposition of large woody tissue, such as stems and large branches, is commonly estimated as the combined volume loss and decrease in specific gravity (Grier, 1978) as

$$m_l = (V_o \times o_o) - (V_{(t)} \times o_{(t)}),$$

where m_l is weight loss, V_o and o_o are volume and specific gravity, respectively, at time zero, and V_t and $o_{(t)}$ are volume and specific gravity at time t, respectively.

The mean residence time and the time required to achieve 95 and 99% of the steady-state forest floor mass can be calculated as $1/k$, $3/k$, and $4.6/k$, respectively (Olson, 1963).

Paul and Clark, 1989). Invertebrates fragment the litter, increasing the surface area and providing greater opportunities for microbial colonization, and incorporate the small pieces of litter into the mineral soil (Swift *et al.*, 1979; Seastedt and Crossley, 1980). Important soil invertebrate groups include nematodes, collembola, mites, earthworms, and termites. Nematodes, collembola, and mites tend to be more prevalent in conifer forests, whereas earthworms are more common in temperate deciduous and tropical forests (Swift *et al.*, 1979; Phillipson *et al.*, 1978). Termites are most prevalent in warm temperate and tropical forests (Gentry and Whitford, 1982). There are very large numbers of soil invertebrates in forest soils, but they comprise a small fraction (generally <5%) of the total organic matter of a forest ecosystem (Ugolini and Edmonds, 1983; Anderson and Domsch, 1980). The biomass of soil invertebrates tends to increase from boreal to warm temperate and tropical forests, except where decomposition is so rapid that little or no forest floor exists (Table 6.4). The effect of invertebrates on subsequent decomposition is greatest for low-quality litter (i.e., high C/N ratio).

During the decomposition process, microorganisms can act as sinks (immobilization) or sources (mineralization) of carbon and nutrients and therefore control the availability of nutrients to vegetation. Bacteria, fungi, and actinomycetes are the most important microorganisms; they can be classified on the basis of numerous structural or functional characteristics, but are often separated into two broad groups—autotrophs and heterotrophs—based on the way they obtain their energy. Autotrophs obtain their energy from either sunlight (photoautotrophs) or from the oxidation of inorganic compounds (chemoautotrophs). Chemoautotrophs are limited to a few species of bacteria, each using a very specific compound (Paul and Clark, 1989), but they play an important role in the cycling of many elements (Chapter 7). Heterotrophs, the organisms responsible for most litter decomposition, use preformed organic matter as a source of energy, which is obtained by cleaving chemical bonds. As in the case of soil invertebrates, the biomass of fungi and bacteria increases from boreal to tropical forests (Table 6.5). Fungi dom-

Table 6.4 Fungi, Bacteria, and Microfauna Biomass (kg ha^{-1}) for Selected Forest Biomes

Biome type	kg ha^{-1}		
	Fungi	Bacteria	Microfauna
Boreal and temperate conifer	836–4620	1–110	84–282
Temperate deciduous	890–1290	1–265	83–786
Warm temperate broad-leafed and tropical	4500	1100	84

Table 6.5 Decomposition Coefficients (k, years) for Foliage and Fine and Coarse Wood for Deciduous and Evergreen Forests in Contrasting Environments

			Tissue Component	
Biome	Leaf habit	Foliage	Fine wood	Coarse wood
Boreal	Decidious	0.39–0.702	0.058–0.120	0.022–0.29
	Evergreen	0.223–0.446	—	—
Cold temperate	Deciduous	0.28–0.85	0.10–0.38	—
	Evergreen	0.140–0.693	—	0.011–0.060
Warm temperate	Deciduous	0.441–2.465	—	0.03–0.27
	Evergreen	0.162–0.751	—	0.04
Tropical	Deciduous	0.62–4.16	—	—
	Evergreen	0.162–2.813	—	0.115–0.461

inate in well-aerated soils, whereas bacteria are more common in anaerobic soils. Fungi also tend to be more prevalent than bacteria in acidic soils because bacteria are less tolerant of low pH (Alexander, 1977).

B. Environmental Controls

Decomposition can be divided into three processes: fragmentation of organic matter, leaching, and microbe-mediated chemical degradation, with the latter process producing CO_2, water, and the energy used by the microorganisms:

$$C_6H_{12}O_6 + 6O_2 \rightarrow 6CO_2 + 6H_2O + \text{energy}. \quad (6.1)$$

Numerous techniques are used to estimate decomposition rates, although no single method provides a complete picture of the factors controlling the decomposition of surface litter and soil organic matter. Methods include measuring the weight loss of the tissue of interest (see Box 6.1), measuring soil CO_2 evolution (discussed below), or measuring ATP, a proxy for microbial activity (Vogt *et al.*, 1980). Each method provides useful information and there are considerable advantages in using combinations of different methods

Table 6.5 summarizes decomposition coefficients for a wide variety of tissues and environmental conditions. In general, decomposition rates are greatest for foliage, slowest for wood, and increase from boreal to tropical forests. Decomposition of organic matter is strongly controlled by environmental conditions that affect the activity of soil invertebrates and microorganisms. As a general rule, microbial activity increases 2.4-fold with a temperature increase of 10°C across the normal temperature range of soils (i.e., decomposition tends to have a Q_{10} of about 2.4; Raich and Schlesinger, 1992). However, extreme temperatures and moisture conditions can decrease the efficiency of most soil microorganisms, resulting in a decrease in decomposition rates and therefore an increase in litter accumulation (Ino and Monsi, 1969).

Meentemeyer (1978) proposed that decomposition could be estimated from annual estimates of evapotranspiration (which he called AET), calculated using the Thornthwaite formulation that incorporates, to some extent, the effects of temperature, precipitation, and growing degree days—an integral of temperature. The success of this relationship, on an annual basis, is illustrated using decomposition data for Scots pine needles (Fig. 6.4). However, the AET model is only a surrogate for the actual regulators of decomposition and is not as useful at the local scale because of a range of other factors that influence decomposition rates. It is important to note that annual averages of AET do not account for seasonal distribution of precipitation and temperature or for extreme temperature and moisture conditions that adversely affect decomposition rates (Fogel and Cromack, 1977; Edmonds, 1979). Whitford *et al.* (1981) noted that the AET–decomposition relationship was a poor predictor of litter decomposition rates in recently clear-cut forests; they suggested that this was because the model did not account for marked changes in microclimate that adversely affect microbial activity. Erikson *et al.* (1985) also noted that suspended woody litter in clear-cut forests decomposed more slowly than woody litter in contact with the soil—presumably because conditions at the soil surface were more moist than in suspended

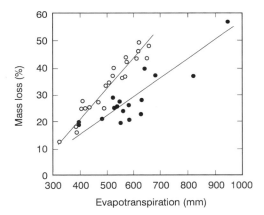

Figure 6.4 First-year mass loss of Scots pine (*Pinus sylvestris*) needles versus actual evapotranspiration. Open symbols represent Scots pine sites in a Scandanavian–NW continental transect and pine sites near the European west coast or exposed to Atlantic coast influence. Solid symbols are for sites around the Mediterranean, in Poland, and the eastern United States [adapted from Coûteaux *et al.* (1995). Data sources include Meentemeyer and Berg (1986) and Berg *et al.* (1993)].

samples. The AET model also cannot account for the presence of soil invertebrates such as termites that hasten decomposition (Whitford *et al.*, 1981; Santos and Whitford, 1982).

A second major factor affecting decomposition rate is litter quality. For example, Tanner (1981) reported that mass loss of foliage in 1 year ranged from 27 to 96% for 15 tree species in a Jamaican montane rainforest and Cuevas and Medina (1986) reported that the time required for 95% of mass loss to occur ranged from 0.4 to 13.6 years for humid tropical tree species in a Venezuelan tropical forest. These large interspecies differences in decomposition rate can be explained by differences in composition of carbon constituents (Table 6.3) and mineral nutrients. It is interesting to note that certain tree species have extremely high concentrations of extractives that have presumably evolved to deter pest and pathogen attack and it is no coincidence that many of these same species are highly valued timber species because of their resistance to rot. The decomposition rate of plant tissue is inversely correlated with the proportion of lignin (Berg *et al.*, 1984; McClaugherty *et al.*, 1985) and is also influenced by the concentration of mineral elements, especially those that limit plant and microbial growth (e.g., N and P). Melillo *et al.* (1982) demonstrated that the ratio of lignin/nitrogen concentration explained 82–89% of the observed variation in decomposition rate for tree species in two contrasting climates (Fig. 6.5). However, some

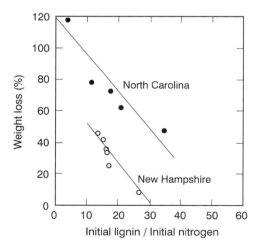

Figure 6.5 Relationship between percentage weight loss during the first year and the lignin/nitrogen ratio of leaf litter for temperate broad-leafed deciduous tree species in North Carolina (closed circles) and New Hampshire (open circles) (redrawn from Melillo *et al.*, 1982).

plant tissues do not decompose at the rate predicted by their size or tissue biochemistry. Fine roots, which tend to have high concentrations of suberin—lignin-like compound laid down as roots mature and become rigid—decompose more slowly than their size or nutrient concentration would suggest. The influence of litter quality on decomposition rates is not uniform across climates; it has the greatest influence in tropical environments and the smallest effect in boreal climates (Meentemeyer, 1978; Baragali et al., 1993).

A third factor influencing decomposition rate is the surface area/volume ratio of the tissue. All other factors being equal, tissue with a small surface area/volume ratio (e.g., stems) will decompose more slowly than tissue with a larger ratio (Berg, 1984; Erickson et al., 1985; Harmon et al., 1986), which emphasizes the importance of soil microinvertebrates as agents for fragmenting litter. It follows that site preparation practices that reduce the size of logging debris should facilitate decomposition and nutrient release from logging residue left on the site.

IV. Carbon Losses from Forest Ecosystems

The efflux of CO_2 from the soil is often referred to as soil respiration or soil surface CO_2 flux; the former term is technically incorrect because the efflux of CO_2 from the soil surface is derived from soil organic matter decomposition and root respiration (Box 6.2). Soil CO_2 flux is commonly the second largest term in forest carbon budgets (Gower et al., 1996c) and of similar importance in the global carbon cycle (Raich and Schlesinger, 1992). The production of CO_2 by microbial and root respiration also plays an important role in nutrient cycling because CO_2 dissolves in water forming carbonic acid (a weak acid) that can dissociate to produce a bicarbonate ion (HCO_3^-) and H^+. Anions such as bicarbonate influence weathering rates and leaching losses of nutrients (see Chapter 7).

Because climate strongly influences decomposition, it is a major factor controlling soil CO_2 flux—a by-product of the oxidation of organic matter. Fung et al. (1987) reported a positive correlation between monthly soil CO_2 flux and air temperature for a diverse group of terrestrial biomes, and annual soil CO_2 flux has also been correlated to mean annual temperature (Raich and Schlesinger, 1992). Using data summarized by Raich and Nadelhoffer (1989), we calculated average annual soil CO_2 fluxes for deciduous and evergreen forests in boreal, temperate, and tropical biomes (Fig. 6.6) and found that, for a given forest type (e.g., deciduous or evergreen), they increased from high to low latitudes. In tropical and temperate regions, soil CO_2 flux was greater in evergreen

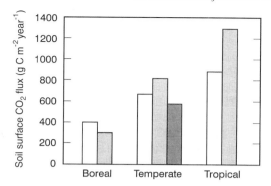

Figure 6.6 Average annual soil surface CO_2 fluxes for different forest biomes. Open columns are deciduous, shaded are evergreen, and the dark shaded column (temperate) is mixed forests (adapted from Raich and Schlesinger, 1992).

than in deciduous forests but the reverse was true for boreal forests, presumably because of the negative effects of low soil temperatures and poor drainage on microbial activity in lowland boreal conifer forests (Flanagan and Van Cleve, 1983; Updengraff et al., 1995).

Correlations between microbial activity and temperature exhibit considerable variation, and attempts to understand the factors that control microbial activity have included field experiments in which microbial activity has been manipulated by varying nutrient and carbon supply in the soil. These studies have shown that microbial biomass is highest in the summer and lowest in the winter—corresponding to seasonal soil CO_2 fluxes—and that the effects of carbon and nutrient additions on respiration are site specific. Strong positive correlations have been established between net primary production and both microbial biomass (Zak et al., 1994) and soil CO_2 flux (Myrold et al., 1989; Raich and Schlesinger, 1992), implying that microbial biomass and activity are dependent on detrital input and may be carbon limited. Microbial activity and biomass are constrained in forests with large organic matter accumulations because of low soil temperatures and/or poor litter quality (Beare et al., 1990; Bridgham and Richardson, 1992; Sugai and Schimel, 1994). Foster et al. (1980) added carbon and nutrients to a boreal pine forest (*P. banksiana*) and concluded that microbial activity was mainly limited by available carbon but also, to a lesser extent, by nitrogen. In warm temperate upland forests where decomposition is more complete and rapid, microbial biomass and soil CO_2 flux appear to be more strongly related to nutrient availability. Soil nitrogen availability commonly limits microbial biomass (Zak et al., 1990; Beare et al., 1990; Wardle, 1992), whereas phosphorus can limit mi-

Box 6.2 Soil CO_2 Flux Measurements

Several techniques are used to measure soil surface CO_2 flux (root mycorrhize respiration + microbial respiration). It is extremely difficult to partition soil surface CO_2 flux into autotrophic and heterotrophic respiration. To date, the most common method used to measure soil surface CO_2 flux was the static alkali absorption technique (Edwards, 1982; Raich and Nadelhoffer, 1989), but there is increasing evidence that this method underestimates soil CO_2 flux, especially at high rates (Ewel et al., 1987, Haynes and Gower, 1995). Because global estimates of soil surface CO_2 flux are largely based on static alkali absorption technique (see Raich and Schlesinger, 1991), data collected in this way may be underestimates. Portable CO_2 analyzers equipped with a special chamber (Norman et al., 1992; Haynes and Gower, 1995) and eddy correlation systems located near the soil surface (Baldocchi et al., 1996) have also been used to estimate soil CO_2 flux. The latter approach has the advantage that it integrates over a large area, but it may be unreliable because of the low wind speeds and poor (aerodynamic) mixing commonly encountered beneath forests canopies, particularly at night.

crobial biomass in some forests (Scheu, 1990). The relative abundance of fungi in relation to bacteria can influence N cycling rates because of the potentially large differences in their C/N ratios (Paul and Clark, 1989).

It is difficult to separate the contribution of microbial (or heterotrophic) respiration and root (or autotrophic) respiration because microbial respiration is influenced by root exudates—labile carbon that leaks from roots. Numerous approaches have been used to partition soil CO_2 flux into the two components, including comparing CO_2 flux from root-free trenched and control plots (Ewel et al., 1987; Haynes and Gower, 1995), comparing soil CO_2 flux from intact forests and recent clear-cuts (Nakane, 1984), $^{14}CO_2$-labeling and tracing studies (Cheng et al., 1993), and process-based models. The contribution of root respiration to total soil CO_2 flux averages 0.45 but is moderately variable, ranging from 0.33 to 0.62 (Table 6.6). The average value is 15% greater than the value assumed by Raich and Schlesinger (1992). There does not appear to be any relationship between the ratio of root/total soil CO_2 flux and climate; however, Raich and Nadelhoffer (1989) reported that annual soil CO_2 flux was positively correlated to annual litterfall. As we noted earlier, they suggested that a carbon balance approach (i.e., carbon loss, as

Table 6.6 The Fraction of Total Soil Surface CO_2 Flux Attributable to Root Respiration

Forest type	Root/total soil CO_2 flux	Source
Cold temperate		
Mixed hardwoods	0.33	Bowden et al. (1993)
Pinus resinosa	0.57	Haynes and Gower (1995)
Pinsu densiflora	0.50	Nakane et al. (1983)
Warm temperate		
Mixed hardwoods	0.35	Edwards and Sollins (1973)
Pinus elliottii	0.58–0.62	Ewel et al. (1987)
Floodplain forest	0.55	Pulliam (1993)
Temperate broad-leaved evergreen		
Nothofagus sup.	0.22	Tate et al. (1993)
Tropical deciduous		
Mixed	0.50	Behera et al. (1990)

reflected by soil CO_2 flux, equals above- plus below-ground detritus input, assuming no change in soil C) can be used to set an upper limit on the total amount of carbon that can be allocated to root turnover, respiration, and exudates (see Chapter 5).

In addition to being lost as carbon dioxide to the atmosphere, carbon can be leached below the rooting zone. This process is particularly important for the soil surface layer, and accumulation in lower horizons is important in soil development, especially for Spodosols (Dawson et al., 1978; Ugolini et al., 1988). The dissolved organic matter consists mostly of high-molecular-weight polymers originating from the canopy and surface detritus layer. Tree species differ greatly with respect to their production of soluble organic substances (Pohlman and McColl, 1988), and concentrations of dissolved organic carbon in the upper soil vary seasonally in conjunction with microbial activity. High dissolved organic carbon concentrations can occur in the spring, coinciding with snowmelt and leaching of organic substances from the forest floor (Antweiler and Drever, 1983). As fulvic and humic acids move downward in the soil, they accumulate in the lower horizons; hypotheses proposed to explain the arrest of these substances in these soil horizons include the metal-fulvate theory (Schnitzer, 1979), flocculation (DeConinck 1980), polymerization or decomposition by microorganisms, or adsorption on the surface of clay minerals (Jardine et al., 1989; Dahlgren and Marrett, 1991). Whatever the mechanism(s) responsible for the arrest of these solutes, their concentrations in the soil solution below the rooting zone of upland forests are low—commonly less than 2 mg liter^{-1} (Sollins and McCorison, 1981; McDowell and Likens, 1988). However, dissolved organic

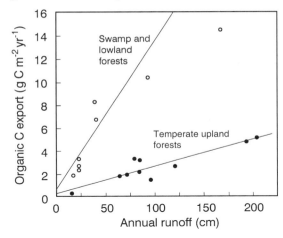

Figure 6.7 Relationship between carbon export and annual runoff for rivers draining lowland boreal and temperate floodplain forests compared to upland forests (redrawn from Mulholland and Kuenzler, 1979).

carbon concentrations are often substantially higher in streams draining lowland boreal and temperate floodplain forests than upland forests (Fig. 6.7).

V. Influence of Forest Management on Soil Carbon Dynamics

Forest management decisions, including "no management"—or simply allowing the forest to develop uninterupted—can affect many of the processes controlling carbon inputs and losses and, hence, soil organic matter content. Carbon and nutrient cycles are tightly coupled (Chapter 7); therefore, any natural or anthropogenic disturbance of the carbon cycle is likely to affect nutrient cycles as well.

Johnson (1992) reviewed the literature and found no discernable effect of harvesting on soil organic matter content; most of the studies he reviewed reported a net change in soil organic matter of less than ±10% (we noted earlier that large spatial variability of the surface litter and soil organic matter content makes it difficult to detect small changes). In general, the more intensive the postharvest site preparation, the greater the loss of soil organic matter (Johnson, 1992). Cultivation is the most extreme site preparation practice and can result in large losses of soil organic matter (Mann, 1986; Detwiler, 1986), although these are generally smaller than in the case of agricultural crops because, in the case of forests, cultivation occurs only once during the stand rotation and the vegetation regrows immediately after planting. Organic matter losses as a consequence of site preparation are caused by increased erosion, accelerated decomposition resulting from higher soil temperatures, improved incorporation of organic matter into the soil facilitating microbial immobilization, and deterioration of soil aggregate structure.

Prescribed burning is an important management tool used to remove logging debris before replanting, reduce fuel loads to minimize the chance of catastrophic fire, and control the invasion of undesirable competing species. In general, prescribed burning appears to have little effect on soil organic matter content (Wells, 1971), but a long-term study has shown that frequent fires cause redistribution of the surface organic matter to the upper mineral soil (Binkley *et al.*, 1994). However, Sands (1983) reported that a prescribed burn decreased soil organic matter by 40–50%. Differences in results obtained from experiments involving prescribed burning may be caused by factors such as fire intensity, the amount and condition of the material in the litter layer, and the composition of that material.

Fertilization generally increases soil carbon storage (Johnson, 1992), although the reasons for this are not fully understood; it may affect the carbon dynamics of roots and soil microorganisms, tending to increase leaf litter production and decrease the relative allocation of carbon to fine root net primary production (see Chapter 5). The varying effects of fertilization on decomposition may be due, in part, to the different types of fertilizer used and their effects on soil microorganisms (Fog, 1988). For example, urea can raise the pH of the soil immediately surrounding the fertilizer pellet to 8.0, whereas ammonium nitrate increases soil acidity. Haynes and Gower (1995) measured soil CO_2 flux in trenched and untrenched plots in control and fertilized red pine plantations in Wisconsin and found that soil CO_2 flux did not differ significantly between trenched plots in control and fertilized stands but was significantly greater in the untrenched control than the fertilized stand for all 3 years of the study. These results suggest that fertilization had little effect on decomposition.

Johnson (1991) cited numerous studies that examined the influence of various forest management practices on soil carbon content, although few of them examined the effects of forest management on the processes that control soil carbon content, such as soil CO_2 flux, humus formation, and decomposition. However, it is necessary to understand how management practices affect the processes so that those practices can be altered, if necessary, before they have detrimental effects on long-term site productivity. The effects of harvesting on soil respiration are inconsistent (Fig. 6.8) but may be explained by the varying impact of harvesting on the structure of forests and, hence, on the environmental conditions af-

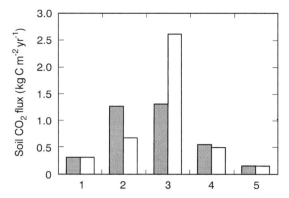

Figure 6.8 Comparison of soil surface carbon dioxide flux for control (hatched columns) and clear-cut (open columns) forests during the first year after harvest. Sources of data are (1) *Populus tremuloides* forest in Alaska (Schlentner and Van Cleve, 1985), (2) *P. densiflora* forest in Japan (Nakane *et al.*, 1984), (3) *P. elliottii* plantation in Florida (Ewel *et al.*, 1987), (4) *Quercus–Carya* forest in Tennessee (Edwards and Ross-Todd, 1983), and (5) mixed broad-leaved deciduous forest in West Virginia (Mattson and Smith, 1993). Soil surface CO_2 flux values for the West Virginia study are based on only 142 days.

fecting organic matter decomposition (Binkley, 1986). It has been variously reported that the decomposition rates of organic matter in clearcuts are slower (Whitford *et al.*, 1981; Erickson *et al.*, 1985), about the same, and faster than in control forests (Gholz *et al.*, 1985b; O'Connell, 1988). In forests where decomposition is strongly temperature limited, such as boreal forests, removal of the canopy increases the solar radiation reaching the soil surface and stimulates decomposition. Thus, the effect of clear-cutting on organic matter decomposition and soil CO_2 flux is site specific. The uncertainties can be resolved—or at least various scenarios can be explored—by the use of process-based decompositon models (Chapter 9).

The replacement of native vegetation by forest plantations is widespread in some regions of the world (Chapter 2), but there is little information about the effects of different tree species on soil organic matter. Altering species or land use practices is likely to influence the soil carbon cycle because detritus production, carbon allocation, and decomposition rates differ among species. Nitrogen-fixing species commonly increase soil organic matter because they lead to enhanced detritus production (Johnson, 1992; see also Chapter 7). As discussed earlier, cultivation decreases soil carbon; therefore, it should not be surprising that reforestation or afforestation increases the soil C content of agricultural or abandoned soils (Table 6.1). The effects of reforestation of previously forested soil are less clear. Alban *et al.* (1982) compared soil properties in four 40-year-old tree plantations established in a common garden design (i.e., adjacent plots on a similar soil) and found that forest floor organic matter content did not differ between the species, but the soil C content did: soil C beneath the aspen plantation was significantly less (31 and 18% at two sites) than under white spruce, red pine, and jack pine. In a similar common garden study design, Gower and Son (1992) found that forest floor biomass differed by 80% (8.7–42.8 mg ha^{-1}) among 30-year-old European larch, red and white pine, red oak, and Norway spruce plantations in southwestern Wisconsin. Few studies have examined the effect of species on the more stable carbon constituents in forest soils.

VI. Role of Forest Soils in the Global Carbon Budget

An estimated 1400×10^{15} g C are sequestered in the form of organic matter in soils across the earth, of which 34% is in forest soils (Post *et al.*, 1982). As discussed earlier, the carbon content of forest soils is subject to large changes caused by changes in land use. Houghton *et al.* (1983) calculated the change in global soil and vegetation carbon content over, approximately, the past century (1860–1980) using land use statistics to calculate agricultural expansion and deforestation. They estimated that the release of carbon in 1980 from land, including soil, was between 1.8 and 4.7×10^{15} g C year^{-1}, compared to 5×10^{15} g C year^{-1} released by the combustion of fossil fuel (Rotty and Masters, 1985). However, as Johnson (1992) points out, the Houghton *et al.* estimate is based on the assumption of 35% carbon loss following deforestation, and empirical data do not support this assumption. These large discrepancies make it difficult to balance the global carbon budget and illustrate the need for better understanding of the effects of land use practices, including forestry, on soil carbon cycling processes.

One likely effect of global warming is accelerated decomposition, resulting in a greater emission of CO_2 into the atmosphere. This would further exacerbate global warming. Experimental soil warming studies have shown that increasing the soil temperature by 5°C above ambient caused annual CO_2 flux from a northern hardwood forest to increase from 712 to 1250 g C m^{-2} year^{-1} (Peterjohn *et al.*, 1994). Jenkinson *et al.* (1992) used the Rothamsted soil carbon cycling model to examine the amount of carbon that could be released under different climate change scenarios. They estimated that an average temperature rise of 0.03°C across the globe would increase CO_2 release by 61×10^{15} g C over the next 60 years—equivalent to about 25% of the CO_2 that will be released from the combustion of fossil fuel if fuel consumption remains unchanged over the next 60 years. Their analysis assumes annual detritus inputs will remain unchanged and that temperature increase will be uniform worldwide—both of which are highly suspect assumptions (see Chapter 5). Nevertheless, such analyses are useful because they force evaluation of the influence of abiotic and biotic factors on soil carbon cycle and its impact on the global C cycle.

VII. Concluding Remarks

Soil organic matter is often the largest carbon pool in forest ecosystems, with many biological, chemical, and physical properties that have beneficial effects on tree growth. Therefore, forest managers should be aware of how forest management practices affect the processes responsible for carbon inputs to and losses from forest soils. Unfortunately, our understanding of the effect of natural factors and management practices on these processes is incomplete, although it is clear that management practices that alter the interrelationships between primary producers and decomposers could alter the long-term productivity of forest soils. There is a great need for better data on soil carbon contents and better understanding of microbial processes as a basis for the development of more realistic soil carbon and nutrient cycling models. Because of the large amount of carbon stored in soils and its susceptibility to climate warming and land use, soil carbon dynamics have significant implications for global ecology.

Recommended Reading

Johnson, D. W. (1992). Effects of forest management on soil carbon storage. *Water Air Soil Pollut.* **64**, 83–120.

Harmon, M. E. *et al.* (1986). The ecology of coarse woody debris in temperate ecosystems. *Adv. Ecol. Res.* **15**, 133–302.

Paul, E., and Clark, (1989). "Soil Microbiology and Biochemistry." Academic Press, San Diego, CA.

Stevenson, F. J. (1982). "Humus Chemistry." Wiley, New York.

Raich, J. W., and Schlesinger, W. H. (1992). The global carbon dioxide flux in soil respiration and its relationship to vegetation and climate. *Tellus* **44B**, 81–99.

7

Nutrient Distribution and Cycling

The soil must be man's most treasured possession: so he who tends the soil wisely and with care is assuredly the foremost among men.
—Sir George Stapledon (1964)

The growth of forests is often limited by mineral nutrient(s), despite the fact that the nutrient content of forest soils is generally many times greater than the annual nutrient requirements of a stand. This imbalance between soil nutrient content and plant requirements suggests that there are rate-limiting processes controlling the transformations of mineral nutrients into forms available to trees, which in turn affect the composition, structure, and function of natural forest communities. As discussed in Chapter 5, nutrition can affect carbon assimilation by leaves and the allocation of carbon to biomass components, both of which affect the harvest index or yield. Studies also suggest that the commonly observed age-related decline of forest growth may be related, in part, to increased nutrient constraints (McMurtrie *et al.*, 1995; Gower *et al.*, 1996c). There is also increasing evidence that nutrient availability or nutrient imbalances can greatly alter the species composition of forests.

Forest nutrition is an important factor that must be considered by management, but despite a plethora of fertilization studies we still lack the understanding that would allow us to consistently predict whether, and how, the addition of nutrients to the soil will affect the growth of forests. Responses depend on the soil type, the kind of fertilizer used, the amount applied, and the timing of the application as well as—perhaps more important—limitations imposed by other resources such as water, temperature, and light (φ_p). Our understanding of the physiochemical and biological processes controlling soil nutrient dynamics and uptake by trees is incomplete and as a result mechanistic models of nutrient uptake are only beginning to emerge (Landsberg *et al.*, 1991).

7. Nutrient Distribution and Cycling

Other forest management practices, such as site preparation, intensive harvesting, clear-felling, and prescribed burning, also alter nutrient cycles of forests, but it is unclear whether these changes significantly affect the long-term productivity of sites. Questions arising about nutrition and its effects on tree growth or the effects of external factors and conditions on biogeochemical cycles of forests for particular sites can rarely be answered using data from empirical studies because, in most cases, appropriate data do not exist. Where there have been pertinent studies, the results tend to be highly site specific and cannot be easily extrapolated to other sites of interest. Furthermore, it is difficult to anticipate how forest ecosystems subject to current environmental conditions and management practices will respond to future global changes such as increasing deposition of atmospheric nitrogen (Aber, 1992). Few empirical approaches have provided reliable indicators of nutrient shortage, susceptibility to leaching losses, sensitivity to acid rain, or other factors affecting the structure and function of forest ecosystems. Therefore, if forest ecosystems are to be managed on a sustainable basis, it is essential to un-

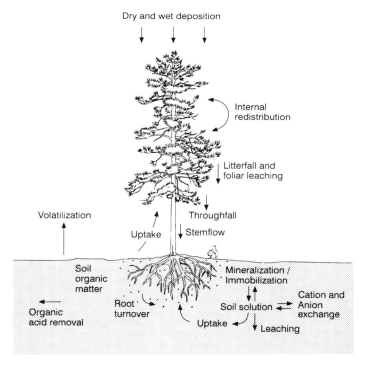

Figure 7.1 Schematic diagram showing the various components of the nutrient cycle of forests.

derstand (i) the natural processes regulating nutrient inputs, (ii) the transformations both within and between the soil and vegetation, and (iii) nutrient losses (Fig. 7.1). We also have to understand how these processes are affected by natural and anthropogenic disturbances.

The purpose of this chapter is to review the basic framework of forest nutrition, including the major sources of nutrient inputs to forests, the transfer of nutrients between the vegetation and soil, the internal recycling or redistribution and use of mobile nutrients (including the feedback mechanisms between the vegetation and soil), and the potential pathways for nutrient losses from forest ecosystems. The first part presents some basic concepts and principles; the remainder of the chapter draws on those concepts and principles to examine how various forest management practices and natural "disturbances" affect the biogeochemical cycles of forest ecosystems. Where possible, we compare contrasting forests ecosystems around the world. We have attempted to provide a balanced presentation, but the focus tends to be on nitrogen and, to a lesser extent, phosporus because these two nutrients have traditionally been perceived as the ones most commonly limiting forest growth, although this viewpoint is changing (Gosz, 1986).

I. The Essential Plant Nutrients and Ion-Exchange Capacity of Soils

A. Essential Plant Nutrients

The mineral elements essential for plant growth can be divided into two categories based on the amounts required by plants (Salisbury and Ross, 1992): macro (carbon, hydrogen, oxygen, nitrogen, phosphorus, sulfur, calcium, magnesium, and potassium) and micronutrients (boron, chlorine, copper, iron, manganese, molybdenum, cobalt, and zinc). Nitrogen (N), sulfur (S), and phosphorus (P) are important constituents of amino acids, proteins, and nucleic acids. Nitrogen is needed to build chlorophyll molecules and phosphorus to construct energy-transfer compounds (ATP and ADP). Potassium (K) differs from other macroelements in that it is not incorporated into structural tissue—it regulates the charge balance across plant membranes. Calcium (Ca) is an important constituent of calcium pectate, which can be thought of as "cell-wall glue." Magnesium (Mg) is required to construct chlorophyll and most micronutrients are used to construct coenzymes in plants. Boron (B), cobalt (Co), and molybdenum (Mo) are essential cofactors in the nitrogen fixation process.

Many methods have been used to determine whether trees have adequate nutrition; however, Stone (1978) pointed out that it is difficult to

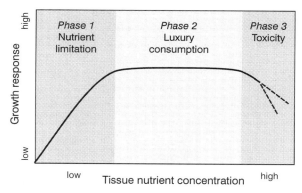

Figure 7.2 Schematic diagram illustrating the general pattern of plant growth responses to nutrient concentrations. The three phases indicate nutrient limitation, where the addition of nutrients is likely to result in increased growth; luxury consumption, where there will be no response to additional nutrients, and toxicity, when additional nutrients have negative effects.

quantify nutrient supply and availability. Figure 7.2 is a schematic illustration of the relationship between nutrient concentrations in plant tissue, usually foliage, and growth responses to added nutrients. In phase I, during which growth is limited by inadequate nutrition, the addition of the limiting nutrient(s) will increase plant growth. The second, so-called luxury consumption phase, is characterized by little or no growth response to increasing nutrient concentrations, and in the third phase nutrients become toxic and growth is likely to decrease in response to increased nutrient concentrations. Although there is little doubt that a general relationship of the type shown in Fig. 7.2 exists for each element, foliar nutrient concentrations are generally unreliable indicators of plant nutrition for several reasons: Nutrient concentrations vary considerably both temporally and spatially in the canopy, wild plants tend to adjust their growth rate to match nutrient supply resulting in relatively conservative foliar nutrient concentrations (Chapin *et al.,* 1986), and nutrient imbalances and multiple limitations can mask the problem. A more promising approach involves the graphical analysis of foliage nutrient content, concentration, and relative foliage mass, as indicated in Fig. 7.3.

Other approaches used to determine nutrient availability, such as visual symptoms, soil chemical analysis, and bioassays, have had limited success. It is worth noting that Wessman *et al.* (1988) reported that the nutrient content of canopies for forests along a nitrogen mineralization gradient in Wisconsin could be estimated by remote sensing, but this approach is far from operational at the time of writing.

I. The Essential Plant Nutrients and Ion-Exchange Capacity of Soils

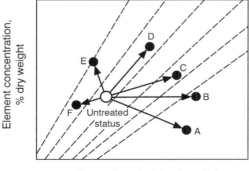

Figure 7.3 A method of graphical analysis to identify nutrient limitation. Following fertilization, a trajectory similar to A indicates decreased nutrient concentration and increased leaf mass and nutrient content, suggesting the specific nutrient was not limiting. Trajectory B indicates no change in nutrient concentration but an increase in nutrient content and leaf mass, implying nutrient transport into the foliage is approximately balanced by leaf growth and/or retranslocation. Trajectory C denotes an increase in nutrient concentration and leaf mass, implying nutrient limitation. Trajectory D occurs when nutrient concentration and content increase but leaf mass remains constant, signifying luxury consumption or nutrient storage. Trajectory E results from decreased leaf mass and increased leaf concentration, indicating nutrient toxicity (adapted from Timmer and Stone, 1978).

B. Ion-Exchange Capacity of Soils

Although the nutrient content of a forest soil is substantially greater than the annual demand by the stand, only a small fraction of the nutrients residing in the soil solution or at exhange sites on the organic matter and clay particles is available for uptake. A thorough treatment of soil chemistry such as is available in Bohn *et al.* (1979) is beyond the scope of this book; instead, we provide a sufficient introduction to the principles regulating nutrient exchange at the surface of mineral and organic soil particles for the reader to understand the basic soil processes regulating nutrient availability in "undisturbed " or disturbed forest ecosystems and to understand why the effects of disturbances (e.g., fire, clear-felling, nitrogen saturation, etc.) on nutrient cycles differ among forests. We also point out the mechanisms that commonly minimize the effects of any perturbation on ecosystem function.

During soil formation (pedogenesis), secondary minerals form from weathered parent material. Weathering can be of two types: physical, such as the freezing and thawing action, or chemical, such as the dissolution of limestone in water. Two types of secondary minerals are widely recognized—2:1 and 1:1 clays—and the relative abundance of each is strongly influenced by the degree of weathering and to a lesser extent parent material (Jenny, 1980). Young boreal and moderately weathered temperate forest soils (e.g., Entisols, Inceptisols, and Alfisols—see Chapter 2) are dominated by 2:1 secondary clays, whereas the more strongly weathered soils (e.g., Ultisols and Oxisols) common to warm temperate, subtropical, and tropical forests are dominated by 1:1 secondary clays. The nomenclature 2:1 and 1:1 is based on the structure of the silicate clays: 2:1 clays are composed of one aluminum octahedral layer between two silica tetrahedral layers, whereas 1:1 clays are composed of one aluminum octahedral and one silica tetrahedral layer. The 2:1 clays, such as montmorillonite and illite, have a net negative surface charge referred to as cation exchange capacity. That is to say these clay particles (colloids) attract and hold positively charged ions (cations) such as the base cations [calcium (Ca^{2+}), potassium (K^+), magnesium (Mg^{2+}), sodium (Na^+)] and other cations [ammonium ($NH4^+$), hydrogen (H^+), and aluminum (Al^{3+})]. The former group is referred to as base cations because they commonly form bases in soil solution [e.g., $Ca^{2+} + 2H_2O \rightarrow Ca(OH)_2$]. Cation-exchange capacity is measured as milliequivalents (meq) per 100 grams of dry soil (or $cmol^+/kg$ of dry soil if SI units are followed). The percentage of the total cation-exchange sites occupied by the base cations potassium, sodium, calcium, and magnesium is referred to as base saturation.

The source of the net negative charge arises from either isomorphic substitution or the dissociation of H from a hydroxide radical (OH). Isomorphic substitution results from the "substitution" of a cation of very similar size but with a smaller valence charge (e.g., Mg^{2+} substitutes for Al^{3+}) inside the lattice of the secondary mineral during formation. This charge is pH independent because it is the result of the structure of the clay and cannot be reversed by covalent bonding. The dissociation of H^+ from exposed hydroxide groups (R-OH) along the edges of clay particles and soil organic matter produces a net negative charge (R-O$^-$) that attracts and holds cations. Unlike isomorphic substitution, this form of cation-exchange capacity is highly pH dependent, increasing as soil pH increases (less acidic), and is especially important for 1:1 secondary clay minerals such as kaolinite (see Table 6.1). Often, the pH-dependent charge of soil organic matter is far more important for sesquioxides and exceeds that of clay minerals. Cation-exchange capacity and percentage base saturation both increase from young (Entisols) to moderately weathered soils (Inceptisols and Alfisols) but decline in heavily weathered soils (Ultisols and Oxisols) (Bockheim, 1980). In the most heavily weathered tropical soils, composed solely of aluminum oxides, the small cation exchange capacity is almost entirely associated with the organic matter.

Unlike most temperate forest soils, many tropical soils are composed of iron and aluminum oxides and hydroxides; these soil colloids have a variable net charge depending on soil pH (Uehara and Gillman, 1981), especially those soils that are derived from volcanic ash (Kinjo and Pratt, 1971). At low soil pH, the surface of iron and aluminum oxides and hydroxides adsorb H^+, resulting in a net positive charge, referred to as anion-exchange capacity, whereas the opposite occurs at high soil pH (i.e., a net negative charge occurs)—hence the name "variable charge." It is important to note that the 2:1 secondary minerals common to temperate forest soils can also have a net positive charge (anion exchange capacity) but it only occurs at pH less than 2.0, making it absent in all natural forest soils (Sposito, 1984). However, heavily weathered temperate forests soils that contain aluminum and iron oxides and hydroxides may have positive anion-exchange capacity (Johnson et al., 1986).

II. Nutrient Distribution

A. Elemental Ratios

It has been known for some time that functional relationships exist between microbial chemistry and soil nutrient cycling processes (Waksman and Starkey, 1931; Redfield, 1958), although we are still learning about the full implications of the relationships between the nutrient requirements of microorganisms and nutrient cycling in terrestrial ecosystems. Redfield (1958) first recognized that plankton have relatively constant carbon, nitrogen, and phosphorus ratios, which could be used to estimate the abundance of these and other elements in the oceans. Although it is now known that the cellular stoichiometry of plankton is simple and does not hold for terrestrial vegetation (Bolin et al., 1983; Reiners, 1986), Redfield's analysis has become the basis for explaining the biogeochemical cycles of terrestrial ecosystems. For example, the stoichiometry of C/N, C/P, and C/S is substantially smaller for decomposers than for vegetation (Table 7.1) because of the high proportion (by weight) of cellulose in plants (see Chapter 6). This is especially the case in trees, which have a large fraction of their total biomass in structural support tissue (i.e., woody tissue). Moreover, plants produce a wide variety of C-based compounds that are difficult to decompose (e.g., lignin, tannins, and suberin). The high C/element ratios for forest vegetation relative to decomposers is further widened by the removal of nutrients prior to tissue senesence (see below). This large disparity in the element stoichiometry of decomposers and vegetation leads to an imbalance between C and essential nutrients for decomposers and causes nutrients to be immobilized by the decomposing organisms, resulting in nu-

Table 7.1 Carbon/Element Ratios of Microorganisms and Angiosperms[a]

Organism group	C: element ratio			
	C/N	C/P	C/S	C/Ca
Microorganisms				
Bacteria	6–7	18–46	>46	46–62
Fungi	8	9	—	145
Angiosperms				
Herbs	10	230	>230	58
Woody	21	1103	221	184

[a]Data from Bowen, 1979; Anderson and Domsch, 1980.

trient limitation for the synthesis of plant tissue. Reiners (1986) summarized the interrelationships between elemental stochiometry of organisms and biogeochemical cycles of terrestrial ecosystems in the form of axioms and theorems.

The stoichiometry of the various parts of trees, especially foliage, can vary. Vitousek *et al.* (1988) reviewed tissue element concentrations in material from major forest biomes and concluded that biomass allometry differences between biomes had little effect on element stochiometry. They did, however, note several systematic C/element ratio patterns: in general, C/N, C/P, and C/S ratios were greater in conifers than in angiosperms; foliage of tropical trees that contain large concentrations of nitrogen-based defense compounds (e.g., alkaloids) have lower C/N and higher N/P ratios than most other forest types, and nitrogen-fixing trees, such as alder, locust, and acacia, also have lower C/N and higher N/P ratios. Conversely, trees with carbon-based defense compounds (e.g., phenolics and terpenes) in the foliage have high C/element ratios (McKey *et al.*, 1978; Bryant *et al.*, 1983). High C/N or C/P ratios are common in forests growing on nitrogen- or phosphorus-deficient soils. The C/N ratios of leaves are higher in colder climates; this has been observed both within (Van Cleve *et al.*, 1983) and among biomes (Vitousek *et al.*, 1988) and may be explained by the adverse effects of low temperatures on decomposition and nutrient mineralization (Chapter 6).

B. Nutrient Distribution in Forests

Table 7.2 provides information on nutrient content and distribution for select forests from many of the major forest biomes. The purpose of the table is not to provide an exhaustive review of the literature, but merely to illustrate several general patterns, although because nutrient content varies greatly among forests within a similar biome it is difficult to make

Table 7.2 Total Nutrient Contents (kg ha^{-1}) of Some Major Forest Biomes and the Distribution of Nutrients within the Biomes[a]

Biome/Forest type	Location	Nitrogen (% of ecosystem)				Phosphorus (% of ecosystem)				Calcium (% of ecosystem)				Source
		Content	Soil	Wood	Leaf	Content	Soil	Wood	Leaf	Content	Soil	Wood	Leaf	
Boreal needle-leaved evergreen														
Picea mariana, lowland	Alaska	689	75	24	1	2	92	5	3	1,692	98	2	<1	1
Pinus banksiana	Ontario, Canada	3,729	96	3	1	6100	100			98,707	99	<1	<1	2
Boreal broad-leaved deciduous														
Betula papyrifera	Alaska	2,879	91	8	1	26	75	21	4	7,546	97	3	<1	1
Temperate needle-leaved evergreen														
Pinus taeda	Tennessee	7,650	95	4	1	960	98	2	<1	930	84	16	1	3
Pinus resinosa	Wisconsin	4,400	91	7	2	107	31	56	13	1,700	74	24	2	4
Temperate broad-leaved deciduous														
Liriodendron tulipifera	Tennessee	3,500	95	4	1	2840	98	2	<1	8,130	93	6	<1	1
Fagus silvatica	Sollings, W. Germany	6,332	94	5	1	3150	98	2	<1	245	49	49	2	1
Temperate broad-leaved evergreen														
Eucalyptus regnans	Victoria, Australia	18,222	97	2	<1	3356	98	1	1	2,149	56	38	1	5
Eucalyptus obliqua	Victoria, Australia	14,097	96	3	<1	1743	98	2	<1	1,952	83	16	1	5
Tropical broad-leaved evergreen														
Montane	New Guinea	19,200	99	1	<1	16	37	63	<1	3,750	71	29	<1	6
Lowland	Puerto Rico	5,480	85	13	2	1410	97	2	<1	13,720	99	<1	<1	7

Note. Sources: 1, Cole and Rapp (1981); 2, Foster and Morrison (1976); 3, Johnson and Todd (1987); 4, Bockheim *et al.* (1986); 5, Feller (1980); 6, Edwards and Grubb (1982); 7, Lugo (1992).

[a] More complete compilations for particular forest types can be found in the source publications.

broad generalizations. For example, it is clear that not all tropical forests contain a large proportion of their total nutrient content in the vegetation, as Stark (1971a,b) found for lowland tropical forests growing on the extremely infertile spodosols in Amazonia.

A large fraction of the nutrients in forest ecosystems is stored in the soil and detritus, with a relatively small percentage (<5%) in the foliage. However, the higher nutrient concentrations of foliage, and the high turnover rates compared to woody tissue, mean that leaf litterfall is the major nutrient input to forest soils. This illustrates the dangers of using nutrient content of either soil or foliage to infer nutrient availability, and this emphasizes the need to understand nutrient transfer processes.

III. Nutrient Cycling

A. General Biogeochemical Cycles for Nitrogen, Phosphorus, Sulfur, and Base Cations

For brevity, we provide a general overview of the cycle of several macronutrients (N, P, S, and Ca) and then discuss in more detail the major nutrient inputs, internal transformations, and losses. More detailed treatments of forest biogeochemical cycles are provided by Binkley (1986) and Schlesinger (1991).

In contrast to that of other nutrients, the nitrogen cycle is strongly controlled by microbially mediated biological processes. Nitrogen fixation, the process of converting atmospheric nitrogen into organic nitrogen by microorganisms in plant roots and free-living organisms, is unique to the nitrogen cycle. Nitrogen can be reduced from NO_3^- to N_2O or N_2 by microorganisms and subsequently lost to the atmosphere. Unlike all other macronutrients, nitrogen is seldom derived from the soil by weathering.

An important geochemical process in the phosphorus cycle is weathering of apatite or calcium phosphate. In contrast to nitrogen, phosphorus does not undergo oxidation or reduction reactions in forest ecosystems (Binkley and Richter, 1987). The exchange of phosphorus between soil solution and the soil particles is controlled by two major mechanisms—anion exchange capacity and ligand exchange with iron and aluminum hydroxides (Bohn et al., 1979). Once phosphorus is taken up by plants, it is incorporated into organic molecules with monoester or diester bonds or remains in the inorganic form. The strong physicochemical reactions of phosphorus with soil or humus result in extremely low leaching losses of phosphorus compared to other anions or cations (Johnson and Cole, 1980).

The sulfur cycle has similarities to both the nitrogen and phosphorus cycles. Oxidation and reduction reactions are important (Johnson, 1984), and in anaerobic conditions SO_4^{-2} can be reduced by microorganisms to produce H_2S, which is lost to the atmosphere:

$$SO_4^{-2} + 10H^+ + 8e^- \rightarrow H_2S + 4H_2O. \tag{7.1}$$

The natural sulfur cycle has been greatly modified by humans. Iron ore smelting and the combustion of fossil fuels release large amounts of SO_2^{-2} into the atmosphere (Oppenheimer *et al.*, 1985). Excellent syntheses of our current understanding of the effects of acid deposition on forest biogeochemical cycles have been provided by Binkley *et al.* (1989) and by Johnson and Lindberg (1992). Some SO_2^{-2} can be directly absorbed by forest canopies (Lindberg *et al.*, 1986), but most of it reacts with water molecules to form HSO_4^- and SO_4^{-2}:

$$SO_2 + H_2O \rightarrow H^+ + HSO_3^- \tag{7.2}$$

$$2HSO_3 + O_2 \rightarrow 2H^+ + 2SO_4^{-2}. \tag{7.3}$$

Once taken up by plants, sulfur may remain in inorganic form but, more commonly, it is incorporated into organic compounds such as amino acids and cysteine. Except for forests experiencing large inputs from sulfur deposition, the strong physicochemical reactions of sulfur with iron and aluminum hydroxides, and to a lesser extent uptake by plants, result in small leaching losses of sulfur (Johnson and Cole, 1980).

In general, the cycling of base cations (e.g., calcium, magnesium, and potassium) in forest ecosystems is simpler than for nitrogen, sulfur, and phosphorus. A major difference is that Ca, Mg, and K occur as cations (positively charged ions), whereas P, S, and N (except when nitrogen is in the NH_4^+ form) occur as anions (negatively charged ions). Weathering and atmospheric deposition can be important sources of input for base cations, which are held on negatively charged exchange sites of clay and humus particles.

B. Nutrient Inputs

Forests receive nutrient inputs from the atmosphere as either wet or dry deposition. The importance of atmospheric deposition as a nutrient input to forests is strongly influenced by geographic location. Annual nitrogen atmospheric inputs in relatively pristine forests in the United States commonly range between 5 and 10 kg ha^{-1} year^{-1} but can reach 27 kg ha^{-1} year^{-1} for forests near large industrial areas (Lovett, 1994). Annual sulfur deposition rates for forests in the United States range from 300 to 2000 eq SO_4^{-2} ha^{-1} year^{-1} (Lindberg, 1992), whereas Bredemeier (1988) reported deposition rates as high as 1800 to 5400 eq SO_4^{-2} ha^{-1} year^{-1} for forests in northeastern Germany. On average, wet deposition of nitrogen and sulfur comprise 40–60% of total deposition (Lindberg et al., 1986; Hicks et al., 1992). Inputs of base cations, such as calcium, magnesium, and potassium, are greatest in moderately dry to arid regions (Young et al., 1988), whereas sodium, sulfur, and chloride are carried as aerosols in sea spray and can be locally important for forests near the coast (Junge and Werby, 1958; Johnson et al., 1977).

Forests modify the chemistry of precipitation as it passes through the canopy (throughfall) and moves down the stems of trees (stemflow). Nutrients and plant-derived organic acids that are leached from the foliage, deposited particles on foliage that are washed off, and the evaporation from the forest canopy of water that contains chemicals all change the chemistry of the water. Although nutrient concentration is higher in stemflow than throughfall, its contribution to total nutrient waterborne input is generally <20%. Throughfall, on average, is the largest pathway of sulfur, potassium, sodium, and magnesium deposition (Parker, 1983). In contrast, carbon content in throughfall supplies only 5% of the total input, the primary source being aboveground detritus. The pH of throughfall is generally higher than that of the precipitation entering the canopy because bicarbonate ion concentrations are increased in throughfall (Parker, 1983), although if appreciable leaching of organic acids occurs the pH of throughfall can be lower than that of incoming precipitation (Edmonds et al., 1991). Throughfall in deciduous hardwoods is generally less acidic and higher in base cations than in coniferous canopies (Tarrant et al., 1968; Henderson et al., 1977; Verry and Timmons, 1977; Cronan and Reiners, 1983) and, except for forests subject to chronic N deposition, the NH_4^+ and NO_3^- concentrations tend to be lower in throughfall water, suggesting that nutrient absorption occurs in the canopy (Feller, 1977; Jordan et al., 1980). Horn et al. (1989) estimated that 15% of the annual N uptake of *Picea abies* stands was met by canopy uptake. However, the nitrogen concentration of throughfall can be higher than that of precipitation in temperate and tropical rain forests that contain a large number of N-fixing epiphytic plants in the canopy (Nadkarni, 1984; Edmonds et al., 1991).

Weathering refers to the release of nutrients from aging parent material or rock. It is described qualitatively by Eq. (7.4):

$$\frac{d[n]}{dt} = f(\text{time, climate, topography, parent material, and biota}), \quad (7.4)$$

where $[n]$ denotes the amount of nutrients weathered from parent material. Quantitative descriptions of nutrient release rates by weathering are not readily available, but several generalizations can be made:

- Weathering rates are positively related to precipitation and temperature (Strakhov, 1967), therefore, the most heavily weathered soils occur in lowland tropical forests, intermediately weathered soils are most common in temperate forests, and the least weathered soils occur in boreal forests.
- Within a similar climate, parent material has a strong influence on weathering rates and the nutrients released as a result of weathering.
- Minerals with a high degree of crystallization and low energy of formation are more resistant to weathering (Bohn *et al.*, 1979).

It is difficult to obtain direct estimates of the rates of nutrient release from weathering rocks because they are normally very low. A common procedure, recognized as the most reliable approach for estimating mineral weathering rates *in situ* (Clayton *et al.*, 1979), is the use of geochemical mass balance studies, which involve constructing watershed hydrology and nutrient budgets (Likens *et al.*, 1977; Henderson *et al.*, 1977; Feller and Kimmins, 1979). A critical assumption of this approach is that the watershed has an underlying layer of impermeable bedrock so that all water and nutrients moving out of the catchment pass the gauging station. Another critical, and often poor, assumption implicit in the watershed approach to estimating weathering is that the content of exchangeable cations remains constant.

Calcium, sodium, and magnesium are commonly released by weathering of silicate rocks, and potassium inputs can be high for forests underlain by mica and feldspar (Table 7.3). Soils derived from serpentine parent material, a highly metamorphosed, ultramafic rock, have unusually high Mg, Fe, and trace elements relative to Ca concentrations. The imbalance between magnesium and calcium has marked effects on species composition and growth of forest communities on these soils (Whittaker *et al.*, 1954; Proctor and Woodell, 1975; Schlesinger *et al.*, 1989) and is a good illustration of how biogeochemical processes can affect forest composition, structure, and function.

Mineralization, which differs from decomposition in that the latter is a generic term describing the breakdown of organic matter, involves the concurrent release of carbon dioxide and the conversion of an element from an organic to inorganic form. The main source of organic matter is above- and below-ground detritus production (see Chapter 6), but leachates, root exudates, fauna carcasses, and feces all contribute. Mineralization is the major mechanism of nutrient input to forests, particularly in the case of nitrogen in temperate forests. Consequently, net primary production (NPP) is often highly correlated to nitrogen mineralization (Fig. 7.4). Mineralization is largely carried out by heterotrophic fungi

Table 7.3 Examples of Nutrient Release Rates as a Result of Weathering Forests Underlain by Different Parent Materials

Parent material	Forest type/location	(kg ha^{-1}year^{-1})			Source
		K	Ca	Mg	
Moraine/gneiss	Northern hardwoods/ New Hampshire	7.1	21.1	3.5	1
Schists	Mixed hardwoods/ Maryland	2.3	1.3	1.1	2
Serpentine	Mixed hardwoods/ Maryland	t	t	34.1	2
Sand (outwash)	Mixed hardwoods/ New York	11.1	24.2	8.4	3
Sand (quartz)	Mixed hardwoods/ New York	0.01	0.04	0.01	4
Dolomite	Western conifer/ California	4	86	52	5
Tuffs/brecia	Western conifer/ California	1.6	47	11.6	6
Alluvial-1000	Lowland tropical/ Venezuela	6	10	1.5	7
Alluvial-5000	Lowland tropical/ Venezuela	1.2	0.2	0.3	7

Note. —, not measured; t, trace. Sources: 1, Likens *et al.* (1977); 2, Cleaves *et al.* (1970); 3, Woodwell and Whittaker (1967); 4, Art *et al.* (1974); 5, Marchland (1971); 6, Fredriksen (1972); 7, Hase and Foelster (1983).

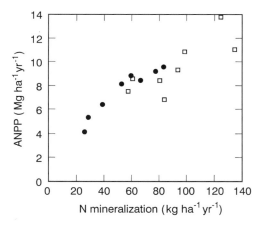

Figure 7.4 Relationship between above-ground net primary production (ANPP) and annual nitrogen mineralization for forests in southern Wisconsin (data from Pastor *et al.*, 1984 (closed symbols); Nadelhoffer *et al.*, 1985 (open symbols)).

and bacteria; therefore, factors, such as anaerobic soil conditions, extreme soil temperatures, high soil acidity, and the solubility of trace elements, which all adversely affect the soil heterotroph populations and their effectiveness (Alexander, 1977), are likely to have parallel effects on mineralization rates.

Meaurements of mineralization and nitrification in many forest ecosystems show large spatial and temporal variability of these processes in forests. Nitrogen mineralization rates differ by factors of two to five among forests in different locations on the landscape in areas of similar climate (Pastor *et al.*, 1984; Zak *et al.*, 1989; Zak and Grigal, 1991; Nadelhoffer *et al.*, 1985). It is difficult to determine the exact causes for these large differences, but important factors include climate, time, relief, parent material, and biota (Jenny, 1980). Nitrogen mineralization rates have been reported to range from 6 kg ha^{-1} year^{-1} for lowland black spruce to 46 kg ha^{-1} year^{-1} for an upland mixed birch–aspen stand in boreal forests (Ruess *et al.*, 1996). Annual nitrogen mineralization rates for temperate forests typically range from 40 to 80 kg ha^{-1} year^{-1} with the higher values often occuring beneath deciduous rather than evergreen forests. Few annual nitrogen mineralization estimates have been reported for tropical forests. Matson *et al.* (1987) reported rates ranging from 588 to 1140 kg ha^{-1} year^{-1} for cleared and early successional fertile moist tropical forests in Coast Rica; these values are some of the highest reported for forests in the literature and probably should not be considered typical but are indicative of the potentially high rates that can occur in fertile, moist tropical forests. The systematic changes in nitrogen availability during stand development, and its causes, are discussed in Chapter 8.

Mineralization does not often occur at the same rate as decomposition because of the large difference in C/element ratios between decomposers and detritus. The C/N ratio of microorganisms is approximately 8/1, whereas the same ratio may be 40/1 to 60/1 for foliage and 200/1 to 400/1 for wood detritus; therefore, we expect organic matter decomposition and element mineralization to occur more rapidly in detritus with a low C/element ratio than in material with a high ratio. Referring back to the discussion of critical elemental ratios, it is apparent that microorganisms must accumulate large amounts of nutrients to maintain satisfactory elemental stoichiometry; their rapid growth and large surface area/volume ratio make them superior competitiors, compared to plants, for available nutrients (Jones and Richards, 1977; Vitousek and Matson, 1984). As a result, nutrient immobilization is commonly observed before mineralization begins. This tends to happen most frequently with limiting elements such as nitrogen and phosphorus, although it can also occur with other elements (Staff and Berg, 1982). The relationship between litter quality and nutrient release by mineralization provides a very strong positive feedback in the ecosystem, sensitive to ecological processes such as stand development and succession (see Chapter 8) and management practices such as fertilization and site preparation. For example, nitrogen is mineralized within months of the fall of deciduous foliage, whereas mineralization and nutrient release from coarse woody detritus may take decades (Grier, 1978; Lambert *et al.*, 1980). Therefore, the management of logging residue, known as slash, affects the nutrient dynamics of the sites and the long-term fertility of forest soils. In addition to a number of other important functions of woody debris in forest ecosystems (see Harmon *et al.*, 1986), it temporarily immobilizes nutrients (e.g., nutrient sink), preventing large nutrient losses [Vitousek and Matson, 1985; Ross *et al.*, 1995; but see Arthur and Fahey (1994)].

A process unique to the nitrogen cycle is biological nitrogen fixation—the conversion of atmospheric N_2 to NH_3^+ [Eq. (7.5)] by organisms that possess the enzyme nitrogenase

$$N_2 + 6H^+ + 6e^- \rightarrow 2\,NH_3^+. \tag{7.5}$$

Two major types of biological nitrogen fixation occur: asymbiotic and symbiotic. Asymbiotic N fixation may be carried out by free-living autotrophic or heterotrophic microbes, most commonly bacteria and blue–green algae. Asymbiotic N fixation is widespread; it is more common in forest soils with high levels of organic matter (Granhall, 1981) but can also occur in coarse woody debris and in epiphytes in the forest canopy. Waide *et al.* (1987) have shown that asymbiotic N fixation rates increased to a maximum of 14–16 kg N ha^{-1} year^{-1} during the first 3–5 years following clear-cutting, presumably because of the large increase in carbon input to the forest floor. Fixation rates can reach 30 kg N ha^{-1} year^{-1} but are typically in the range 0.1–4 kg N ha^{-1} year^{-1} in boreal forests, 0.1–5 kg N ha^{-1} year^{-1} in temperate coniferous forests, 0.1–6 kg ha^{-1} year^{-1} in temperate deciduous forests, and 2–20 kg N ha^{-1} year^{-1} in tropical forests (Boring *et al.*, 1988).

Symbiotic N fixation involves a relationship between bacteria or actinomycetes and the roots of plants. Examples of symbiotic N-fixing associations include the bacteria *Rhizobium* and the host tree black locust (*Robinia pseudoacacia*) in the southeastern United States and the actinomycete *Frankia* and red alder (*Alnus rubra*) in the U.S. Pacific Northwest. They appear to be common in many tropical forests (Boring *et al.*, 1988). Nitrogen-fixing understory shrubs, such as *Ceanothus* spp. in the western United States, *Myrica* spp. in the eastern United States, *Acacia* spp. in the eucalypt forests of Australia, and the legume *Lupinus arboreaus* in New Zealand, can also be responsible for substantial N inputs in managed and native forests. Nitrogen inputs from symbiotic N fixation are often several orders of magnitude greater than asymbiotic N fixation, ranging from 30 to over 300 kg N ha^{-1} year^{-1} (see review by Binkley, 1983; Boring *et al.*, 1989). It seems that high rates can occur in all forest biomes; Van Cleve *et al.* (1981) reported annual N-fixation rates of 156–362 kg N ha^{-1} by *Alnus incana* in boreal forests in Alaska, whereas Dommergues *et al.* (1984) measured N-fixation rates of 250–500 kg N ha^{-1} by *Casuarina equisetifolia* and *Leucaena leucocephala* forests in the tropics. In temperate forests, nitrogen-fixing species are most common in early successional forests. Factors that reduce the carbon gain of the host plant, such as water and nutrient limitations, shading, and limitation of trace elements (e.g., Co, Mo, and B), which are essential cofactors in the N-fixation process, restrict N fixation in plants.

Because nitrogen-fixing plants do not rely on nitrogen from the soil, they accelerate primary succession by adding critical organic matter and nutrients required by other plants; the inclusion of N-fixing tree species with other trees may also increase the growth of the companion species. Binkley *et. al.* (1984) compared the nitrogen and carbon cycles for Douglas fir grown with and without sitka alder (*Alnus*) and found higher soil nitrogen and carbon concentration, foliage N concentration, and foliage N content in the Douglas fir grown with, rather than without, alder (Table 7.4). The greater availability of nitrogen increased foliage biomass, which presumably was responsible for the greater above-ground net primary production of Douglas fir stands grown with alder. However, maintaining pure stands of N-fixing species on a site can lead to nitrate leaching, a slow decline in base cations, and an increase in soil acidity (Van Miegroet and Cole, 1985). When introduced as exotics to naturally

Table 7.4 Summary of Selected Ecosystem Characteristics for Douglas Fir Forests Growing with and without Sitka Alder[a]

Ecosystem characteristic	Douglas fir	
	Without alder	With alder
Soil nitrogen (%)	0.09	0.16
Soil carbon (%)	2.05	3.34
Forest floor N (kg/ha^{-1})	36	282
Litterfall N (kg/ha^{-1}/year^{-1})	16	112
Foliage N–Douglas fir (%)	0.93	1.10
Foliage mass (kg/ha^{-1})	9.6	11.7
Aboveground NPP (kg/ha^{-1}/year^{-1})	8.5	14.4

[a] Adapted from Binkley *et al.* (1984).

infertile soils, N-fixing species can substantially alter the composition, structure, and function of native ecosystems (Vitousek et al., 1989).

Given the benefit of N fixation to the mineral economy of plants and the common nitrogen limitations of forest ecosystems, it is puzzling why symbiotic N fixation is not more prevalent. One possible explanation is that a shortage of the micronutrients required to construct the nitrogenase enzyme may inhibit the growth of nitrogen-fixing species—for example, the availability of molybdenum may limit nitrogen fixers in forests in the Pacific Northwest (Silvester, 1989). A second factor may be related to the carbon balance of the host: N-fixation rates of symbiotic plants appear to be closely related to their C balance (Bormann and Gordon, 1984). Analysis of the energy costs of fixing atmospheric nitrogen compared to nitrogen uptake from the soil suggests that the two processes have equivalent costs (Gutschick, 1981), which may explain why N fixers are typically early successional species and are more common in tropical than other forest ecosystems. Also, N-fixing plants often use water less efficiently than non-N-fixing tree species.

C. Nutrient Transfers

For many of the essential nutrients, detritus is the most important means of nutrient input to forests. Annual above- and below-ground detritus production transfers organic matter and nutrients to forest soils: The organic matter is an important carbon source for heterotrophs and the precursor of humus; the nutrients are converted to inorganic forms, making them available for plant uptake. Therefore, in order to manage forests on a sustainable basis, balance the nutrient budget, or understand the linkages between the nutrient cycles of adjacent terrestrial and aquatic ecosystems, a thorough understanding of the factors influencing nutrient inputs via detritus is essential.

The nitrogen and phosphorus content of above-ground fine litterfall generally decreases from typical values of about 100 kg N ha^{-1} year^{-1} and 5 kg P ha^{-1} year^{-1} in tropical forests to 15 kg N ha^{-1} year^{-1} and 5 kg P ha^{-1} year^{-1} in boreal forests (Fig. 7.5), but the amounts of N and P in above-ground detritus can vary enormously within a forest biome, as shown by the large variation around the mean for each forest biome. The large intrabiome variability of litterfall nutrient content is probably caused by differences in water and nutrient availability. For instance, Vitousek et al. (1982) reported a strong positive relationship between soil nitrogen mineralization rate and litterfall nitrogen content for temperate forests. Within major climatic regions, there seems to be no consistent pattern in the N and P contents of above-ground detritus in evergreen and deciduous forests: Gower and Son (1992) found no differences between the N and P content of litterfall from 30-year-old evergreen and

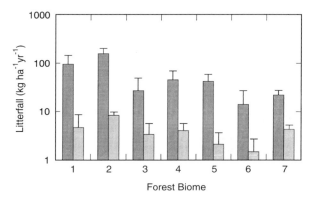

Figure 7.5 Annual above-ground nitrogen (dark bar) and phosphorus (light bar) litterfall inputs for major forest biomes of the world. 1, tropical broad-leaved semideciduous and deciduous (3,2); 2, tropical broad-leaved evergreen (119,122); 3, temperate broad-leaved deciduous (50,27); 4, temperate needle-leaved evergreen (71,50); 5, temperate broad-leaved evergreen (26,31); 6, boreal broad-leaved deciduous (10,7); and 7, boreal needle-leaved evergreen (19,15). Values in parentheses are sample size for nitrogen and phosphorus, respectively.

deciduous tree plantations in adjacent plots, but their results do not correspond to those obtained from earlier studies in which litterfall nutrient content was found to be greater in deciduous than evergreen forests (Bray and Gorham, 1964; Rodin and Bazilevich, 1967).

The addition of fertilizer to conifer forests often decreases litterfall nutrient content during the first year after fertilization because of greater needle retention, which perhaps serves as a mechanism to store nutrients. Continued fertilization, however, increases the nitrogen content of leaf litter (Turner, 1977; Gholz *et al.*, 1991; Raison *et al.*, 1992; Gower *et al.*, 1996c). Severe droughts can cause premature sensence of foliage, which may result in higher than normal nutrient content in litterfall (Raison *et al.*, 1992; Gower and Son, 1992).

Fine root turnover is an important nutrient input in forest ecosystems, comprising, on average, more than 50% of the total (above- and below-ground) detritus nitrogen content (Table 7.5) although there is, inevitably, considerable variation between forest types within a biome. For example, using fine root and litterfall nitrogen content data from Cuevas and Medina (1986, 1988), we estimated that the nitrogen release attributable to fine root turnover in caatinga and tierra firme tropical forests in Venezuela ranged from 12 to 256 kg N ha^{-1} year^{-1} and comprised 26–68% of the total nitrogen content of detritus. On an absolute basis, nitrogen input from fine root turnover appears to increase from boreal

Table 7.5 Annual Ratio Above- and Below-ground Nitrogen Input and Below-ground: Total Nitrogen Input Ratio for Evergreen and Deciduous Forest in Contrasting Biomes[a]

Biome leaf habit	Above-ground (kg N Ha^{-1} year^{-1})	Below-ground (kg N Ha^{-1} year^{-1})	Below-ground: total ratio
Boreal			
Evergreen	3–10	14–27	0.74[b] (0.58–0.86)
Deciduous	14–32	26–44	0.62 (0.54–0.74)
Cold temperate			
Evergreen	14–41	20–125	0.64 (0.37–0.88)
Deciduous	24–66	24–79	0.53 (0.27–0.64)
Warm temperate			
Deciduous	29–55	22–68	0.52 (0.28–0.69)
Tropical			
Evergreen	12–120	4–256	0.40 (0.05–0.68)

[a] Data are from Reuss et al. (1995), Nadelhoffer et al. (1985), Vogt et al. (1986b), and Cuevas and Medina (1986). Studies were only included if corresponding above- and below-ground nitrogen data were available. Nitrogen turnover for the tropical forests was calculated by multiplying fine root N concentration by fine root NPP. This calculation assumes fine root biomass is in steady state, with no retranslocation of nitrogen from senescing fine roots.
[b] Average.

to tropical forests, but on a proportional basis (i.e. fine root/total above- plus belowground turnover) it decreases from tropical to boreal forests.

Woody litter plays an important role in the biogeochemical cycle of forests despite the fact that the annual nutrient input in this form is low; because of its high C/element ratio relative to leaves and fine roots, woody litter can immobilize nutrients for decades before net mineralization occurs (Grier, 1978; Lambert et al., 1980). There are few data showing the nutrient content of woody litter, but it appears that the nutrient content in coarse woody detritus can be 10–20% of that for fine aboveground detritus (Vogt et al., 1986a). The change in coarse woody input during stand development is discussed in Chapter 8.

Nutrient uptake is governed by soil processes regulating the amount of nutrients in soil solution and the amount and distribution of fine roots and mycorrhizae. Uptake by trees is achieved by three processes: mass flow, diffusion, and root interception. Mass flow is the passive movement of nutrients through the soil along water potential gradients maintained by transpiration (Chapter 4). Diffusion refers to the movement of ions from zones of high to low nutrient concentration; extraction of nutrients from the rhizosphere by fine roots and mycorrhizae creates zones of nutrient depletion relative to the surrounding soil solution. Root interception refers to the uptake of nutrients that results from the growth of fine roots and mycorrhizae through the soil into regions where nutrient concentrations are high. The relative importance of each of these processes varies depending on the nutrients concerned and the form in which they occur in the soil, and also on tree species, soil type, and climate. However, several generalizations can be made. Most cations move through the soil matrix by mass flow, whereas phosphorus, potassium, and nitrate move largely by diffusion (Mengel and Kirkby, 1979). Regardless of the process of nutrient acquisition, the most time-limiting step is ion transport to the root membrane, not movement across the membrane and into the root symplast (Clarkson, 1985), from which we can assume that there is little selective advantage for trees, and plants in general, in supporting energy-expensive enzymes to expedite ion tranport across root membrances (Chapin, 1980).

We have previously described mycorrhizae and their relative abundance in forest ecosystems (see Chapter 5) and therefore will restrict this discussion to the role mycorrhizae play in nutrient uptake. Mycorrhizae benefit the nutrition of trees through several different mechanisms. Most notable is the fact that mycorrhizal roots have substantially greater surface area/biomass ratios than nonmycorrhizal roots. Simulations suggest that increasing this ratio increases nutrient absorption more than can be achieved by increasing other morphological or physiological characteristics by corresponding amounts (Clarkson, 1985). This feature is particularly important in the uptake of nutrients such as phosphorus, which do not move easily or rapidly through the soil by diffusion or mass flow.

There is also an increasing amount of information suggesting that mycorrhizae and roots can increase the availability of nutrients by altering the chemistry of the rhizosphere. For example, it has been suggested that fungal symbionts can increase the nutrients available for uptake by

- exudation of organic acids that hasten the weathering of parent material (Boyle and Voigt, 1973; Boyle *et al.,* 1974; Bolan *et al.,* 1984) or solubilize phosphorus from compounds unavailable for uptake (Rovira, 1969; Grierson and Attiwill, 1989; Fox *et al.,* 1990);
- releasing enzymes that stimulate decomposition rates of organic matter (Dodd *et al.,* 1987);
- storage of phosphorus as polyphosphates (Martin *et al.,* 1983);
- increasing phosphatase activity at the mycorrhizal surface (Dighton, 1983; Dodd *et al.,* 1987). It has also been suggested that mycorrhizae and tree roots release soluble carbon-based exudates that prime microbial activity in the rhizosphere, thereby increasing nutrient mineralization rates.

There is some evidence that mycorrhizae can also benefit the nitrogen nutrition of trees by preferential uptake of NH_4^+ (as opposed to NO_3^-) and subsequent assimilation into glutamine by the fungi. Because electroneutrality must be maintained in both the soils and the plant, the uptake of NH_4^+ results in the exudation of H^+, which increases NH_4^+ in the rhizosphere (Smith and Smith, 1990). Subsequent research has shown that the glutamine can be stored as a soluble pool of nitrogen within the fungi and released to the plant as needed (Chalot et al., 1991).

Our incomplete understanding of factors regulating carbon allocation to fine roots and the paucity of data available on the quantity of nutrients needed to construct new roots to replace senescing roots (Meier et al., 1985; Nadelhoffer et al., 1985; Reuss et al., 1996) makes it difficult to calculate the annual nutrient uptake rates of forests or plantations directly. More often, scientists estimate annual nutrient "uptake" as the sum of the annual increment in nutrient mass of new tissues. Most of the values reported are probably underestimates because of the exclusion of belowground biomass components (but see Meier et al., 1985). Cole and Rapp (1981) summarized annual nutrient requirement, uptake, and retranslocation for 32 boreal and temperate forests studied during the International Biological Programme. There is little such information available for tropical forests. As can be seen in Table 7.6, annual uptake rates for nitrogen and calcium range from a low of 3–8 kg ha^{-1} year^{-1} for boreal forests to a high of 43–169 kg ha^{-1} year^{-1} for temperate deciduous forests. Annual nitrogen uptake rates can, however, exceed 400 kg ha^{-1} year^{-1} in fertilized forests (Johnson, 1992a). Cole and Rapp found that annual nutrient uptake rates were generally greater for temperate than boreal forests and were greater for deciduous than evergreen forests. However, it is not clear whether this latter pattern is a consequence of the fact that deciduous forests generally occur on more fertile soils than evergreen forests (Vitousek et al., 1982) or is a result of greater physiological nutrient demand by the deciduous forests. Son and Gower (1991) calculated annual nitrogen and phosphorus uptake rates for five tree species grown in adjacent plots and found that the greater nutrient requirements of deciduous relative to coniferous species are attributable to their greater physiological demands. However, the deciduous species had modified soil nitrogen mineralization rates after only 30 years (Gower and Son, 1992).

Plants can also meet their annual nutrient demands by removing nutrients from aging or senescing tissue. This process is commonly referred to as retranslocation. Nutrient retranslocation within trees is a dynamic process that varies with age and growing conditions, and it appears that the likelihood that particular nutrients will be retranslocated is strongly associated with their physiological role. For example, nitrogen, sulfur, phosphorus, and potassium are required for dynamic physiological processes, are mobile, and are often retranslocated; calcium is a constituent of calcium pectate, or the cell-wall glue, and is immobile. Foliage is the

Table 7.6 Annual Nutrient Uptake Rates (kg ha^{-1} year^{-1}) for Boreal, Temperate, and Subtropical[a] Forests

Biome/forest type	Number of stands	Nitrogen		Phosphorus		Calcium		Potassium	
		Average	Range	Average	Range	Average	Range	Average	Range
Boreal									
Evergreen, needle-leaved	3	5	3–7	1	1–2	6	3–8	2	1–3
Deciduous, broad-leaved	1	25	—	6	—	39	—	13	—
Temperate									
Evergreen, needle-leaved	13	47	24–88	6	28–74	45	28–74	33	15–48
Evergreen, broad-leaved				3	1–7				
Deciduous	14	75	43–115	6	48–169	85	48–169	51	40–59
Sub-tropical									
Evergreen, broad-leaved	1	107	—	7	—	—	—	55	—

[a] Adapted from Cole and Rapp (1981).

most important source of nutrients for retranslocation; typical retranslocation rates for nitrogen and phosphorus from senescing foliage range from 50 to 60% (Chapin and Kedrowski, 1983; Reich *et al.*, 1992), but can be as high as 90% for *Larix* (Gower and Richards, 1990). Nutrients can also be removed from aging wood (Cowling and Merrill, 1966). Whittaker *et al.* (1979) reported only 3–10% withdrawal of nitrogen and phosphorus from stemwood, which appears to be low compared to estimates of about 50% reported by others (Meier *et al.*, 1985; Son and Gower, 1991). Retranslocation of nutrients from roots has been reported for perennial grasses (Woodmansee *et al.*, 1981), but little is known about its importance in trees. McClaugherty *et al.* (1982) suggested that retranslocation of nutrients from fine roots of forests is small, but Meier *et al.* (1985) concluded that retranslocation supplies a major portion of the annual nitrogen and phosphorus requirements for *Abies amabilis* stands in Washington.

On a stand basis, retranslocation of nutrients, especially nitrogen and phosphorus, supplies a significant portion of the annual nutrient requirement. Cole and Rapp (1981) reported that retranslocation contributed little to the annual nutrient requirement of evergreen conifer forests; however, this does not appear to be the case. Meier *et al.* (1985) estimated that 37–58% of nitrogen and 48–59% of the annual phosphorus requirements of a subalpine *A. amabilis* stand were met by retranslocation. Turner and Lambert (1986) estimated that 50–60% of the annual phosphorus requirement of *P. radiata* plantations were met by retranslocation. Son and Gower (1991) compared nitrogen and phosphorus budgets for evergreen and deciduous tree plantations and found that uptake accounted for 72–74% of the annual nitrogen requirement of these evergreen conifers, whereas retranslocation was the predominant source (76–77%) for annual nitrogen requirement of the two deciduous species.

There has been long-standing interest in understanding the relationship between leaf habit, or more correctly leaf longevity, and nutrient use efficiency, in the hope of explaining the distribution and physiological competitiveness of species (Chapin *et al.*, 1986; Gower and Richards, 1990; Sheriff *et al.*, 1995). Numerous definitions of nutrient use efficiency exist, ranging from instantaneous, leaf-level indices to annual, stand-level use efficiency indices; few studies have considered below-ground nutrient use (see Meier *et al.*, 1985). Important physiological factors that influence nutrient use efficiency include carbon gain per unit nutrient invested in photosynthetic tissue, efficient removal or retranslocation of nutrients from senescing tissue, and greater nutrient absorption rates per unit of root or mycorrhizae.

It has been suggested that evergreens cycle nutrients internally more efficiently than deciduous trees because evergreens typically occur on more infertile soil (Monk, 1966); however, support for this hypothesis is mixed. Chapin and Kedrowski (1983) concluded that evergreens retranslocated a greater proportion of foliage nitrogen than deciduous species, but Reich et al. (1992) found no relationship between leaf longevity and retranslocation of nitrogen from foliage. A species that provides support for the argument that the deciduous habit leads to more efficient nutrient use than the evergreen habit is the deciduous conifer, *Larix* (Gower and Richards, 1990; Gower et al., 1995). This genus occurs in montane, subalpine, and boreal forests in the northern hemisphere and is an important cover type in Siberia. A combination of morphological and chemical leaf characteristics enables the species to maintain higher photosynthetic and nutrient retranslocation rates than most other tree genera.

Vitousek (1982) examined nutrient use efficiency patterns in world forests and reported a strong inverse relationship between nitrogen use efficiency and litterfall nitrogen content (Fig. 7.6) A similar analysis for tropical forests suggested that the same inverse relationship exists for phosphorus.

Fertilization studies have shown that fertilized trees commonly do not retranslocate as much nitrogen from senescing needles as unfertilized trees (Turner, 1977; Miller, 1984; Birk and Vitousek, 1986). However, generalizations about this can be misleading. Detailed studies by Nambiar and Fife (1991) showed that the nutrient content of *P. radiata* needles fluctuated in cyclic patterns: phases of retranslocation often coincided with new shoot growth in spring and summer—when conditions were also suitable to soil nutrient mineralization—whereas accumulation in needles occurred in autumn and winter when growth was slow. Nambiar and Fife showed (in a series of papers reviewed in the reference cited)

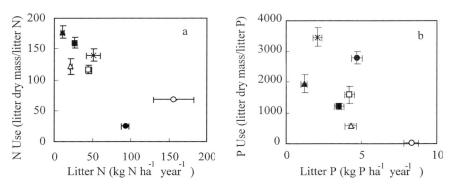

Figure 7.6 The relationship between the ratio litter mass/litter N content and litterfall nutrient content (from Vitousek, 1982).

that there is a direct link between shoot growth and nutrient retranslocation. It is also increasingly recognized that slow growth rates and storage of nutrients are important physiological traits that ensure the survival of plants in harsh environments, but may not result in extraordinarily high nutrient use efficiency (Chapin et al., 1986)

Foliar leaching affects nutrient use efficiency (Gray and Schlesinger, 1989) and increases nutrient transfer to the soil. Leaching rates from foliage are nutrient and species dependent (Parker, 1983), with potassium being readily leached, compared to calcium, because potassium is not incorporated into structural tissue. N and P are moderately susceptible to leaching (Tukey, 1970). On an annual basis, deciduous foliage is more susceptible to leaching losses than evergreen foliage with thick cuticles, but leaching losses over the lifespan of the foliage may be comparable for evergreen and deciduous species (Chapin, 1980). Overall, foliar leaching losses are minor fluxes in forest ecosystems (Miller, 1984), except for forests subject to chronic acid deposition, which can cause lesions in the cuticle, thus accelerating leaching.

D. Nutrient Losses from Forests

From both the forest nutrition and the water quality viewpoints, nutrient losses from forests by leaching and surface drainage are of great interest to forest managers. Large losses will reduce the fertility of the soil and may reduce water quality in nearby streams. These concerns were initially highlighted by the results from a now-famous watershed experiment—the Hubbard Brook clear-felling study (Bormann and Likens, 1979). Today, there is also strong interest in the effects that leaching losses associated with chronic low-level nitrogen deposition may have on terrestrial ecosystems.

Cole and Gessel (1965) first suggested that the availability of mobile anions is a major factor determining leaching rates in forest ecosystems and, in several excellent papers, Johnson and colleagues outlined the sources and mobility of the dominant anions in forest soils (Johnson et al., 1977, 1986; Johnson and Cole, 1980). Much of the discussion below is based on these papers. Potentially important anions in forest soils include nitrate (NO_3^-), bicarbonate (HCO_3^-), phosphate ($H_2PO_4^-$ or HPO_4^{-2}), sulfate (SO_4^{2-}), organic anions (R-O^-), fluoride (Fl^-), and chloride (Cl^-). Their relative importance is influenced by species composition, climate, and geographic location. Below, we briefly review the sources and mobility of these anions.

Bicarbonate concentration in the soil solution is determined by CO_2 pressure (pCO_2) and pH, which is the negative logarithm of hydrogen ion concentration. In the presence of water CO_2, the end product of microbial (Chapter 6) and root respiration (Chapter 5) dissolves, producing carbonic acid (H_2CO_3):

$$H_2O + CO_2 \longleftrightarrow H_2CO_3, \tag{7.6}$$

which in turn can dissociate to form bicarbonate and a hydrogen ion:

$$H_2CO_3 \longleftrightarrow H^+ + HCO_3^-. \tag{7.7}$$

The reaction depicted in Eq. (7.7) is strongly pH dependent and is only important in forest soils with pH greater than 4.5. The bicarbonate anion (HCO_3^-) is present in all forests but is most important in tropical forests and least important in cold temperate and boreal forests (Fig. 7.7). The HCO_3^- anion is less important in colder forests for two reasons. First, low temperatures adversely affect root and microbial respiration rates (Chapters 5 and 6), decreasing the production of CO_2 and pCO_2 in the soil atmosphere. Second, low soil temperatures prevent complete decomposition, resulting in high concentrations of organic acids (Cronan, 1978), which lowers soil pH, suppressing the dissolution of carbonic acid [Eq. (7.7)].

Sulfate, derived from the dissociation of sulfuric acid,

$$H_2SO_4 \longleftrightarrow H^+ + HSO_4^- \tag{7.8}$$

$$HSO_4^- \longleftrightarrow H^+ + SO_4^{-2}, \tag{7.9}$$

is only important for forests near major industrial centers where combustion of fossil fuel is high or in soils with pyrite (FeS) in parent material. Phosphorus anions ($H_2PO_4^{-1}$ and HPO_4^{-2}) are important in the forest floor, and result from phosphorus mineralization (Yanai, 1991); however, because these anions are strongly adsorbed and undergo physicochemical reactions with aluminum and iron sesquioxides (Johnson and Cole, 1980), they comprise a negligible fraction of the total anions in the soil solution below the rooting zone, even after harvest (McColl, 1978; Yanai, 1991).

The processes regulating the production and consumption of organic acids are extremely complex and poorly understood (Stevenson, 1980). They are thought to be the result of incomplete humification of organic matter or produced as leachates from the canopy and forest floor of coniferous forests (Ugolini et al., 1977; Dawson et al., 1978). Organic acids are likely to be the dominant anions in forests where decomposition is incomplete. Johnson et al. (1977) estimated that organic anions comprised 30–50% of the total anions in the soil solution for a subalpine conifer (A. amabilis) forest (Fig. 7.7). Similar patterns have been observed for subalpine forests in Maine (Cronan, 1979).

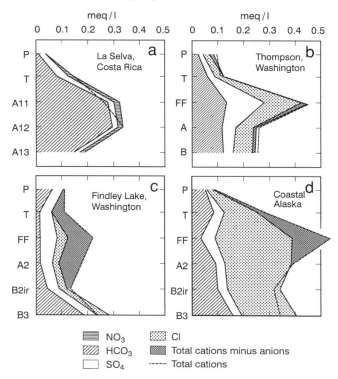

Figure 7.7 Comparison of the relative occurrence of mobile anions within and among four contrasting forest ecosystems. The four sites include (a) a lowland broad-leaved evergreen tropical forest at La Selva Costa Rica, (b) a low-elevation, needle-leaved evergreen conifer forest (Douglas fir, *Pseudotsuga menziesii*), (c) a high-elevation needle-leaved, evergreen conifer (Pacific silver fir, *Abies amabilis*) temperate forest in western Washington, and (d) coastal, boreal needle-leaved, evergreen forest in Alaska. The letters on the y axes indicate soil horizons (adapted from Johnson *et al.*, 1977). Total cations minus anions is assumed to equal organic acids.

Much of the information about the production and consumption of nitrate, which is better understood than in the case of most other anions, was obtained from studies that examined the mechanisms regulating nitrate leaching losses following timber harvesting (see Section IV,c). There are two dominant sources of nitrate production, with their relative importance being dependent on geographic location and soil fertility. Atmospheric deposition is important in forests located near large industrial regions. The most important source of NO_3 in nonpolluted forests is a two-step, microbially mediated conversion of NH_4^+ to NO_3^-, commonly referred to as nitrification:

$$2NH_4^+ + 3O_2 \rightarrow 2NO_2^- + 4H^+ + 2H_2O \qquad (7.10)$$

$$2NO_2^- + O_2 \rightarrow 2NO_3^-. \qquad (7.11)$$

The oxidation of NH_4^+ to NO_2^- is commonly carried out by *Nitrosomonas*, a chemoautotroph that uses O as an electron donor, whereas the oxidation of NO_2^- to NO_3^- is performed by *Nitrobacter*, also a chemoautotroph.

Chloride and flouride typically constitute only a small fraction of the total anions, except for forests located near an ocean (Hingston and Gailitis, 1976; Johnson *et al.*, 1977). Some reduction of chloride is typically observed in lower soil horizons (Fig. 7.7), which may be a result of plant uptake or absorption by sesquioxides.

In addition to leaching losses, certain nutrients can be lost to the atmosphere. The microbially mediated conversion of inorganic forms to gaseous form, which is lost to the atmosphere, is unique to the nitrogen and sulfur cycles. Denitrification is the conversion of NO_3^- to N_2O or N_2 [Eq. (7.12)] by facultative anaerobic bacteria, commonly of the genus *Pseudomonas* (Knowles, 1982). Nitrate is the terminal electron acceptor in the absence of oxygen. The ratio of N_2O/N_2 (nitrous oxide to molecular N) produced during denitrification is not constant; increased concentrations of nitrate, nitrite, molecular O_2, and soil acidity in the presence of NO_3^- enhance N_2O production relative to molecular N (Firestone *et al.*, 1980):

$$5CH_2O + 4H^+ + 4NO_3^- \rightarrow 2N_2 + 5CO_2 + 7H_2O. \qquad (7.12)$$

Equation (7.12) shows that denitrification rates are dependent on an adequate supply of labile carbohydrates and NO_3^-: denitrification in a nitrate-rich, cleared tropical forest was limited by carbon availability (Matson *et al.*, 1987), whereas denitrification in many cleared temperate forests is limited by nitrate availaibility (Robertson *et al.*, 1987). Denitrification was originally thought to be an important process only in flooded soils, but studies have shown that the slow diffusion of oxygen into soil microaggregates allows anaerobic zones to form even in upland forest soils (Tiedje *et al.*, 1984; Sexstone *et al.*, 1985). It is these microsites that make the process highly variable both spatially and temporally. For example, Robertson *et al.* (1988) reported the coefficient of variation for denitrification rates for an agricultural field in Michigan to be 275. However, nitrogen losses from upland forests by denitrification are small—typically <2 kg N ha^{-1} year^{-1} (Robertson and Tiedje, 1984; Matson *et al.*, 1987)—and even fertilized forests do not appear to sustain large denitrification losses (Matson *et al.*, 1992). Extrapolating fluxes, with such extreme variability, from a soil sample to a stand or landscape is problematic; however, incorporating landscape-scale attributes, such as soil texture and drainage class, into empirical models leads to better landscape-scale estimates of denitrification (Groffman and Tiedje, 1989).

IV. Impacts of Natural and Anthropogenic "Disturbances" on Nutrient Cycles

A. Species Effects on Biogeochemical Cycles

It has long been known that plants affect soil nutrient status (Müller, 1879, 1884; cited in Miles, 1985); in fact, some species may affect pedogenic processes so dramatically that they alter the trajectory of succession (Miles, 1985). Plants create feedbacks to nutrient cycles through nutrient uptake, use, and detritus production (both quantity and quality), which collectively influence microbial popoulations, and they in turn influence nutrient availability (Gosz, 1981; Hobbie, 1992). This issue is of increasing importance as afforestation and reforestation with exotic species become common worldwide (Chapter 2). Perhaps the best example of species influence on nutrient cycles is the effect, discussed earlier, of nitrogen-fixing species on the nitrogen cycle (see Table 7.4); there is also increasing evidence that non-N-fixing trees can have substantial effects on forest nutrient cycles. Recent summaries of various components of forest nutrient cycles have shown that nitrogen mineralization (Vitousek *et al.*, 1982) and nitrogen content of above-ground detritus (Vogt *et al.*, 1986a) are greater for deciduous than evergreen forests. Such comparative studies provide valuable insights into forest nutrient cycles, but they cannot be used to infer cause and effect because, in most cases, other important factors that affect soil development and nutrient availability are not held constant. Gosz (1981) provided one of the first conceptual models illustrating the feedbacks between vegetation and soils: The quantity and quality of litter entering the forest floor regulate microbial populations, which in turn regulate litter decomposition and nutrient mineralization rates and strongly influence net primary production and litter chemistry. Many of these concepts have been incorporated into forest ecosystem process models that simulate soil carbon and nitrogen dynamics (Chapter 9).

The differential effects of species on the nutrient content of vegetation and soil (Alban *et al.*, 1982; Son and Gower, 1992; Lugo, 1992) and on soil chemistry (Binkley and Valentine, 1991) have been demonstrated in numerous studies, but few have identified the processes responsible for the changes. Gower and Son (1992) found that nitrogen mineralization rates varied by more than a factor of two in plantations of five tree species only 29 years after planting; 80% of the variation was explained by the lignin/nitrogen ratio of litterfall. There is little information available on the importance of below-ground processes in relation to interspecies differences in forest nutrient cycles, but data for grassland ecosystems suggest that it is important (Wedin and Tilman, 1990). It is worth noting the potential effect that particular mycorrhizal associations—vesicular arbuscular (VA) versus ectomycorrhizae—can have on nutrient cycling. Read (1991) speculated that mycorrhizal associations replace each other along environmental gradients because of their different ability to utilize various nutrient fractions. He suggested ericoid mycorrhizae predominate in cold wet heathlands, where decomposition is slow, because they have the greatest capacity for direct uptake of organic residues. Ectomycorrhizae predominate in boreal and temperate forests and have intermediate capacity to utilize the organic nutrient pools, and VA or endomycorrhizae predominate in warm temperate and tropical forests because of their capacity to utilize various organic and inorganic phosphorus pools.

Interspecies differences in nitrogen use and cycling have important implications for forest management, particularly as reforestation and afforestation in the tropics increase. Houghton *et al.* (1991) estimated that between 313 and 412 × 10^6 ha have been converted from natural forests to other land uses since 1850. Today, much of this land is degraded and no longer suitable for crops or pasture. There is great interest in reforesting these areas to minimize erosion and perhaps even ameliorate the soils. Fisher (1995) compared soil chemical and physical properties before and after planting 11 (8 native and 3 exotic) tropical tree species in replicated plantations in Costa Rica. He found that in only 4 years the bulk density decreased beneath 8 of the 11 species, organic carbon increased significantly beneath 3 species, and extractable potassium and phosphorus increased significantly beneath 4 species. Increases in nutrient availability were commonly associated with tree species with deep roots that accumulate nutrients from a large soil volume and redistribute them, as litterfall, in a small volume near the soil surface. In a study by Lugo (1992) on the effects of plantations of *P. caribaea* and natural forests in Puerto Rico, the mean residence times of nitrogen and phosphorus at four sites ranged from 1 to 4 years in the plantations but, under natural forest, exceeded 1 year at only one site. The much higher residence time of the nutrients in the pine plantations was attributed to the poor litter quality of the pines compared to the natural forests. The natural forests also had greater root density and biomass, further enhancing the differences in nutrient cycles between them and the pines.

Forest biogeochemical cycles are affected not only by plants but also by vertebrates (Pastor *et al.,* 1988), phytophagous insects (Mattson and Addy,

1975; Schowalter *et al.*, 1986), and herbivores, which can indirectly stimulate or retard nutrient cycling and hence the availability of nutrients. Periodic, short-lived outbreaks of folivores often stimulate nutrient cycling rates by generating large amounts of prematurely senesced foliage, insect tissue, and frass, all of which increase nutrient transfer to the forest floor (Schowalter, 1981; Schowalter and Crossley, 1984; Swank *et al.*, 1981; Seastedt and Crossley, 1984). For example, Swank *et al.* reported a greater population of nitrifying organisms and elevated NO_3^- export for a mixed hardwood forests defoliated by a fall cankerworm (*Alspphila pometaria*). In some circumstances, severe episodes or chronic low levels of defoliation can adversely affect nutrient cycling rates because plants produce facultative or inducible defense compounds to deter herbivory (Coley *et al.*, 1985). These are usually phenolics and tannins—carbon-based compounds that play a major role in chemical defense of many forest trees. Polyphenolics (e.g., tannins and lignins) form recalcitrant complexes with carbonaceous constituents, such as cellulose and starch (Davies *et al.*, 1964; Benoit and Starkey, 1968), that adversely affect decomposition and nitrification rates (Handley, 1961; Baldwin *et al.*, 1983).

B. Fire

Wildfires are a natural component of some terrestrial ecosystems, although they are often viewed by humans as destructive and, as far as possible, prevented. This is a controversial policy because fire suppression over long periods has, in some cases, led to large changes in forest species composition or loss of species that depend on fire to regenerate (Kilgore and Taylor, 1979; Lorimer, 1990). Fire suppression policies also tend to lead to unforseen consequences because it is almost impossible to prevent wildfires in perpetuity, so that when fires do occur in areas where fire has been excluded they are usually much worse, and have a greater affect on forest biogeochemical cycles, than they would have been if left to occur with "natural" frequency—whatever that may be. Vegetation that is adapted to relatively frequent, usually low-intensity fires may be killed by the high-intensity wildfires that occur when forests accumulate heavy fuel loads and finally burn in catastrophic wildfires. There are many examples of such wildfires that have followed long periods of fire suppression: the most spectacular in recent years have been the Ash Wednesday fires that raged through southern Australia in 1983 and the fires in Yellowstone National Park, in the United States, in 1988.

In Australia, temperate evergreen forests of the type known as dry schlerophyll, which may be in flammable condition for long periods in most years, can be expected to burn relatively frequently, but wet schlerophyll forests are likely to burn much less frequently. Average fire frequencies for major forest types in North America include 30–60 years for coastal plain pine forests in the southeast, 70–100 years for pine forest in the Great Lakes region and boreal forests, 200–400 years for oak–hickory forests in the eastern United States and dry conifer forests in Rocky Mountains, greater than 400 years for coastal conifer forests in the Pacific Northwest, and greater than 1000 years for northern hardwoods (Aber and Melillo, 1991). Boreal forests provide an excellent example of ecosystems in which the structure and function of the forests is very strongly influenced by fire. They occur in inaccessible areas with sparse human populations and are usually of low commercial value; fire suppression is almost nonexistent and single fires can burn unhindered across very large areas.

In addition to naturally occuring fires, fire has been used by humans for centuries. Prescribed fires—or controlled burns—are used to control competing vegetation, reduce fuel load and minimize the likelihood of catastrophic fires, improve wildlife habitat, and reduce logging slash or debris from the previously harvested forests. In the southeastern United States alone, approximately 10^6 ha of loblolly pine plantations are deliberately burned annually (Richter *et al.*, 1982). The intensity of prescribed fires is likely to be of the order of a few thousand watts per meter at the fire front.

The biological and chemical effects of fire differ widely among forests, depending on vegetation condition, fuel quantity and quality, and fire intensity, duration, and frequency. Soil temperatures greater than 60°C for several minutes kill most soil biota and temperatures greater than 120°C sterilize the soil (Warcup, 1981). The intensity of prescribed burns is typically much smaller than that of wildfires and slash removal. Only about 5% of the total energy released during a fire is transferred into the soil (Packham, 1970). Major factors that influence the effect of fire on soil temperature are fire intensity and two physical soil properties: soil heat capacity and thermal conductivity. The thermal conductivity of air, which comprises 50% of the soil volume, is several orders of magnitude lower than organic matter or mineral particles so that heat generated from fire does not penetrate very deeply into forest soils. For example, Humphreys and Craig (1981) reported that soil temperature did not exceed 25°C at depths of about 3, 5.5, and 7 cm for prescribed burn, wildfire, and heavy slash fire, respectively; however, soil temperatures exceeded 200°C down to 10 cm beneath a log pile fire. In general, the surface layers of dry soils reach higher temperatures than those of wet soils, but the temperature increases are propagated to greater depth in wet soil.

The major pathway for nutrient loss during, or as a consequence of, fire is volatilization. During prescribed burns, temperatures at the soil surface seldom rise high enough to lead to volatilization of large amounts of nutrients, but in wildfires and large debris fires they may do so (Grier, 1975). The threshold for elemental volatilization varies among nutrients but generally follows the pattern: C = N < S < P < monovalent cations < divalent cations (Fig. 7.8). The low combustion temperature for nitrogen results in a strong positive relationship between the combustion of organic matter and nitrogen volatilized (Raison *et al.*, 1985) that is almost 1:1.

Fire is commonly used to get rid of unwanted slash (debris) after a stand is harvested for timber. Weston and Attiwill (1990) compared soil nitrogen dynamics in unburnt forests with forests burnt by wildfires of varying intensity (surface and crown fire) and with a logged and slash fire treatment. They found that

- total inorganic N in the 0- to 5-cm soil layer increased with fire intensity;
- NO_3^- concentration in the soil solution at 10 cm increased with fire intensity ranging from 0.6 mg ml^{-1} in the unburnt forest to 70 mg ml^{-1} in the logged and burned forest;

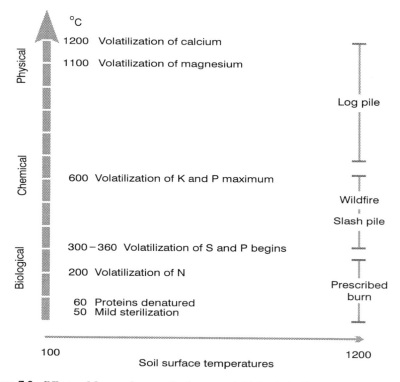

Figure 7.8 Effects of fire on forest soils. In general, biological effects occur at temperatures less than 120°C, most chemical effects occur by 600°C (except for the volatilization of divalent cations), and physical changes occur above 600°C. For reference, the temperature ranges likely to be generated by fires of different types are indicated (modified from Walker *et al.*, 1983).

- cations in the soil solution were positively correlated to NO_3^-, suggesting it was the major transport anion.

Total inorganic nitrogen concentration in the soil and NO_3^- in the soil solution returned to concentrations similar to the unburnt forests after 205 days in the surface fire plots and Weston and Attiwill (1990) suggested that microbial immobilization was a key process in limiting NO_3^- leaching losses. Others have reported that prescribed burns have little or no effect on soil nitrogen content (McKee 1982, Waldrop et al., 1987; Schoch and Binkley, 1986; Boyer and Miller, 1994), foliage nitrogen concentration of remaining trees (Binkley et al., 1992; Boyer and Miller, 1994), and soil nitrogen mineralization rates (Bell and Binkley, 1989; but see Vance and Henderson, 1984).

Numerous studies have examined the effect of one to several prescribed fires on biogeochemical cycles of temperate forests, but few data are available from long-term studies. One exception is the study conducted by Binkley et al. (1992) who examined the 30-year cumulative effect of prescribed fire intervals of 1, 2, 3, and 4 years for a loblolly pine—longleaf pine (*Pinus taeda* and *Pinus palustris*) forest in South Carolina. (Fig. 7.9). They found that the effects of prescribed burns on soil fertility were restricted to the forest floor and that the total nutrient content in the forest floor + upper 20 cm of mineral soil was not significantly affected by fire frequency, although there was a general decline in nutrient availability in the forests that were burned annually.

Losses of nutrients resulting from fire are likely to be offset by increased

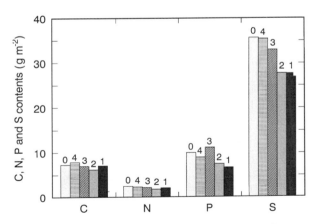

Figure 7.9 Carbon (C), nitrogen (N), phosphorus (P), and sulfur (S) content (g m^{-2}, except for nitrogen which is g m$^{-2} \times 10^{-2}$) for forest floor + upper 20 cm of mineral soil for a loblolly/long-leaf pine forest in South Carolina subjected to different burning intervals: control (0), every fourth (4), third (3), and second (2) year, and annually (1). P content is the sum of labile inorganic + organic forms (adapted from Binkley et al., 1992).

nutrient mineralization, which is probably caused by increases in soil temperature and base cations; both increase the activity and efficiency of microorganisms (Paul and Clark, 1989). Increased availability of base cations decreases soil acidity and may enhance nitrogen fixation (Jorgensen and Wells, 1980). Periodic light fires may also volatilize terpenoids or polyphenols that inhibit nitrogen mineralization and nitrification (White, 1986, 1988).

Fire is also used to convert tropical forests to agriculture—also referred to as slash and burn—which is a common component of the shifting cultivation practiced in the tropics. Although nitrogen volatilization losses can reach 300–700 kg ha^{-1} (Nye and Greenland, 1960; Ewel *et al.*, 1981), additional unexplained losses of 500–2000 kg N ha^{-1} from the soil have been attributed to slash and burn practices (Nye and Greenland, 1960; Brinkmann and de Nascimento, 1973; Sanchez *et al.*, 1982). There is a very poor understanding of whether these losses result in long-term decreases in soil fertility. Matson *et al.* (1987) reported that felling and burning of a relatively fertile tropical forest in Costa Rica significantly increased nitrogen mineralization rates, but the increase was short-lived: Concentrations similar to those of the secondary intact forest were reestablished after less than 6 months. They also suggested that substantial amounts of nitrogen mineralized by fire may be retained as exchangeably bound nitrate deep in the soil and that this nitrogen would become available to the trees during reforestation. (As we discussed earlier, this mechanism of anion retention occurs only in forest soils with significant anion exchange capacity.)

During the past two decades, the frequency and extent of fires in forest ecosystems has increased dramatically in relation to long-term trends in Canada (Van Wagner, 1988), the United States (USDA Forest Service, 1992), and the Commonwealth of Independent States (Krankina, 1992) (previously the Soviet Union). There has also been increasing burning in tropical and subtropical countries (see Chapter 2). The overall result is accelerated release of CO_2 and other greenhouse gases to the atmosphere, contributing to the greenhouse effect and climate change. Clearly, the effects of fire, both natural and prescribed, should be incorporated into global carbon budgets. According to simulated carbon budgets, the increased extent of fires is responsible for a total release of 5.5 Pg—or 43% of the total CO_2 released to the atmosphere annually (Auclair and Carter, 1993). Matson *et al.* (1987) point out that nitrogen losses from clear-felling and burning of tropical forests alone may be roughly equal to half of the industrial nitrogen fixed globally.

C. Timber Harvesting

Bormann and Likens (1979) reported extremely high nutrient losses from a watershed that was clear-felled, relative to a control watershed

IV. Impacts of Natural and Anthropogenic "Disturbances" on Nutrient Cycles

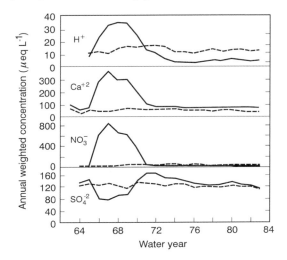

Figure 7.10 Schematic diagram of annual nutrient losses from control (dotted line) and treated (clear-felled + herbicide; solid line) watersheds at the Hubbard Brook Experimental Forest in New Hampshire. The large losses of nitrogen in the form of NO_3^- are caused by high nitrification rates, which explains the concomitant increase in H^+ ion production. Leaching losses of base cations were also high as a result of maintaining electroneutrality in the soil. It is important to note that the experiment is not representative of standard forest management practice because continuous application of herbicide prevented all regrowth.

(Fig. 7.10), in the Hubbard Brook study. Their results prompted a series of experimental and comparative studies and the development of process-based vegetation–soil N-cycling models to examine nutrient budgets and processes regulating nutrient leaching losses from forests. These efforts continue today and have resulted in a much better understanding of vegetation and soil processes that regulate nitrogen mineralization, transformation, abiotic and biotic immobilization, and losses in forest soils.

Vitousek *et al.* (1982) used trenched plots—large blocks of soil surrounded by trenches lined with plastic and back filled with soil, maintained free of vegetation by weeding—and laboratory incubation studies in a variety of North American forest ecosystems to examine the controls on nitrate leaching in temperate forest ecosystems. Nitrate concentrations collected from lysimeters placed below the rooting zone in control and trenched plots did not differ greatly in many of the forests, suggesting there are inherent processes that, at least temporarily, prevent large leaching losses when plant uptake is not important. Long-term laboratory incubation studies demonstrated that nitrification occurred for many of the forests. The work of Vitousek *et al.* demonstrated that NH_4^+ and NO_3^- uptake by regenerating vegetation is critical for the prevention of large nutrient losses.

Competition among heterotrophic organisms, plants, and nitrifiers—the microorganisms reponsible for converting NH_4^+ to NO_3^-—for nitrogen plays a major role in determining NO_3^- leaching losses (Riha et al., 1986). Until recently, it was widely assumed that heterotrophs are better competitors for NH_4^+ than nitrifiers (Jones and Richards, 1977; Vitousek et al., 1982), which would imply that the nitrifying population is strongly regulated by the rate heterotrophs mineralize organic matter and release NH_4^+. ^{15}N-labeling studies have shown that heterotrophs are a very important immobilization sink for NH_4^+ and NO_3^- (Vitousek and Matson, 1984, 1985; Matson et al., 1987; Zak et al., 1994). The results have direct implications for residue management following natural disturbances (windstorms) or timber harvests. Harvesting and site preparation practices that leave woody residue (high C/element ratio) on the site should increase heterotrophic competition for NH_4^+, resulting in reduced net nitrification until vegetation demand for nitrogen becomes reestablished (Ross et al., 1995). However, Davidson et al. (1990), working in a grassland soil, reported that a significant proportion (12–46%) of the mineralized N was converted to NO_3^- by nitrifiers even though soil NO_3^- pools were very low. They argued that these data suggest that nitrifiers may be more effective competitors for NH_4^+ than has been assumed.

Once NH_4^+ is nitrified, the major mechanisms for retaining NO_3^- in the soil include plant uptake, microbial immobilization, and NO_3^- accumulation. Nitrate sorption on the exchange surface of clay particles or humus is unlikely in temperate and boreal forests because these soils rarely have a significant anion-exchange capacity. Denitrification may reduce NO_3^- leaching losses but still results in a net loss of nitrogen and causes other environmental concerns. Therefore, in forests in which the capacity of vegetation to absosrb nitrogen is reduced or eliminated and microbial immobilization potential is saturated, NO_3^- leaching losses are likely to occur. The exceptions are likely to be forests in which nitrogen availability is extremely low; this can result in plants producing low-quality litter-containing polyphenols or other compounds that inhibit nitrifying bacteria (Lamb, 1975; Olson, 1963; White, 1988).

Timber harvesting may affect the long-term fertility of sites by direct removal of significant amounts of nutrients in the biomass, leading to adverse effects on long-term forest productivity. The removal of aboveground biomass from tree plantations managed on relatively short rotations removes nutrients that would require many decades to be replaced by atmospheric inputs (Table 7.7). This is particularly the case with intensive systems: During the "oil crisis" of the 1980s there was great interest in plantations for energy from biomass, but Hanson and Baker (1979) showed that intensive, short-rotation coppiced plantations are not sus-

Table 7.7 Nutrients Removed by Conventional (Stem Only) and Whole Tree (Total Above-ground) Harvests from Selected Forests[a]

Species/location	Stand age (years)	Treatment	N (kg ha^{-1}) Removed in harvest	N (kg ha^{-1}) Rotation balance	P (kg ha^{-1}) Removed in harvest	P (kg ha^{-1}) Rotation balance	Ca (kg ha^{-1}) Removed in harvest	Ca (kg ha^{-1}) Rotation balance
Populus deltoides/ Alabama	7.5	WT	185	133	26	−22	348	−314
Eucalyptus grandis/ New South Wales, Australia	10	C	267	−247	10	−8	613	−513
		WT	416	−396	16	−14	730	−630
Mixed deciduous/ Tennessee	38	C	110	152	7	14	410	−235
		WT	315	−53	22	−1	1090	−915
Picea abies/ Exp. Forest, Wisconsin	28	C	140	140	16	−2	NA	NA
		WT	687	−407	97	−83	NA	NA
Larix decidua/ Exp. Forest, Wisconsin	28	C	160	120	20	−6	NA	NA
		WT	208	72	40	−26	NA	NA

Note. Data sources are White (1974), Son and Gower (1992), and Johnson *et al.* (1982). C, conventional; WT, whole tree; NA, not available.
[a] Rotation balance is defined as (nutrient removed/annual atmospheric nutrient input) multiplied by stand age. The selected ages do not necessarily correspond to rotation length stands, but illustrate potential nutrient losses.

tainable without large additions of fertilizers. The impact of more traditional forest management regimes is less clear: Empirical studies suggest that whole-tree harvesting (e.g., removal of stem, branches, and foliage) increases the rate of loss of nutrients by several-fold relative to a conventional harvest (e.g., stem only) in which, usually, only a small fraction of the total nutrients of the ecosystem is removed (see Table 7.2). Because nutrient requirements and within-tree allocation patterns differ among species, it should not be surprising that the impact of timber harvesting on nutrient removal is species dependent. C/element ratios are often lower in broad-leaved deciduous species than in coniferous species; therefore, removal of an amount of biomass from a deciduous forest will result in a greater nutrient removal than the same amount of biomass from a coniferous forest (Son and Gower, 1992). More nutrients are removed when old forests are harvested than in younger stands, but the longer rotation lengths of the older forests provide more opportunity for nutrients to be replaced by natural inputs. The actual impact of nutrient removal is site specific and dependent on annual nutrient inputs by weathering and atmospheric deposition. Turner and Lambert (1986) concluded that at least four rotations would be needed before a detectable decrease in forest floor nutrient content would occur, although this may not always be true for base cations, which may be more plentiful in woody tissue than in the soil (Table 7.2). Credible analyses of this problem of nutrient removal by tree harvesting require a complete understanding of the direct and indirect effects of timber harvesting on all the processes that control nutrient availability. Process-based models (Aber *et al.*, 1979a; Dewar, 1996) provide a useful management tool for examining the long-term implications of the variables associated with timber harvesting on the long-term productivity of soils.

Another potentially adverse effect of timber harvesting on long-term site productivity is increased erosion. Erosion of the nutrient-rich surface detritus and upper mineral soil can cause significant nutrient losses from the site, increasing the turbidity of streams and, among other effects, possibly degrading fish spawning habitat (MacDonald *et al.*, 1991). Erosion is likely to increase in proportion to the amount of disturbance to the surface; therefore, excessive road building and intensive site preparation often lead to greater nutrient losses through this process. Pye and Vitousek (1985) reported greater nitrogen and phosphorus losses associated with more intensive site preparation in a moderately sloping loblolly pine (*P. taeda*) forest; maximum nitrogen and phosphorus losses were 254 and 61 kg ha^{-1}, respectively, for prepared areas of windrow. It is unlikely that erosion losses can be totally avoided, although several studies have shown that the use of "best management practices" can minimize, and in some cases, completely avoid them (Lee, 1980).

D. Nitrogen Saturation

A shortage of nitrogen is widely recognized as one of the major factors limiting forest productivity; however, there is increasing concern that, in some areas, chronic nitrogen deposition may reverse this pattern and cause forest decline and excessive leaching of nitrogen into ground and surface waters (Aber *et al.*, 1989; Aber, 1992). The phenomenon of continuous low levels of atmospheric nitrogen deposition and its effects on forest ecosystem pattern and processes is referred to as "nitrogen saturation." There is no agreement, as yet, on its definition. Proposed definitions, in increasing order of severity, include (i) lack of a growth response to added nitrogen (Nilsson, 1986), (ii) nitrogen loss exceeds "normal" leaching losses (Christ *et al.*, 1995), (iii) nitrogen leaching loss equals or exceeds input (Ågren and Bosatta, 1988), and (iv) nitrogen deposition induces physiological damage or species change (van Breenan and van Dijk, 1988). The common thread linking these definitions is the gradual decline in the capacity of an ecosystem to retain nitrogen as increased nitrification results in greater losses of NO_3^- and accompanying base cations. Aber *et al.* (1989) proposed a conceptual model linking nitrogen deposition and changes in forest ecosystem structure and function (Fig. 7.11). We note that the influence of nitrogen deposition on some ecosystem characteristics may differ among forests depending upon whether N is the most limiting resource.

Few data are available to test the hypotheses represented in Fig. 7.11. The mechanism responsible for nitrogen retention is not well understood; however, it is necessary to characterize critical thresholds at which forest ecosystem integrity becomes severely impaired. Several comparative studies have been carried out in the United States and in Europe but, like the acid rain studies of the 1970s and 1980s, they lack adequate control treatments. Johnson (1992) used nitrogen cycling data from the Integrated Forest Study (Johnson and Lindberg, 1991) to examine the relationship between ecosystem (vegetation plus soil) retention capacity and atmospheric N deposition. He found that

- ecosystem retention of atmospherically deposited nitrogen ranged from +99% to −266% and was not correlated with atmospheric N-deposition rates;
- in 19 of the 24 forests studied, nearly all of the nitrogen retained by the ecosystem was in the vegetation; retention in the soil was low.

Experimental studies have been started. Christ *et al.* (1995) added ammonium sulfate [$(NH_4)_2SO_4$] at three levels (40, 160, and 560 kg N ha^{-1} year^{-1}) to a beech-dominated forest at Hubbard Brook Experimental Forest in New Hampshire and measured soil and vegetation nitrogen dynamics. They found that all three levels of nitrogen addition increased

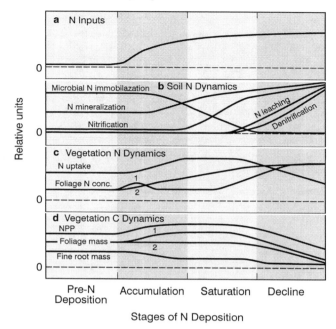

Figure 7.11 Conceptual model illustrating hypothesized changes in (b) soil and (c) vegetation nitrogen dynamics and (d) vegetation carbon dynamics in relation to (a) chronic nitrogen atmospheric deposition (modified from Aber *et al.*, © 1989 American Institute of Biological Sciences).

nitrogen leaching in the upper soil horizon, but only the highest N treatment caused significant N leaching loss in the form of NH_4^+ rather than the expected NO_3^-. The results from this study corroborate those of another experimental study (Aber *et al.*, 1992) but are in direct contrast to comparative studies that found that nitrification rates were positively correlated to N deposition in forests along nitrogen deposition gradients (Nilsson *et al.*, 1988; McNulty *et al.*, 1991). They are also inconsistent with the conceptual model proposed by Aber *et al.* (1989).

V. Concluding Remarks

Our understanding of the effects of nutrition on forest growth and biomass allocation has increased greatly over the past few decades, although much of the research during that period has focused on nitrogen and, to a lesser extent, phosphorus. Areas of research related to forest nutrition from which we can expect productive results in the near future include studies on the influence of the availability of elements other than N and P on forest growth and studies on the interaction of elements and of nutrient and water interactions. The development and testing of mechanistic models of nutrient uptake by and distribution within trees must be a high priority: There have been empirical studies in these areas in agriculture for many years, but because of the complexity of the soil–plant system our ability to predict the effects of various levels of nutrition on the growth of crops has not improved greatly. The situation with regard to forests is worse. Trees are long lived and forest ecosystems less homogeneous than agricultural systems. Empirical research will provide local information, which will often be of limited value; therefore, understanding and being able to quantify the mechanisms is essential.

Nutrient and carbon cycles are tightly linked so that anything that alters the litter quality (i.e., C/element ratios or the composition of carbon constituents) is likely to cause feedbacks affecting nutrient availability, net primary production, and carbon allocation. Numerous empirical studies have shown that fauna—ranging from folivores to large vertebrates such as moose—and flora can affect nutrient cycles of forest ecosystems. The interactions of these organisms with nutrient cycles is poorly understood. Future research in this area must increase our understanding of the relationship between biodiversity and critical ecosystem properties such as nutrient cycles, and if forests are to be managed on a sustainable basis studies are needed to elucidate the mechanisms responsible for the inherent stability of forest ecosystems. Concurrent with experimental studies, process-based models should be used to determine whether forest management practices are sustainable. Many empirical studies have demonstrated that current controversial forest management practices (e.g., clear-felling, fire, etc.) may not seriously threaten the long-term sustainability of forest ecosystems if they are carefully controlled. Their effects should also be monitored. The effects of many disturbances, both natural and anthropogenic, are very site specific and are often complicated by complex interactions. Important mechanisms that minimize nutrient leaching losses from sites include plant and microbial immobilization of nutrients and, in some tropical forests, adsorption of anions such as nitrate on variable-charged aluminum and iron hydroxides. Nonbiological reactions of nitrogen with soil organic matter following fertilization and clear-felling, or for forests subject to atmospheric nitrogen deposition, may be important processes that result in nitrogen storage in the soil as opposed to nitrate leaching; however, it is clear that our understanding of the relative importance of heterotrophs, vegetation, nitrifiers, and abiotic immobilization is incomplete.

We note, in conclusion, that humans have been able to exploit forest systems for millenia by simply moving to new areas when the biogeochemical cycles of terrestrial and aquatic ecosystems they were occupying become too severely altered This option is no longer available.

Recommended Reading

Aber, J. A. (1992). Nitrogen cycling and nitrogen saturation in temperate forest ecosystems. *Tree* **7**, 220–224.

Attiwill, P. M., and Adams, M. A. (1993). Nutrient cycling in forests. *New Phytol.* **124**, 561–582.

Binkley, D., Richter, D., David, M. B., and Caldwell, B. (1992). Soil chemistry in a loblolly/longleaf pine forest with interval burning. *Ecol. Appl.* **2**, 157–164.

Boring, L. R., Swank, W. T., Waide, J. B., and Henderson, G. S. (1988). Sources, fates, and impacts of nitrogen inputs to terrestrial ecosystems: Review and synthesis. *Biogeochemistry* **6**, 119–159.

Johnson, D. W. (1992). Nitrogen retention in forest soils. *J. Environ. Qual.* **21**, 1–12.

Johnson, D. W., and Cole, D. W. (1980). Anion mobility in soils: Relevance to nutrient transport from forest ecosystems. *Environ. Int.* **3**, 79–90.

Parker, G. G. (1983). Throughfall and stemflow in the forest nutrient cycle. *Adv. Ecol. Res.* **13**, 57–113.

Vitousek, P. M., Gosz, J. R., Grier, C. C., Melillo, J. M., and Reiners, W. A. (1982). A comparative analysis of potential nitrification and nitrate mobility in forest ecosystems. *Ecol. Monogr.* **52**, 155–177.

Vogt, K. A., Grier, C. C., and Vogt, D. J. (1986b). Production, turnover, and nutrient dynamics of above- and belowground detritus of world forests. *Adv. Ecol. Res.* **15**, 303–377.

8

Changes in Ecosystem Structure and Function during Stand Development

Forest succession—the changes in species composition, structure, and function that occur as forests age—has long been a subject of speculation, prompting the formulation of numerous theories relating to the driving factors and predictability of the process (McIntosh, 1981). Studies on succcession have generally placed more emphasis on changes in species composition than on understanding how and why the structure and function of forest ecosystems change during the process. However, in addition to changes in plant communities, the structure and function of the communities also change. The focus in this chapter is mainly on the changes in carbon and nutrient cycles that are a consequence of, and contribute to, the changes in forest structure that occur as stands progress through the various stages and states that together constitute succession. We are not concerned here with the long-term ecological patterns of succession that may take place over many growth cycles and generations of forests. These are subject not only to the unpredictable consequences of the direct activities of humans but also to the consequences of changes in climate that appear likely as an indirect effect of those activities.

It is clear that if forests are to be managed on a sustainable basis we must understand how natural and anthropogenic disturbances affect succession. Many of the early hypotheses pertaining to structural and functional changes in forest ecosystems during natural succession (Odum, 1969) are, as yet, untested and the causes of these changes are still poorly understood. Such changes are relevant to global ecology as well as to forest management: The simulation models currently being used to evaluate the global carbon balance (e.g., Melillo *et al.*, 1993; Potter *et al.*, 1993; Ruimy *et al.*, 1994) do not account for the functional changes that occur during forest succession. In particular, the well-documented age-related decline in forest productivity and, hence, possibly the rate of carbon sequestration, is not adequately modeled or is ignored. The disturbances caused by humans, which include land management practices such as grazing, fire suppression or controlled burning, flood control, the application of herbicides, as well as logging, can greatly affect the state of forests and therefore the successional processes that determine their future structure and function. The effects of these actions can be studied directly by monitoring the same stand over time but cannot be predicted unless the initial state of the forest is specified and the results of the disturbance, in terms of changes to the forest structure, are described precisely. It is then necessary to understand how these changes will affect the trajectory, or direction, of future change.

In the following sections, we provide an outline of succession theory, of ideas about the relationships between species diversity and succession, and of the way stand functional characteristics, such as the fluxes of carbon, nutrients, and water, change as forest stands change in structure during stand development. Changes in functional characteristics are, inevitably, associated with changes in stand and ecosystem productivity: Some of the hypotheses about causes of changing above-ground net primary production (ANPP) are discussed and the probability of systematic changes in net ecosystem productivity (NEP) is considered. A major focus of this chapter is on the way structural and functional characteristics change during stand development; it is increasingly clear that these changes are interrelated.

I. General Succession Theory

There are two basic types of succession:

- Primary succession refers to the development of forest communities in areas that have not previously supported vegetation, such as newly exposed substrate resulting from retreating glaciers, sand dunes, and lava flows. Such areas are of considerable interest in ecological terms but are small as a proportion of the forested, or potentially forested, areas of the globe—in most regions they do not exist at all; therefore, primary succession *per se* is generally of minor concern.
- Secondary forest succession refers to the reforestation of an area following destruction of the existing stand. Primary and secondary succession are thought to follow the same patterns, except that the temporal scale is significantly longer for primary than secondary succession because of the length of time required for soil organic matter to accumulate in primary succession (Fig. 8.1).

Differences in soil carbon accumulation rates between primary and secondary succession were presented in Table 6.1.

Succession is a continuous process and does not really consist of distinct stages, but it is useful to identify major stages in stand development

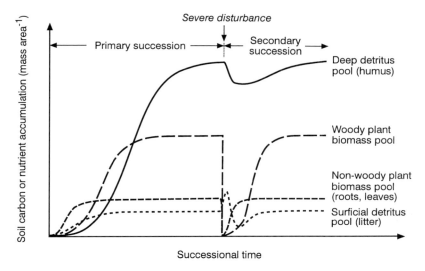

Figure 8.1 Schematic diagram comparing carbon and nutrient accumulation patterns in primary and secondary successional forests (adapted from Reiners, 1981).

and their general attributes. This procedure underlies classification schemes proposed by Bormann and Likens (1979) and by Oliver (1981) for studying succession in forests redeveloping after catastrophic natural disturbances, such as wildfires or hurricanes, or after clear-cutting—a human-induced disturbance that produces similar results (Fig. 8.2a). Disturbances that destroy stands completely tend to result in even-aged stands. Depending on disturbance frequency and intensity, stands may remain even-aged or may slowly convert to uneven-aged. Catastrophic disturbances that remove the overstory but leave seedlings and saplings in the understory intact (e.g., windthrow and selective tree removal) may not result in even-aged stands because of the great variation in the age of survivors (Henry and Swan, 1974). Borman and Likens identified four stages: (i) reorganization, (ii) aggradation, (iii) transition, and (iv) steady state; these are roughly equivalent to the stand reinitiation, stem exclusion, understory reinitiation, and old-growth stages of stand development proposed by Oliver (1981).

Stand reorganization marks the beginning of a new stand. It is characterized by high availability of resources (i.e., light, nutrients, and water) and, consequently, low competition. The aggradation stage is characterized by increased tree mortality (stem exclusion) caused by inter- and intraspecific competition for resources. Thinning, or the selective removal of trees to reduce competition for resources, is typically carried out in managed forests shortly after the onset of this stage. The third stage, referred to as the transition or understory reinitiation phase, reflects the

232   8. Changes in Ecosystem Structure and Function during Stand Development

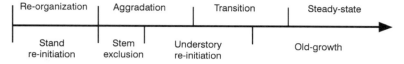

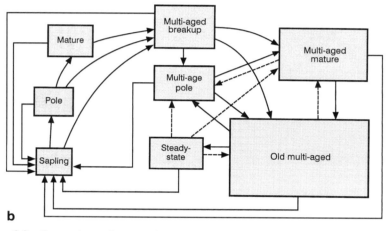

Figure 8.2 Comparison of stages of stand development for (a) forests subject to a catastrophic disturbance creating an even-aged stand [adapted from Bormann and Likens (top) (1979) and Oliver (bottom) (1981)] and (b) forests subject to frequent low to moderate intensity canopy disturbances and highly infrequent catastrophic disturbances creating a multi-aged stand (adapted from Frelich and Lorimer, 1991a).

renewed importance of understory as gaps develop caused by tree mortality. The fourth stage, the steady-state or old-growth phase, corresponds to the climax sere of forest succession and is characterized by large accumulations of living biomass and coarse woody debris. Managed forests seldom reach this stage of stand development because it is long past the period of maximum annual biomass increment—in fact, biomass increment can actually be negative at this stage (Grier and Logan, 1977; Gries, 1995). However, forests in the fourth stage have numerous desirable attributes, such as high stand structural diversity, that must be considered when managing forest landscapes.

Some mesic temperate forests, such as the northern hardwood forests in the Lake States region of the United States and broad-leaved evergreen forests in New Zealand, rarely experience catastrophic disturbances. Instead, noncatastrophic disturbances, such as selective logging, windthrow, or deaths caused by competition, which typically involve the death or removal of single or small groups of trees (Canham and Loucks, 1984;

Frelich and Lorimer, 1991a; Lusk and Ogden, 1992), are the norm. Such disturbances may occur at any stage of succession, although the probability of tree mortality increases with tree age (Dahir, 1994). Frequent noncatastrophic disturbances produce highly diverse, uneven-age stands, the development of which can be described by a complex web of changes (Fig. 8.2b) rather than the linear transition as shown in Fig. 8.2a. It is interesting to note that, when done correctly, selective tree removal (not to be confused with "high-grading"—the removal of the best-quality trees) mimics this type of disturbance and produces an uneven-aged stand.

II. Changes in Species Composition

The question of how species diversity changes during stand development has intrigued ecologists for centuries (McIntosh, 1981). This once "academic" issue is now recognized as having global importance because of the unprecedented rate of change of landscapes, and with it the extirpation of species, all of which are traceable, directly or indirectly, to the exponential growth of the human population (Ehrlich and Ehrlich, 1981). Many of the forest biomes of the world have already been subject to several centuries of human impact (see Chapter 2); therefore, for many forest types it is difficult to know exactly what pristine forests were like. We cannot currently answer the question, "how many species are enough to ensure ecosystem function and stability?" because of our very poor understanding of the relationship between biodiversity and ecosystem function (Schulze and Mooney, 1993; Johnson *et al.*, 1996). Given the growing demand for forest products, it is unlikely that forest use will decline; therefore, the critical issue that ecologists, foresters, and policymakers now face is how to manage forests on a sustainable basis without further decreasing biodiversity.

Multiple physiological constraints prevent any one species from monopolizing all competitive traits for all combinations of constraints (Huston and Smith, 1987; Tilman, 1993). Consequently, plants have evolved compromise mechanisms that enable them to deal with various environmental constraints. For example, greater allocation of carbon to foliage and stem growth makes a species a better competitor for light, whereas greater allocation of carbon to fine roots enhances the capacity to compete for nutrients and water. These two strategies are, to some extent, incompatible and would be optimum in different circumstances; they may not only vary between species but also within species under different conditions (Chapter 5).

There are few empirical data available that can be used to test theories on the relationship between species diversity and succession. The species diversity of a particular stand at a particular stage of development depends on the number of species that have the suite of ecophysiological characteristics best adapted to the environmental constraints. Tilman's (1993) equilibrium theory predicts that species diversity should follow a sigmoidal curve with lowest diversity where the availability of resources, such as nutrients and light, is low and highest diversity where the resources are not in short supply but competition between species results in equal limitations to growth by light and nutrients. Denslow (1980) concluded that the greater the spatial extent of a habitat and the longer the period over which succession has been able to operate without major disturbance, the greater will be the number of species. The intermediate disturbance hypothesis proposed by Connell (1978) predicts that maximum species diversity occurs at intermediate disturbance frequencies. Frequent disturbance allows only species that colonize rapidly to become established, whereas long periods without disturbance exclude inferior species, thereby reducing diversity. These two theories are not mutually exclusive because nutrient and light availability change as a function of disturbance (see below).

III. Stand Functional Characteristics

We can obtain insights into the cause(s) of changes in functional characteristics, such as carbon, nutrient, and water fluxes, during stand development if we understand the changes that take place in forest structure during the natural progression of successional stages. In fact, the so-called natural progression is seldom undisturbed: Superimposed on it are stochastic natural disturbances, such as lightning strikes, windstorms, and pest or pathogen outbreaks, that alter forest structure. Management practices, such as thinning, prescribed burning, and tree harvesting ranging from selective removal to clear-felling, also alter the natural trajectory of stand structure during development. As noted earlier, all these disturbances tend to produce uneven-aged stands. Other management perturbations, such as the effects of fire suppression or wildlife management practices, can be subtle in their effects on species composition and stand structure (Kilgore and Taylor, 1979; Marquis, 1981; Pastor et al., 1988; Lorimer, 1992). However, the remainder of this chapter focuses on changes in the structure and function of forest ecosystems during natural succession.

As forests age, the accumulation of woody biomass can be described by an asymptotic curve. Not surprisingly, the rate of accumulation is strongly influenced by environmental conditions: Severe modification of the soil thermal regime or soil fertility can accelerate the attainment of maximum aboveground woody biomass, although in some cases these changes may cause reversion to vegetation communities of lower stature

(Yarie and Van Cleve, 1983). It is less clear how the living fraction of woody biomass, or respiring tissue, changes during stand development (Sprugel, 1984). This issue may be critical to understanding why NPP decreases during stand development, because stem respiration rates are positively correlated to stem sapwood volume (Ryan, 1991; Ryan et al., 1995). Sprugel (1984) argued that sapwood volume need not necessarily be proportional to total woody biomass, but the empirical data needed to rigorously test his hypothesis are not currently available.

Many of the functional changes in forests during succession appear to be directly or indirectly related to changes in foliage mass or leaf area. In general, foliage mass increases during early stand development (stand reorganization), reaches a maximum during the aggradation stage, and remains constant or decreases in older forests (Gower et al., 1994; Ryan et al., 1996). Foliage mass reaches an asymptote most rapidly in tropical forests and relatively slowly in boreal forests. The cause of the decline in foliage mass remains speculative, but empirical data suggest that nutrient limitation may be partly responsible (Gholz and Fisher, 1982; Flanagan and Van Cleve, 1983; Binkley et al., 1995). The fact that the foliage mass in some forests does not decline with age may be explained by the replacement of shade-intolerant species by shade-tolerant species, with the latter trees supporting a greater leaf area (Peet, 1981).

Few data are available to characterize root biomass accumulation during stand development. Coarse roots follow an accumulation pattern similar to that of above-ground woody biomass (Gholz and Fisher, 1982), but less is known about fine roots and no clear relationship has been observed between total live fine root mass and stand development. Vogt et al. (1987) found that total fine root mass did not differ among different-aged Douglas fir stands in Washington, but other scientists have found that live fine root mass is greater in older than in younger forests (Grier et al., 1981, Gholz et al., 1986). Vogt et al. (1987) reported that the ratio of understory/overstory fine root mass is bimodal in conifers, with the highest peaks during the stand reinitiation and understory reinitiation stages.

In general, understory biomass and species diversity are greatest in the stand reorganization stage, decrease in the aggradation stage, and increase in the transition stage as canopy gaps form as a result of tree mortality. The rapid turnover and high quality (in terms of nutrients) of understory detritus makes its role in carbon and nutrient cycling during succession more important than its small contribution to the organic matter or nutrient content of the forest would imply (MacLean and Wien, 1978; Attiwill et al., 1983). Lockaby et al. (1995) reported that excluding understory by applying herbicide significantly decreased nutrient availability in a loblolly pine (*Pinus taeda*) forest in Alabama.

Coarse woody debris is an important, yet until recently largely ignored,

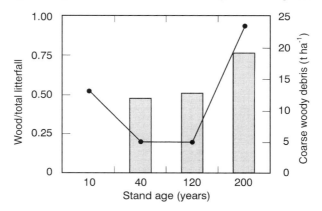

Figure 8.3 The ratio of coarse/wood (wood and foliage) detritus production (bars) and coarse woody debris mass for an *Acer saccharum*-dominated northern hardwood chronosequence. Approximate time periods elapsed since catastrophic disturbance for the four stages are 10, 40, 120, and >200 years, respectively (adapted from Gries, 1995).

component of forest ecosystems. It has a very large carbon/nitrogen (C/N) ratio and small surface area/volume ratio and is therefore a significant factor in nutrient immobilization and storage in forest ecosystems (see Chapter 6). The spatial distribution of coarse woody debris in forests is highly variable, making it difficult to detect patterns associated with stand age. Harmon *et al.* (1986) suggested that coarse woody debris biomass should exhibit a bimodal pattern during stand development, with peaks occurring immediately after disturbance—observed by Gore and Patterson (1986) and by McCarthy and Bailey (1994)—and late in stand development, when there is increased mortality of large trees (Tyrrell and Crow, 1994; Spies *et al.*, 1988; Dahir, 1994). Gries (1995) observed a U-shaped pattern for the northern hardwood age sequence with a maximum occurring in the old-growth stand (Fig. 8.3). The accumulation of coarse woody debris in mature forests and the increasing fraction of aboveground detritus composed of this material will lead to a gradual decline in nutrient availability during stand development (see below).

IV. Forest and Ecosystem Productivity

The first hypotheses describing changes in functional characteristics such as carbon and nutrient flow during succession were published almost 30 years ago (see Odum, 1969) but, until recently, there have been few tests of these hypotheses. There has also been virtually no research into the structural and physiological changes underlying the functional changes. In keeping with the major focus of this book, we will largely restrict the discussion to carbon and nutrient cycling, and their interrelationships, during succession.

Gross primary production (GPP)—or the sum of NPP plus autotrophic respiration (R_a) (see Chapter 5)—is difficult to measure in large-stature forests; therefore few empirical data are available, especially for a complete forest chronosequence. However, forest ecosystem process models have been used to examine relationships between GPP and stand development and, although the results will reflect limitations or shortcomings in the models, they provide the best information we have. In general, GPP is very low immediately after disturbance and increases in proportion to leaf area index during the early years of stand development. GPP is predicted to reach a maximum near canopy closure and then decline, partly as a result of the decline in leaf area. An interesting point is that the decline in GPP does not appear to be as great as the decline in ANPP.

There is increasing evidence from both the ecological and forestry literature that aboveground NPP decreases as stands age, irrespective of forest type (Table 8.1). Kira and Shidei (1967) first speculated that the age-related decrease in ANPP was caused by increased autotrophic respiration. Foliage has the highest maintenance respiration rates of tree tissues (Ryan *et al.*, 1994), but it is unlikely that foliage respiration can explain the decline in ANPP because leaf area remains stable or decreases with stand age. Phloem has high maintenance respiration costs, but the surface area of branches increases while stem surface area increases only slightly (Whittaker and Woodwell, 1967; Sprugel, 1984). Sapwood is the tissue most commonly considered responsible for the decline in ANPP with stand age (Yoda *et al.*, 1965; Whittaker and Woodwell, 1967; Kira and Shidei, 1967), but the assertion that increasing sapwood respiration is the cause of this decline has been undermined by recent observations that it represents only a small percentage (5–10%) of the annual stand carbon budget (Ryan *et al.*, 1995). Few attempts have been made to estimate R_a for forest stands of different age; however, the results from all the studies made so far suggest that R_a is constant, or increases slightly, but cannot account for the large decreases in ANPP (Moller *et al.*, 1954; Ryan and Waring, 1992; Murty *et al.*, 1996). Ryan and Waring estimated that maintenance respiration for woody tissue increased by 18 g C m^{-2} year^{-1} between ages 40 and 245 years for a lodgepole pine (*Pinus contorta*) age sequence, whereas wood production and associated construction respiration decreased by 164 g C m^{-2} year^{-1}.

A second explanation for the decline in ANPP with stand age is decreased nutrient availability. The availability of nitrogen is strongly controlled by litter decomposition that, in turn, is controlled by environ-

Table 8.1 Above-ground Net Primary Production (ANPP, t ha^{-1} year^{-1}) for Forest Age Chronosequences[a]

Biome/species	Location	Age range (year)	ANPP Maximum	ANPP Mature	% Change[b]	Source
Boreal						
Larix gmelina	Yakutsk, Siberia	50–380 (3)	4.9	2.4	−51	1
Picea abies	Russia	22–138 (10)	6.2	2.6	−58	2
Cold temperate						
Abies balsamea	New York	0–60 (6)	3.2	1.1	−66	3
Acer saccarhum[c]	Michigan	10–200 (4)	5.7	3.1	−46	4
Pinus contorta	Colorado	40–245 (3)	2.1	0.5	−76	5
Pinus densiflora	Mt. Mino, Japan	18–390 (7)	16.1	7.4	−54	6
Populus tremuloides	Wisconsin	8–63 (5)	11.1	10.7	−4	7
Populus grandidentata	Michigan	10–70[d]	4.6	3.5	−24	8
Pseudotsuga menziesii	Washington	22–73 (4)	9.3	5.1	−45	9
Warm temperate						
Pinus elliottii	Florida	2–34 (6)	13.2	8.7	−34	10
Pinus radiata	Puruki, New Zealand					
	Tahi	2–6 (5)	28.5	28.5	0	11
	Rua	2–7 (6)	29.2	23.5	−20	11
	Toru	2–8 (7)	31.1	31.1	0	11
Tropical						
Pinus caribaea	Afaka, Nigeria	5–15 (4)	19.2	18.5	−4	12
Pinus kesiya	Meghalaya, India	1–22 (9)	30.1	20.1	−33	13
Tropical rain forest	Amazonia	1–200 (8)	13.2	7.2	−45	14

[a]From Gower et al., 1996; [b]minus sign denotes a decrease in ANPP; [c]Gries, 1995; [d]continuous measurements.

mental conditions and the chemical and physical characteristics of litter (Chapter 6). Several scientists have noted that the fraction wood/total (wood and foliage) litter increases during stand development for boreal (Van Cleve and Noonan, 1975), temperate evergreen (Turner and Long, 1975), and temperate deciduous forests (Gries, 1995). Gries (1995) reported that this ratio increased from near zero for a 10-year-old to 0.73 for an old-growth northern hardwood stand (Fig. 8.3). Also, the major components of the woody litter change from twigs and branches to large stems during succession, which is important because the larger woody material has a significantly lower surface area/volume ratio, thereby adversely affecting the rate of microbial attack on the wood and, hence, the decomposition rate (see Chapter 6). The slow decomposition rates and very high C/N ratio of large logs cause coarse woody debris to be a major nutrient immobilization source for decades (Grier, 1978; Foster and Lang ,1982; Lambert et al., 1980).

If there is a positive feedback between the vegetation and soil, this could further exacerbate the decline in nitrogen availability during stand development (Gosz, 1981). Retranslocation of nitrogen from senescing foliage increases the C/N ratio of leaf detritus, lowering litter quality and further increasing N limitation. Lower litter quality during stand development increases N immobilization during litter decomposition, which in turn decreases net N mineralization as stands age (Davidson et al., 1992; Hart et al., 1994).

Nutrient limitation, especially nitrogen, adversely affects leaf photosynthesis and L^*. Reduction in L^* tends to reduce light interception and GPP (see Chapter 3; McMurtrie et al., 1994). Nitrogen limitation also shifts biomass allocation from above-ground components to fine roots and mycorrhizae (Keyes and Grier, 1981; Gower et al., 1992; Haynes and Gower, 1995), but no forest age sequence studies have, to our knowledge, included fine root and mycorrhizal NPP so that such shifts in biomass allocation during succession have not been detected.

A third explanation for the decline in ANPP during stand development is stomatal constraint. Zimmerman (1983) pioneered the study of the hydraulic architecture of trees and the ascent of water to the canopy. Much of his research focused on the ecological compromises resulting from the need to maximize water transport to the canopy while minimizing the probability of permanent damage to the xylem caused by cavitation during periods of drought or very rapid transpiration (dynamic water stress). Zimmerman suggested that trees avoid permanent xylem dysfunction by decreasing leaf-specific conductivity (i.e., flow rate of water in the xylem per unit pressure gradient causing the flow) toward the apices, thereby increasing the nodal resistance and confining cavitation and embolism to branches that play a less important role in carbon assimilation. This hydraulic architecture would imply that resistance to water transport should be greater for large branches of older trees than for small branches of younger trees. The importance of this adaptation appears to be greater than originally thought: Findings by Tyree and Sperry (1988) suggest that all trees operate near the point of catastrophic xylem dysfunction caused by dynamic water stress.

Because water vapor loss and CO_2 uptake both occur through the stomates, conservative hydraulic architecture affects carbon assimilation. Ryan and Waring (1992) noted that a 15% reduction in canopy photosynthetic rate would account for the large discrepancy between modeled and measured ANPP. Mattson-Djos (1981) noted reduced stomatal conductance in older trees, as did Yoder *et al.* (1994), who also found that photosynthesis rates in the foliage of older lodgepole pine and ponderosa pine trees were 14–30% lower than in the foliage of younger trees. These results support the point made by Yoder *et al.* (1994) and are supported by carbon isotope ratios measured by Ryan and Waring in young and old ponderosa and lodgepole pine. [Differences in the ratios of the stable carbon isotopes $^{13}C/^{12}C$ in foliage and stem tissues reflect differences in average stomatal conductance because the carboxylating enzymes involved in photosynthesis discriminate against the heavier carbon isotope (^{13}C), with discrimination decreasing as the leaf mesophyll CO_2 concentration decreases (Farquhar *et al.*, 1982). Therefore, the $^{13}C/^{12}C$ ratio will fall when photosynthesis is limited more by stomatal conductance than by carboxylating enzyme activity.] On a diurnal basis, delayed opening and earlier closing of stomata have been related to reduced hydraulic conductance (Borghetti *et al.*, 1989; Sperry *et al.*, 1993).

A. Net Ecosystem Productivity

NEP, or the net change in carbon storage in the ecosystem, includes fluxes from vegetation, detritus, and mineral soil and is an important descriptor of the overall functioning of ecosystems. NEP quantifies carbon accumulation or loss and, because carbon and nutrient cycles are strongly coupled, allows inferences to be drawn about the loss of essential nutrients from forests. Few estimates of NEP exist for forests (see Chapter 5), with fewer still for similar forest types at different stages of succession; therefore, the points made in this section are more speculative than those in the sections on changes in ecosystem structure and function. However, despite the paucity of data, a reasonable picture can be pieced together based on information about changes in each of the major carbon fluxes during succession. Odum (1969) first suggested that NEP should approach zero for steady-state or old-growth forests, and the argument we develop here supports this hypothesis, although it has never been validated empirically.

Catastrophic disturbances, such as wildfire and clear-felling, reduce leaf area to near zero, causing NPP to approach zero and setting the scene for secondary succession. The removal of the forest canopy results in elevated soil temperatures (in summer) and changes in soil moisture status, both of which tend to lead to increases in heterotrophic respiration (R_h) immediately after stand reinitiation. We can therefore expect NEP to be negative at the start of a secondary succession cycle or during the stand reorganization phase (Fig. 8.4, top). The length of time that NEP remains negative is likely to vary among ecosystems. Forests in which L^* recovers rapidly after disturbance (e.g., tropical forests) probably experience very short periods of negative NEP, whereas forests in which recovery in L^* is slow (e.g., boreal forests) are likely to experience negative NEP for longer periods. Disturbance intensity may also affect the magnitude and duration of negative NEP—the more severe the disturbance, the longer NEP is likely to remain negative. As leaf area recovers, net carbon accumulation by autotrophs increases rapidly and exceeds R_h, resulting in positive NEP; NEP for aggrading forests during early succession ranges from 0.9 to 5.0 ha^{-1} year^{-1} (see Chapter 5).

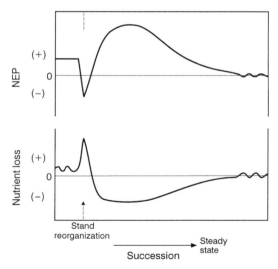

Figure 8.4 Relationship between net ecosystem production (NEP), nutrient loss, and their interaction during forest succession. Immediately following disturbance, NEP goes negative (i.e., the ecosystem is losing carbon to the atmosphere) because heterotrophic respiration exceeds net primary production. During this period, nutrient demand by the vegetation is low and microbes become a source instead of a sink for nutrients as the dominant process changes from immobilization to mobilization. Modified from Vitousek and Reiners (1975) to include a negative NEP pulse immediately after disturbance.

Gross primary production follows L^*, which reaches a maximum during the early aggradation stage and then oscillates around a "quasi steady-state."

It is important to note that small-scale disturbances can result in small patches of early successional forests embedded in a steady-state forest landscape (Frelich and Lorimer 1991b). At any one time, some patches may be accumulating carbon while others are losing carbon, but the net result is a near-zero carbon accumulation for old-growth forests.

V. Nutrient Cycling

It is not surprising that soil biogeochemical cycles change during succession because nutrients, especially nitrogen, are so tightly coupled to carbon cycling (Chapter 6 and 7). It is therefore possible to explain decreased nitrification rates during stand development in terms of changes in environmental and litter quality conditions. Nutrient mineralization rates will be affected by the type of disturbance: They may be relatively high if a substantial portion of debris with a high C/N ratio is removed, but if woody debris is left on a site it helps to immobilize nutrients and reduce nutrient loss immediately following disturbance (Vitousek and Matson, 1985). Rice (1974) has argued strongly that late successional species produce allelochemicals that suppress nitrification, thereby increasing nutrient retention. If organic matter decomposition rates are reduced, this will also result in slower mineralization of essential nutrients, especially those that are tightly coupled to organic matter (Davidson et al., 1992; Hart et al., 1994). The low C/N ratios of forbs and understory species that first colonize disturbed sites contribute to high mineralization rates during stand reinitiation: Mineralization rates are often highest during this stage and decrease during the remainder of stand development (Robertson and Vitousek, 1981; Vitousek et al., 1989).

The varying availability of nutrients during stand succession influences nutrient use and recycling by the vegetation. Nutrient requirements and uptake reach a peak about the time of canopy closure (Turner, 1975; Gholz et al., 1985a; Fig. 8.5), which coincides with the period of maximum tissue production, especially nutrient-rich tissue such as foliage. From this point on in succession, retranslocation of nutrients from foliage to woody tissues increases and a smaller proportion of the annual nutrient requirement is met by uptake. As a result, litter quality decreases, decreasing nutrient availability.

During stand development, nutrient use efficiency appears to increase as soil nutrient availability decreases (Fig. 8.6), but the positive feedback mechanism between litter quality and nutrient availability in later stages

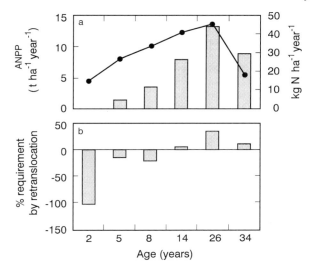

Figure 8.5 Relationship between (a) above-ground net primary production (solid dots connected by line), nitrogen requirement (bars), and (b) percentage of nitrogen requirement supplied by internal retranslocation for a slash pine (*Pinus elliottii*) age chronosequence in north-central Florida (adapted from Gholz *et al.*, 1985a).

led Odum (1969) to speculate that forest biogeochemical cycles get tighter, or more closed, as forests age.

Vitousek and Reiners (1975) suggested that Odum's (1969) hypothesis was too simplistic: They proposed, on the basis of the strong microbial controls on decomposition and nutrient mineralization, that the overall "tightness" of forest nutrient budgets is closely coupled to NEP (Fig. 8.4, bottom). Following disturbance, when NEP is < 0, nutrient loss occurs

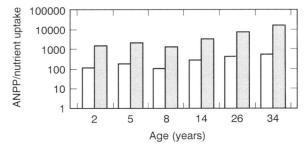

Figure 8.6 Nitrogen (open columns) and phosphorus (shaded columns) use efficiency for a slash pine (*Pinus elliottii*) age chronosequence (adapted from Gholz *et al.*, 1985a). Note logarithmic scale on *y* axis.

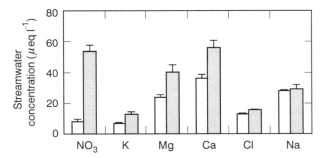

Figure 8.7 Comparison of average growing season (June 1–September 30, 1973 and 1974) stream water concentration of essential (N, K, Mg, and Ca) and nonessential (Cl and Na) nutrients from successional (open columns) and mature (shaded columns) spruce–fir *(Picea rubens–Abies balsamea)* forests in the White Mountains of New Hampshire (adapted from Vitousek and Reiners, 1975).

unless nutrients are immobilized by microorganisms (Chapter 6). The loss of each element is proportional to the demand by the regrowing vegetation; therefore, as L^* recovers, nutrient uptake should parallel NEP. However, as NPP and NEP decline in older stands (Chapter 5), nutrient storage also decreases. This hypothesis is supported by the average growing season streamflow nutrient concentration data from aggrading and old-growth spruce–fir forests in the White Mountains in New Hampshire, where the outflow of nutrients essential for plant growth was consistently higher from mature than from successional stands (Fig. 8.7). However, there appears to be little difference in streamflow concentrations of nonessential nutrients such as Na between successional and mature stands. The hypothesis leads to the prediction that storage of essential nutrients should approach zero as NEP reaches zero in old-growth or climax forest. This has never been tested.

VI. Concluding Remarks

Major points that emerge from this chapter are that species composition and structural and functional characteristics are inextricably linked and change during succession.Therefore, any forest management practice that alters the natural successional trajectory is likely to affect other ecosystem characteristics. It is important that forest managers recognize that changes will occur after disturbance and are able to assess their likely consequences in terms of productivity and sustainability. Given that information about those changes, and knowledge of the soil and weather, we can use some of the ideas discussed here to make projections about the likely course of changes in forest structure and function. We can make reasonable predictions about changes in stem populations (see Chapter 3) and the time course of leaf area development, about nutrient mineralization rates, and, on the basis of all that, about the fluxes of carbon and water.

With increasing concern about biodiversity and its maintenance, it is becoming essential that the objectives of forest managers, in this respect, should be clear. The question of scale is pertinent in relation to management for biodiversity. Biodiversity will tend to be low if forests are reduced to a series of even-aged stands that are not allowed to progress to the point at which they contain old trees and the gaps associated with old-growth stands: The number of niches, with a range of conditions that would be conducive to species with different adaptive mechanisms, is reduced in such even-aged stands. However, if the stands are small, interspersed between areas of very different characteristics and growth patterns, the range of habitats will be increased and we can expect species diversity for the region as a whole to be high.

On a global scale, we need to seek predictable changes in ecosystem function during development and the various stages of succession; these will allow us to improve our capacity to predict change in global carbon balance. We note, in passing, that although we stated in the introductory section that we are not concerned with long-term succession, the current rapidly changing global atmospheric CO_2 concentrations will almost certainly affect forest succession through effects on growth and biomass accumulation rates, particularly in the early stages (reorganization and aggradation) and, hence, on the patterns of forest development.

It is clear that the improvement of theories of forest succession, in the general sense used here, which encompasses changes in structure and function as well as changes in species, is an area that warrants a great deal more research attention in the future.

Recommended Reading

Bormann, F. H., and Likens, G. E. (1979). "Pattern and Process in a Forested Ecosystem." Springer-Verlag, New York.
Frelich, L. E., and Lorimer, C. G. (1991). A simulation of landscape-level stand dynamics in the northern hardwood region. *J. Ecol.* **79,** 223–233.
Gorham, E., Vitousek, P. M., and Reiners, W. A. (1979). The regulation of chemical budgets over the course of terrestrial ecosystem succession. *Annu. Rev. Ecol. Syst.* **10,** 53–84.
Gower, S. T., McMurtrie, R. E., and Murty, D. (1996). Aboveground net primary production decline with stand age: potential causes. *Trends Ecol. Evol.* **11,** 378–382.
Odum, E. P. (1969). The strategy of ecosystem development. *Science* **164,** 262–270.

Sprugel, D. G. (1984). Density, biomass, productivity and nutrient-cycling changes during stand development in wave-regenerated balsam fir forests. *Ecol. Monogr.* **54**, 165–186.

Vitousek, P. M., Van Cleve, K., and Matson, P. A. (1989). Nitrogen availability and nitrification, primary, secondary, and old-field seres. *Plant Soil* **115**, 229–239.

Yoder, B. J., Ryan, M. G., Waring, R. H., Schoettle, A. W., and Kaufmann, M. R. (1994). Evidence of reduced photosynthetic rates in old trees. *For. Sci.* **40**, 513–526.

9

Ecosystem Process Models

Scientists have different ideas about what is meant by a "model," in the context of biology and ecology. Landsberg (1986, p. 2) defined a model as a "formal and precise statement, or set of statements, embodying our current knowledge or hypotheses about the workings of a particular system and its responses to stimuli." Models therefore provide an organized way of stating hypotheses and of focusing on the questions that must be asked at the beginning of a research program. They may also be practical tools to be used by forest managers to estimate stand growth rates (productivity) and to predict the effects of management actions or of insect or pathogen attack. Ecologists may develop and use models to calculate carbon, nutrient, and water flow through ecosystems or to investigate species dynamics in ecosystems. On a wider scale, concern about increasing atmospheric CO_2 concentrations and the probable resulting changes in global climate, and about factors such as atmospheric pollution, make it essential that we have models capable of predicting the probable effects of these phenomena (Landsberg *et al.*, 1991).

This chapter is concerned with process-based models, i.e., models that are based on the processes, or mechanisms, underlying the responses to change (stimuli) of the system under study. The objective of process-based forest growth, or ecosystem, models is to simulate the growth of stands in terms of the underlying physiological processes that determine growth and the way the stands are affected by the physical conditions to which the trees are subject and with which they interact. This means, for example, that forest growth may be described in terms of radiation interception, photosynthesis, and carbon allocation rather than by empirical equations developed from statistical analysis of measurements made on trees and stands (see Section I). Process-based models have the potential to be far more flexible than empirical relationships and can be used in a heuristic sense to evaluate the consequences of change and the likely effects of stimuli. However, they are not without their problems, the most important of which is that they tend to require many parameter values that are often difficult to obtain. Furthermore, the fact that a model is composed of (mathematical) descriptions of various physiological processes does not mean it contains no empirical relationships; at some point, all the relationships used are empirical—the justification for process-based models rests on the fact that "lower-level" empiricism translates into more flexible and realistic responses to change than relationships that subsume the processes responsible for the responses.

There are, clearly, various levels at which process-based models can be written. In general, it is not practical to attempt to model a system across more than two levels of organization; that is, to attempt to explain an observation in terms of processes more than one level down. Landsberg (1986, pp. 13–20) has discussed this in some detail.

Empirical models are based on relationships between outputs and inputs—for example, site index curves giving stand yield estimates based on stem diameter and tree height measurements. Such models are very widely used in forestry, but they are site specific, strongly influenced by the conditions pertaining during the period when the measurements on which they are based were made, and they cannot be used to examine the consequences of any significant departure from the conditions pertaining over that period. In particular, empirical models cannot simulate the consequences of changes in such factors as atmospheric CO_2 concentrations, temperature, or water regimes (Landsberg *et al.*, 1991). However, from the point of view of the forest manager, simple empirical models may be of more immediate practical use than process-based models, most of which have not yet been brought to the point at which they can be said to be useful practical tools.

The position was stated unequivocally by Leech (1985), who made it clear that forest managers are concerned with robust, pragamatic models, that they are likely to be more concerned with local politics, economics, and wood flow than precise predictions of the probable response to conditions in a particular stand, and that they are unlikely to use complex physiological models. When they are aware of them at all, managers tend to regard such models as esoteric and abstract, requiring values for parameters they have no knowledge of and no way of measuring. The question therefore arises: Are we (scientists) justified in our pursuit of knowledge about systems, without concern for short-term utility but with the underlying assumption that this knowledge will somehow filter through to those who might actually use it, or should we consciously orient our science, and the understanding it gives rise to, toward practical objectives? Without becoming involved in this (interesting) philosophical argument, we state here our view that the future must lie with process-based models, but we accept that these must be produced in forms relevant to forest managers and decision- and policymakers at other levels, and that the outputs must be practically useful. We note that forest ecosystem models are already at the stage at which they can contribute to debates about ecosystem sustainability, global carbon balance, and biomass productivity, but in the area of management for wood production some advances and changes will be needed to compete with the conventional forest growth models. A useful discussion of the considerations involved in this question of modeling for prediction or analysis of the behavior of ecological systems is given by Sands (1988).

We discuss models developed as research tools, which can be used to collate our knowledge of systems and to explore the consequences of our hypotheses, and we consider the possibility of using those models, or the principles embodied in them, as the basis for management tools. These tools must be relatively simple, reasonably reliable, and produce answers in the form the managers need: CO_2 uptake or biomass estimates would not generally be considered useful. Therefore, included among the advances needed there must be some procedure for converting biomass calculations into wood production because, although the emphasis in natural forest management is shifting away from the historical focus on wood production toward sustainability of the ecosystems as a whole, the same types of model can be used to simulate and analyze the growth of plantations and the required end result from them is certainly wood.

This chapter provides brief reviews and assessments of some of the more important and well established of the current generation of process-based forest ecosystem models and, in the last section, indicates how these models can and should be used to assist in the decisions made in the course of practical forest management. The models considered in this chapter are at several levels, which will become apparent in the course of the discussions. First, however, we provide an outline of some conventional forest models and the approach used in their development.

I. Forestry Models

Conventional forestry models, as we noted earlier, are essentially statistically derived stand growth curves based on measurements that provide information about the progressive changes in stem volume of trees on particular sites or in particular regions. These measurements commonly include tree number (stand density), stem diameters, average or mean dominant height, and, in some cases, stem taper functions. The models derived from them do not generally invoke any physiological processes or have any mechanistic basis for simulating response to changing conditions. The growth curves may be related to site indices, where the site index is an arbitrary measure of the productivity of a site, usually expressed in terms of average dominant tree height at 20 or 50 years. Site indices reflect the influence of soil fertility, climate and microclimate, tree species, and genetics. The logic is arguably completely circular: We cannot say anything about the productivity of a site until we have grown trees there and measured them, at which point obviously, we know how well they have grown. Furthermore, the site index for a particular location could vary from one period to another depending on whether there were more years with good growing conditions than bad or *vice versa*, whether there were pest attacks that may not have been identified and were almost certainly not quantified, and so on. Nevertheless, curves representing the growth to be expected from particular species growing on sites with specified indices are standard tools for forest yield estimation and will remain so until something better is available. Clearly, these curves cannot be used with any confidence for areas and forest types other than those for which they were derived.

Being statistical descriptions, forestry models are based on measurements made on sample plots. Ideally, tree populations in these plots should be monitored and the dimensional measurements repeated at regular intervals. The location of measurement plots that can be said to represent a forest is a matter of considerable importance. A book by Vanclay (1994) provides a good overview and treatment of the techniques and considerations involved in modeling forest growth and yield using the conventional, or traditional, approaches.

Forest growth models tend to be either stand models, which project the growth and development of entire stands, or individual tree-based models, which have the advantage that age structure, spacing, and individual tree attributes can be included. The model STANDSIM (Opie, 1972; Campbell *et al.*, 1979), developed in Australia using data from *Eucalyptus regnans* forests, mainly 15 years of age or older, is a relatively simple stand model requiring as input the site index of a stand and its stocking density. The model contains functions to predict from this information the basal area of the stand at 15 years of age and the frequency distribution of the diameters at breast height of the trees in the stand. It also calculates annual basal area increment as a function of stand age, basal area, and site index. Other functions distribute this basal area increment among the individual trees of the stand and predict mortality.

A good example of a complex model of forest dynamics is the model of Ek and Monserud (1974; see also Ek and Monserud, 1979), which is an individual tree-based simulator that consists of separately estimated component models for overstory tree height growth, diameter growth, and survival. It also has separately estimated components for reproduction and understory tree height growth and survival. A competition index, calculated at the start of every growth period, is computed for every tree by considering all other trees to be potential competitors in terms of crown overlap. Enormous amounts of data are required to develop such models. Ek and Monserud (1979) evaluated the performance of this model for Lake States (United States) northern hardwood stands for a range of ages and harvest conditions. The model performed well (see also Hasse and Ek, 1981).

Dale *et al.* (1985) reviewed tree growth models and divided them into forest growth and community dynamics models. Forest growth models tend to be site specific, whereas community dynamics models are species specific. Competition is a major component of both types and may be described in terms of tree size, number (stand density), or water relations; in the community dynamics models—often called gap models—the approach to simulating tree growth is based on extremely simplistic representations of the physiological processes that determine tree growth. Nevertheless, these community dynamics models are more transportable than the forest growth models (see Shugart, 1984; see also Section II,H).

A stand growth model, based on descriptions of individual tree growth, which uses physiological processes as the basis for quantifying competition, was produced by Mäkelä and Hari (1986). It is more soundly based, mechanistically, than any conventional forestry model or extant community dynamics model. Instead of defining a competition index, the concept of photosynthetic light ratio is applied. This is the ratio between actual photosynthesis produced under shade from other trees and the potential, or unshaded, photosynthesis. Stand structure determines the degree of shade cast on any particular tree and the individual tree model is based on the carbon metabolism of the trees, with the assumption that water and nutrient availability do not change over time. (The model was parameterized using data from trees in southern Finland. For many other areas, the assumption about water and nutrient supplies would not necessarily hold, but restrictions caused by these factors could be written into the model.) The model calculates stem biomass growth; tree death is determined by shortfalls in the rate of carbon supply. The Mäkelä and Hari model contains many high-level empirical components, including tree form equations and the carbon allocation routine, which determines the growth patterns. It has not (to our knowlege) been generalized or widely used but contains ideas that should be followed up, developed, and generalized.

II. Current Process-Based Models

We present, in the following sections, relatively detailed discussion of a number of the best known and most widely used of the current "family" of forest ecosystem models. Some of these, as well as other models not discussed here, have been thoroughly reviewed in a useful paper by Ågren et al. (1991). Many models that may have justifiable claims to be important, or that contain novel and useful routines, are not considered, but we believe that those discussed cover the spectrum of approaches to the problem of forest productivity modeling. Most of the models do not deal with the range of ecosystem processes from nutrient cycling to plant competition and species relationships and thus cannot strictly be called ecosystem models. They are generally carbon flow models dealing with different levels of detail and different time scales. The descriptions and discussion given here are based on the latest papers in the literature at the time of writing. They will soon be out of date, either because the models fall into disuse and disappear or because the models are developed, changed, and improved. Nevertheless, we are confident that the principles involved in these models are now well established, and such change and development is unlikely to involve any radical new approaches.

The treatment moves from detailed, short time-step models to simpler, longer time-step models designed for stand, forest, and regional simulation. To set them in context, the most important properties of the models discussed are outlined below; they are then discussed in some detail. Figure 9.1 provides a schematic illustration of the way these models re-

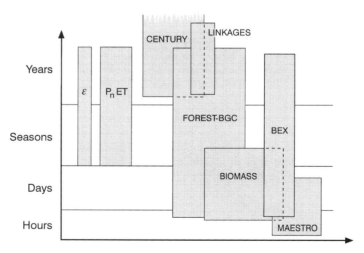

Figure 9.1 Schematic illustration of the complexity and operational time scales of the models discussed. Cross-hatching indicates common elements, e.g., the soil carbon decomposition routines in FOREST-BGC are similar to those in CENTURY; the water balance and some physiological process routines are similar in BIOMASS and FOREST-BGC.

late to one another. The diagram shows that, to deal with processes such as canopy photosynthesis in detail over short periods tends to require highly parameterized models; as the time scales expand, and the level of detail decreases, the number of parameters required also decreases. It can be taken as axiomatic that simulations concerned with large diverse areas, for which calculations may have to be done for a number of subunits (see Chapter 10) or long periods, require relatively simple models with few parameters. These should, however, be firmly based on physical and physiological principles, rather than on empirical input–output relationships, and should be consistent with the more detailed models. One of the uses of detailed, process-based models is to provide the basis for simpler models. We return to this point later.

The first model considered—MAESTRO—is a research tool, with short time steps, very detailed descriptions of canopy structure, and detailed procedures for calculating photosynthesis and transpiration rates. The next three models can be said to belong to the same group. They are all essentially mechanistic carbon balance models based on radiation interception and photosynthesis routines. BIOMASS, aimed at much longer-term simulations than MAESTRO, has far less detailed radiation interception and photosynthesis routines but includes respiration, the allocation of carbon to tree components, and a water balance module, none of which are in MAESTRO. [At the time of writing, the carbon allocation routines in BIOMASS were based on observed time series of allometric ratios (see Chapter 5, Eq. (5.22)), and not on physiological mechanisms, and are therefore empirical.] FOREST-BGC (and its derivative BIOME-BGC) has many of the same structural characteristics as BIOMASS but was designed for larger-scale (spatial) simulations. It includes a more mechanistic attempt to partition carbon, based on more detailed treatment of soil water and nitrogen availability. BEX, a model developed specifically to simulate fluxes in boreal forests, has many of the same characteristics as BIOMASS and FOREST-BGC.

The next level "down" (in terms of reduced complexity) is represented by PnET, an empirical, lumped-parameter model of the carbon and water balances of temperate and boreal forests that represents a compromise between physiological process-based and empirical models. PnET is very different in its design from BIOMASS and FOREST-BGC. The empiricism is at the level of canopy processes, carbon allocation, and stand water balances, and the model works on a monthly time step. Its development was driven by the realization that if models are to be applied extensively (over wide areas) they must require only a few widely available parameters as inputs.

Of the models we consider, LINKAGES is nearest to a genuine ecosystem model. It is among the most developed—in a mechanistic sense—of the family of "gap" models (Botkin et al., 1972; Shugart, 1984), although it uses simple, entirely empirical relationships to describe the interactions between the biological processes of reproduction, growth, and death and the effects of physical factors such as water, temperature, and nitrogen on growth.

CENTURY is a soil-based model, with a relatively crude net primary production module, based on the assumption that growth must be proportional to available nitrogen (N). CENTURY is primarily concerned with soil–organic matter decomposition and turnover.

The "end point" in the process of simplification is the ϵ model—a simple quasimechanistic (ϵ can be justified in terms of physiological processes) equation suitable for the wide-scale estimation of net primary production (NPP) (see Chapter 5). It requires very few parameters, but the values of ϵ used are of paramount importance and determine the output of the model. This model uses a monthly to seasonal time step and is amenable for use with measurements from satellites. There is currently no tested version that includes carbon allocation, although this will be relatively simple to build into the model (see Section III). We provide somewhat more extensive discussion of the ϵ model than of the others considered because we believe that it offers important opportunities for utilizing ecophysiological principles in forest management and decision making.

A. MAESTRO

MAESTRO, developed by Ying Ping Wang and Paul Jarvis (1990 a,b), is a model of an array of trees in a stand. It deals with each individual tree in considerable detail and calculates net photosynthesis and transpiration rates in canopies. MAESTRO does not include stand respiration and is essentially a research tool that allows exploration of the effects of various conditions (e.g., climate, canopy architecture) on canopy photosynthesis.

The positions of all trees are specified by their x, y, and z coordinates. Each is described by the radii of its crown in the x and y directions, crown length, measured as the distance from the top of the tree to the lowest live whorl (MAESTRO was developed for conifers), height from the ground to the crown base, and total area of leaves within the crown. The positions of leaves in the vertical and radial directions are defined by functions describing the leaf area density (LAD) distribution. The slope of the ground in the x and y directions, and the orientation of the x-axis, are also specified to take account of possible horizontal limitations caused by topography.

The spatial scale for MAESTRO for radiation absorption is a point; for photosynthesis and transpiration it is individual leaves. The point represents a canopy subvolume, within which conditions are assumed to be homogeneous. The physical and physiological properties of the leaves within each subvolume have to be specified. The model partitions incoming radiation into direct beam and diffuse photosynthetically active radiation (PAR), near infrared, and thermal radiation. MAESTRO works on an hourly time step.

The leaves within the crowns are classified into three different age classes; if information on the spatial distributions of leaves within tree crowns is not available, the leaf area density is treated as uniform or random within crowns and only the proportion of leaves in each category would be needed as inputs. MAESTRO can be run with a range of leaf angle distributions (see Wang *et al.*, 1992). Photosynthesis is calculated using the Farquhar and von Caemmerer (1982) model. This model, and the calculation of transpiration, requires stomatal conductance values (see Chapters 3–5), which are calculated using the submodel described by Jarvis (1976). [Recent versions may incorporate the model proposed by Ball *et al.* (1987) and modified by Leuning (1990, 1995)]. Leaf dark respiration rates are assumed to be functions of leaf temperature; MAESTRO does not include a solution of the leaf energy balance for leaf temperature, which is assumed uniform through the canopy, as is ambient CO_2 concentration. Transpiration is calculated using the Penman–Monteith equation (see Chapter 4), which requires net radiation flux density at leaf surfaces, the vapor pressure deficit of the air and boundary layer, as well as stomatal conductances.

MAESTRO requires a great deal of input data, including soil surface temperatures, crown structure, leaf transmittance and reflectance values for PAR, leaf age class/density distributions, and a range of physiological parameters for each age class. The inputs to the radiation partitioning model include the zenith angle of the sun and the incident radiation flux densities above the canopy. Radiation extinction coefficients are derived internally from leaf angle distribution data.

Validation of a model such as MAESTRO is difficult. The photosynthetic characteristics of the foliage are obtained from laboratory measurements, but there are few measurements of CO_2 uptake by whole trees, or canopies, that provide suitable test data. Micrometeorological (flux) measurements provide information on net ecosystem productivity (NEP), not canopy photosynthesis. In their original paper, Wang and Jarvis (1990a) tested the model in terms of radiation interception and transmittance demonstrating, for two Sitka spruce sites in Scotland, that predicted mean hourly transmittances corresponded closely to those measured by arrays of sensors beneath the canopies. McMurtrie and Wang (1993; see also Wang *et al.*, 1992) compared canopy photosynthesis rates calculated by the model BIOMASS (see Section II,B) and MAESTRO and found that they corresponded remarkably well (see Fig. 9.2). MAESTRO is more suited to running simulations for short periods, such as a single day, or at most a season. BIOMASS, with its simpler canopy

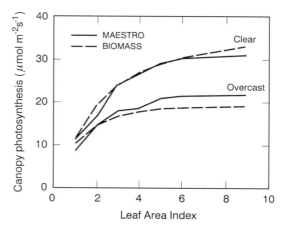

Figure 9.2 Daytime canopy photosynthesis rates calculated using MAESTRO (solid lines) and BIOMASS (dashed lines) for different values of L^*. It is clear that the less complex model gives reliable results (redrawn from Wang et al., 1992).

structure descriptions and radiation interception routines, can readily be used to simulate photosynthesis for periods ranging from days to decades. Therefore, although the comparisons carried out by Wang et al. (1992) cannot be called rigorous tests, the correspondence of the outputs from the two models reinforces confidence in both. The greater physiological detail in MAESTRO is likely to provide more accurate simulation of reality in terms of rates of canopy photosynthesis over short periods under well-specified conditions.

The obvious disadvantage of MAESTRO is the very large number of parameter values, and amount of information, required. It is not likely to be useful, or even interesting, to anyone concerned with the growth of trees and stands and the management of forests. It also has limitations in that it contains no submodel dealing with water stress—although this could be incorporated through the stomatal conductance submodel—and it is nondynamic in terms of the canopy: The trees do not grow. However, MAESTRO is a valuable research tool that can be used to evaluate the probable effects of nutrition on canopy photosynthesis (given a relationship between CO_2 assimilation rates and leaf nutritional status; see Chapter 5) to study the influence of beam and diffuse radiation on PAR absorption, photosynthesis, and transpiration (Wang and Jarvis, 1990 b) or the effects of canopy structure on photosynthesis (Wang et al., 1992). The complete canopy description contained in MAESTRO could be exploited to evaluate the consequences of management options such as various spacings in plantations

Although it may not be possible to test MAESTRO against canopy CO_2 flux measurements, the model can be used to disaggregate such data by providing reliable values for the photosynthesis term, hence allowing estimation of the total respiratory flux from a forest (NEP = photosynthesis − respiration), particularly if the understory is a minor factor. It also provides (one of) the "baseline" models that can be used to test and evaluate simpler models.

MAESTRO is completely general, it is not calibrated for any particular stand, species, or region, and it can be used anywhere in the world provided the necessary input information is available.

B. BIOMASS

BIOMASS, developed by Ross McMurtrie, is a process-based, or mechanistic, model consisting of a series of submodels that describe the operation of various physiological processes involved in the growth of trees (McMurtrie *et al.*, 1990; McMurtrie and Landsberg, 1992). The basic processes modeled are radiation interception by the foliage, carbon fixation by photosynthesis and losses by respiration, the allocation of dry matter to the component parts of the trees and, hence, the growth patterns of the trees (see earlier comment), and the effects of water availability on the growth of stands. BIOMASS now incorporates phenology and development (McMurtrie *et al.*, 1994).

BIOMASS is essentially a stand-level model consisting of a number of submodels that calculate the carbon balance of the canopy from a radiation interception model that uses information about the canopy structure and foliage photosynthetic characteristics. Tree crown shape is represented by geometrical constructions (ellipsoids, cones, etc.) and the plant community by a randomly spaced array of trees. The foliage is divided into three horizontal layers and different photosynthetic parameters can be specified for the foliage of each layer. The radiation interception model that provides the basis for the calculation of canopy net photosynthesis estimates the interception of direct and diffuse radiation by foliage (see Chapter 3). BIOMASS is less demanding computationally than MAESTRO, with more detailed descriptions of the distribution of radiation within canopies, but tests against such models indicate that the resulting errors are minor (Wang *et al.*, 1992).

In the original version of BIOMASS (McMurtrie *et al.*, 1990), leaf photosynthetic properties were described in terms of an empirical equation (the Blackman function), but in later versions (McMurtrie *et al.*, 1992a, b) they were formulated in terms of Farquhar and von Caemmerer's (1982) biochemically based model of photosynthesis in C_3 plants and the stomatal model of Ball *et al.* (1987) (see Chapter 5). Photosynthesis is calculated for both sunlit and shaded foliage.

The carbon balance of the canopy is updated daily after the calculation of maintenance respiration of above-ground tissues, which is dependent on temperature and assumed to be proportional to tissue nitrogen concentration (Ryan, 1991). In BIOMASS, empirical partitioning coefficients, which vary over the annual cycle and are linear functions of foliar nitrogen concentration, are used to allocate carbohydrates to leaves, stems, branches, and roots. Litterfall is subtracted from each component with rates allowed to vary over an annual cycle (McMurtrie and Landsberg, 1992) (see Fig. 9.3).

Stand water use and, hence, the water balance, is calculated from the Penman–Monteith equation (see Chapter 3); biological control of the process is exerted through the stomatal response model. Stomatal response is assumed to decrease in reponse to soil water deficits or overnight frost as well as (atmospheric) vapor pressure deficits.

BIOMASS requires many parameters that must be supplied by the user such as canopy architecture, the values for the parameters of the foliage photosynthetic response relationships, the stomatal conductance model, rooting depth and soil moisture retention characteristics, rainfall interception model parameters, as well as a number of physical parameters (leaf reflectance, air density, psychrometric constant, etc.) The minimum meteorological data set needed is daily precipitation and maximum and minimum air temperatures. Other information needed, such as daily total shortwave energy income, from which PAR can be estimated, can be read from files or generated from formulae provided within the simulation package. The model operates with a daily time step, but because of the nonlinear dependence of photosynthesis and transpiration on meteorological conditions the diurnal course of several variables is required to calculate daily totals.

Much of the early development of BIOMASS was in conjunction with a field experiment on *P. radiata* that included a range of irrigation and fertilization treatments (see Benson *et al.*, 1992). The parameters of the model are therefore well established for this species, and its performance in simulating the growth of *P. radiata* is encouraging. McMurtrie *et al.* (1994) have also shown that the model performs well for a range of pine species (*P. resinosa, P. sylvestris,* and *P. elliiottii*) growing in Wisconsin, Sweden, and Florida, as well as *P. radiata* growing in Australia and New Zealand (Fig. 9.3a). BIOMASS has also been successfully tested with *E. globulus* in Western Australia (Hingston *et al.*, 1995), although to get a good fit with observed growth data it was necessary to adjust the carbohydrate allocation coefficients. The values used to allocate carbon to stem growth ranged from 0.55 to 0.75 of carbon fixed, which emphasizes the need for better understanding of the factors controlling carbon allocation.

The BIOMASS model suffers from the shortcomings that it does not include nutritional dynamics and that the carbon allocation coefficients are empirical. A model developed from BIOMASS, called G'DAY (McMurtrie *et al.*, 1992a; Comins and McMurtrie, 1993), includes soil nitrogen dynamics based on litter decay processes and nitrogen mineralization rates calculated in a manner similar to those for CENTURY (Parton *et al.*, 1987; see Section II,G). Photosynthetic productivity and the dynamic behavior of G'DAY are strongly dependent on C/N pools and ratios, but the model is primarily aimed at evaluation of long-term processes and the attainment of equilibria. The physiological processes are greatly simplified (see McMurtrie *et al.*, 1992; Comins and McMurtrie, 1993).

C. FOREST-BGC

FOREST-BGC (BioGeoChemical), developed mainly by Steve Running (Running and Coughlan, 1988), is an ecosystem model designed primarily to provide estimates of carbon, nitrogen, and water cycling across forested landscapes. The key property of the model is its dependence on L^* to define canopies; the model requires no information about canopy properties such as stand density or canopy architecture. Requiring only L^*, which can be estimated by remote sensing, allows FOREST-BGC to be driven by remote sensing. Key climatic variables required to execute the model include daily maximum and minimum temperature, average relative humidity, daily total precipitation, and shortwave radiation. FOREST-BGC is associated with a group of models [Regional Ecosystem Simulation System (RESSys)] (Running *et al.*, 1989) that perform climatological simulations, on the basis of topography and prevailing weather, to provide the input data to FOREST-BGC.

FOREST-BGC simulates the carbon cycle by calculating photosynthesis, respiration, above- and below-ground net primary production, litterfall, decomposition, and leaf area index. The model uses "green sponge" canopy architecture, i.e., the canopy is treated as a homogeneous mass of thickness L^*. Because L^* is the key structural attribute of the modeled canopies, and because the model is intended for use over long time periods, the carbon available after respiration must be partitioned to leaf, root, and stem growth. Running and Gower (1991) describe the version of the model in which this partitioning is introduced: leaf, stem, and fine root growth are controlled by leaf water status and nitrogen availability. The soil carbon and nitrogen cycling subroutine in FOREST-BGC is patterned after CENTURY (see below). Litterfall and fine root turnover are controlled by functions that link the soil carbon and nitrogen cycles.

The model calculates complete stand water balances (canopy interception, evaporation, transpiration, and drainage). It has a dual time step;

canopy CO_2 and water exchange are simulated on a daily basis and summed for the year and C allocation, decomposition, and N circulation are simulated on an annual time step.

FOREST-BGC has 20 state variables and requires 41 parameter values. The state variables in the model, which describe the C and N distribution in the vegetation and soil, and the ecophysiological attributes, such as specific leaf area, maximum, stomatal conductance, leaf turnover, leaf lignin concentration, and respiration coefficients, can be modified to match the information available about particular species. Complete lists of state variables and parameters used in the static and dynamic carbon allocation versions of FOREST-BGC are provided in Running and Coughlan (1988) and Running and Gower (1991), respectively.

Comparisons between measured and calculated aboveground net primary production for FOREST-BGC are shown in Fig. 9.3b.

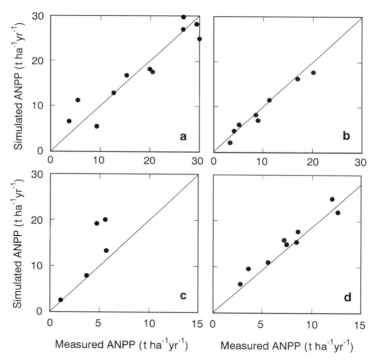

Figure 9.3 Simulated versus measured above-ground net primary production (ANPP) for (a) control, fertilized, and/or irrigated pine forests in Australia, New Zealand, Sweden, and the United States using BIOMASS (data from McMurtrie *et al.*, 1994); (b) control and fertilized conifer forests in Montana, Wisconsin, Florida, and Oregon using FOREST-BGC (data from Gower *et al.*, 1996); (c) evergreen and deciduous boreal forests in Alaska using BEX (data from Bonan, 1991); and (d) evergreen and deciduous temperate forests in the eastern United States using PnET (data from Aber and Federer, 1992).

D. BEX

The primary aim of BEX, developed by Gordon Bonan (Bonan, 1990a, 1991b), was to simulate energy, water, and carbon fluxes in boreal forests. The model has several attributes that are necessary to simulate processes in boreal forests accurately; namely, incorporation of permafrost and the explicit modeling of energy, water, and C budgets of the moss layer (see Chapter 2). No other forest ecosystem model attempts to model these two critical components of boreal forests.

The forest energy budget (see Bonan, 1991a) is computed for a multilayered forest canopy using a daily time step. Daily estimates of ground surface temperature, evapotranspiration, and snowmelt are used to determine soil moisture and soil temperature. The energy balance of n canopy layers is computed as a function of n unknown surface temperatures and the equations are solved for these temperatures using a multiple nonlinear Newton–Raphson iteration (Press et al., 1986). Environmental constraints on tree photosynthesis are mediated through mesophyll resistance, thereby including rate limitations imposed by CO_2 diffusion within the plant cells and biochemical processes of CO_2 fixation. The actual algorithms used to simulate the irradiance and leaf N concentration effects are identical to those used in FOREST-BGC. Carbon allocation is empirically derived and assumed constant. Decomposition is modeled as a function of soil moisture and temperature using empirical constants that are scaled according to forest floor N concentration.

Climate data required to execute BEX include daily averages of air temperature, relative humidity, air pressure, wind speed, and cloudiness. Twenty-seven physiological parameters that define photosynthesis and respiration are needed. Results from 200 Monte Carlo simulations for which all parameters were selected randomly suggested that 12 parameters define environmental influences on stomata and mesophyll resistance and 15 parameters define tree photosynthesis.

BEX has been rigorously tested for evergreen and deciduous boreal forests in interior Alaska (Bonan, 1991b) and used in sensitivity analyses of boreal forest ecosystems to climate change (Bonan et al., 1991). Measured and simulated ANPP did not differ for each of five forests types in Alaska (Bonan, 1991b), although BEX tended to overestimate ANPP for birch and balsam poplar forests (Fig. 9.3c).

E. PnET

PnET, developed by John Aber and Anthony Federer (1992), represents an interesting compromise between empirical models such as gap models (see LINKAGES) and process-based or physiological models. It is a simple, lumped parameter, dual time step model that simulates C and water circulation. PnET attempts to avoid the ecophysiological complexities inevitably encountered when dealing with numerous species, such as typically occur in the forests of the eastern United States; Aber and Federer (p. 464) state that "an effort has been made to retain only as much structural complexity as is required to capture the major dynamics in the systems."

PnET consists of four modules or computational components: climate, canopy photosynthesis, water balance, and carbon allocation. The model has five compartments (foliage, wood, and fine roots, soil water, and snow), three annual C fluxes (net canopy photosynthesis, and fine root and wood production), and eight water fluxes (precipitation, interception, snowmelt, snow–rain division, uptake, evapotranspiration, drainage, and fast-flow drainage).

PnET operates on a monthly time step. Input data requirements are substantially smaller than those for the models discussed previously, and there are fewer parameters, largely because of the simplistic approach to photosynthesis and transpiration calculations (see below). Only three soil parameters are needed and only one (soil water-holding capacity) is varied from stand to stand. Latitude, month, average maximum and minimum air temperature, and insolation are used to calculate vapor pressure deficit, day length, average day- and nighttime temperatures, and daily radiation. Twenty-two vegetation structure and ecophysiological variables are required, although many are generic relationships for forest ecosystems (i.e., fraction of C in tissue mass, fraction of C allocated to growth or maintenance respiration, and temperature parameters for photosynthesis). Species-specific variables include characteristics such as leaf %N, specific leaf weight, leaf longevity, vapor pressure deficit effects on photosynthesis, and transpiration.

Leaf CO_2 and water fluxes are based on two first-order ecophysiological relationships: (i) maximum net photosynthesis ($A_{n \cdot max}$), calculated as a linear function of foliar N concentration (percentage by weight); and (ii) dry mass produced per unit water transpired, known as the water use efficiency (WUE, denoted W_E), which provides a link between carbon gain and transpiration. Based on observed relationships between rates of net photosynthesis and stomatal conductance for C_3 plants, and the assumption that atmospheric CO_2 concentration can be taken as constant, WUE can be calculated from the simple relationship: $W_E = c_E/D$, where c_E is a constant and D is vapor pressure deficit.

PnET, like FOREST-BGC, uses "green sponge" canopy architecture and Beer's law to calculate light absorption by the canopy. Assuming base-rate respiration of foliage to be 10% of the maximum net photosynthesis rate, gross leaf photosynthesis (A, i.e., carbon uptake) is calculated as ($A = 1.1\ A_{n \cdot max}$). Day- and nighttime respiration are calculated as 0.1 $A_{n \cdot max}$ and modified by a Q_{10} of 2, using average monthly day- and nighttime temperatures. Maximum annual canopy photosynthesis is calculated by summing monthly canopy photosynthesis and respiration and then adjusted downward based on temperature, water availability, and vapor pressure deficit. A monthly available water balance is calculated by assuming that a constant fraction of precipitation (0.1) is intercepted and/or evaporated and a constant fraction (0.1) drains rapidly through the soil and is unavailable for plant uptake. Potential transpiration is calculated as (potential gross photosynthesis/W_E) and adjusted downward using a water-deficit scalar. Carbon is allocated to leaves, fine roots, and wood using the following algorithms: leaf C = C_f = L^*/σ_f (where σ_f is specfic leaf area; see Chapter 3); fine root C = C_{fr} = 130 + 1.92 C_f (cf. Raich and Nadelhoffer, 1989); and wood C is the difference between total C accumulated and C_f + C_{fr}.

Net primary production and water balance calculations using PnET have been validated in a diverse range of boreal and temperate evergreen and deciduous forests. A strong near 1:1 relationship was observed between measured and simulated net primary production (Fig. 9.3d) and reasonable agreement was obtained between simulated and measured stream discharge from two temperate deciduous forests (Aber and Federer, 1992).

PnET is unique among the models discussed in that it represents the first attempt to condense the numerous ecophysiological attributes of tree species to some simple "physiological principles." We believe the incorporation of such principles (e.g., relationships between leaf N and leaf longevity, and σ_f and leaf longevity; see Chapter 7) for different functional groups with a simple, but mechanistic treatment of CO_2 and water flux at the leaf level may be a useful direction for ecological modeling to follow. The empirical approach to the estimation of carbon allocation to fine roots is probably satisfactory for global estimates; however, it is unsuitable for stands that are not in steady state and is likely to provide unreliable estimates for individual stands (see Gower *et al.*, 1996a). Despite the greater temporal time step of PnET relative to other models (months compared to hours or days), aggregation of climate data at the monthly time step did not appear to have an undesirable effect on NPP estimates, although stream discharge estimates may suffer from the longer time step. PnET has no specified spatial dimension and has been applied at the stand to small watershed scale.

F. LINKAGES

LINKAGES, developed by John Pastor and Wilfred Post (Pastor and Post (1986) is one of numerous forest stand growth simulator models of the FORET family (Shugart and West, 1977) that evolved from JABOWA (Botkin *et al.*, 1972). FORET models, commonly referred to as gap models, are different from all other models discussed in this chapter because they simulate the birth, growth, and death of individual tree species based on autecological characteristics of tree species. Although now somewhat outdated, Shugart's (1984) book provides a summary of various FORET models. LINKAGES is unique among these because it links carbon and nitrogen cycles with the population dynamics of overstory tree species.

LINKAGES uses subroutines called BIRTH, GROW, and KILL to simulate the birth, growth, and death, respectively, of trees in stands that contain a number of species. All the subroutines are stochastic functions. BIRTH determines the number of individuals of each species that become established in a plot based on the autoecology of each species and four variables: availability of mineral soil nutrients, the presence of leaf litter, relative temperature, and susceptibility to mammalian herbivory. GROW is an empirical tree growth simulator, and KILL simulates tree mortality based on the maximum age reported in the literature of each tree species and average growth for the past 2 years.

The equation used to calculate tree volume growth in optimal conditions is an empirical, asymptotic relationship between leaf area per tree and diameter at breast height, with the asymptote determined by species-specific maximum values for tree height and diameter. The model incorporates the interaction of trees with one another by shading: Because tree heights are computed explicitly, the radiation that reaches a given tree can be calculated in terms of the incident radiation and the radiation absorbed by the leaf areas of taller trees. Growth multipliers are calculated in subroutine GMULT based on degree days, soil moisture, and soil nitrogen availability. The degree day multiplier is a symmetric parabolic function between minimum and maximum degree days recorded for the entire range of each species. The soil moisture growth multiplier is calculated from the fraction of growing season days in which soil moisture is less than the "wilting point" and the site water budget is calculated on a monthly time step using the water budget method of Thornthwaite and Mather (1947). The nitrogen growth multiplier is taken from Aber et al. (1979a). A decomposition subroutine (DECOMP) calculates weight loss, N immobilization, lignin decay, and CO_2 loss from decomposing litter cohorts and N mineralization, weight loss, and CO_2 loss from humus. Weight loss is calculated as a function of lignin/N ratio and actual evapotranspiration (E_a). Nitrogen mineralization is calculated as a function of litter C/N ratio and is multiplied by a decomposition scalar and an evaporation rate scalar based on degree days, soil moisture, and soil nitrogen availability.

LINKAGES contains parameters for 72 upland tree species of the eastern United States. Other data required to run the model include latitude, start and end day of the growing season, average and standard deviation of monthly temperature and precipitation, soil moisture field capacity and wilting point, and initial soil organic matter and nitrogen contents.

G. CENTURY

CENTURY, developed by Bill Parton and colleagues, (1987, 1988), is a generic plant–soil ecosystem model that simulates the water, carbon, and selected nutrient (N, P, and S) budgets of grasslands. Later versions include plant production models for forests, crops, and savannas (Parton *et al.*, 1993). CENTURY uses different plant production submodels for each of the major vegetation types and a common soil organic matter (SOM) submodel. We refer readers to Sanford *et al.* (1991) for a more detailed description of the forest version of CENTURY.

The SOM submodel divides the soil carbon into three pools: active, slow, and passive. These pools correspond roughly to microbes and microbial products, resistant plant material such as lignin and lignin-like components, and physically and chemically stabilized SOM, respectively. Carbon flow between soil pools is controlled by an inherent maximum turnover coefficient and scaled downward based on water- and temperature-controlled functions. The temperature scaling function is based on monthly soil temperature at the surface and the water availability scalar is calculated as the ratio of stored water (0- to 30-cm depth) plus current monthly precipitation to potential evapotranspiration. Microbial respiration occurs for each soil carbon transformation (active to slow to passive) and with the partitioning between CO_2 and SOM determined by soil texture. Turnover coefficients for live tissue are based on empirical monthly constants. Leaf and fine root detritus are transferred into surface and fine root residue pools and allocated to structural and metabolic residue based on the lignin/nitrogen ratio of residue material. The woody tissue components have specific decay rates and the lignin and nonlignin components are transferred to the slow and active SOM pools, respectively.

The water budget, which is used to modify SOM dynamics and plant production, is a simplified monthly time step submodel that calculates evaporation and transpiration, soil water content, snow water content, and flow of water between soil layers. Soil water-holding capacity for each soil layer is calculated as a function of bulk density, soil texture, and SOM content using the equation developed by Gupta and Larson (1979). Water loss by transpiration is derived empirically as a function of live leaf mass, rainfall, and potential evapotranspiration.

The forest production model divides trees into five components: leaves, fine branches, large wood, and fine and coarse roots. Carbon allocation is constant and empirically controlled. Like the SOM submodel, the forest production submodel runs on a monthly time step. Maximum monthly production is calculated as a function of available water and temperature. Monthly maintenance respiration for each tree component is calculated using modified (daily to monthly time step) equations in FOREST-BGC (Running and Coughlan, 1988). A Q_{10} of 2.3 is used in the model.

The SOM submodel of CENTURY has been widely used, with little or no modification, in many of the ecosystem process models that simulate soil organic matter. This submodel has been tested for a wide variety of terrestrial ecosystems and land management practices. It is particularly well suited for exploring the effects of vegetation change and land use effects on soil biogeochemistry dynamics. The grassland version of CENTURY has been coupled to geographical information systems (GIS) and used to simulate regional grassland dynamics (Burke et al., 1989), although we are not familar with the application of the forest version to large spatial scales. The long time step (monthly), empirical treatment of the water budget, and forest production submodels are less desirable than the more mechanistic treatments of forest water and production budgets found in many other models.

H. The Radiation Utilization Efficiency (e) Model

To estimate forest growth over large areas and long time intervals, we need a simple model with a small number of parameters, preferably with robust, conservative values. The detailed, process-based and generally highly parameterized models that we have discussed so far are generally not suitable for widescale use. If they are applied to large areas, it is inevitable that considerable simplifications will have to be made and that parameter values will often be approximated by very general mean values. All this tends to reduce the value of having a detailed model while probably doing little to reduce the computational requirements and the time that must be spent setting the model up. Nevertheless, the simple model needed for practical applications and estimates of CO_2 sequestration by forests over large areas must be soundly based on physical and physiological processes.

Such a model exists. It is essentially a linear relationship between the photosynthetically active solar radiation absorbed by forest canopies and the production of dry mass by forests, i.e.,

$$\text{NPP} = \epsilon \, \Sigma \varphi_{p \cdot a}, \tag{9.1}$$

where NPP denotes net primary dry mass production (see Chapter 5) and $\varphi_{p \cdot a}$ denotes absorbed photosynthetically active radiation over some interval that, in the case of forests, may be a year, although there is a good case for working with seasons. If $\varphi_{p \cdot a}$ is expressed in MJ m^{-2}, and NPP is dry mass, ϵ will have units of mass MJ^{-1}. It is usually expressed in g MJ^{-1}. There is a range of values in the literature (see Russell et al.,

1989; Prince, 1991), some of which are expressed in different units; one of the more common sources of confusion is the use of incident or absorbed solar radiation rather than absorbed photosynthetically active radiation. The problem is discussed in some detail by Landsberg et al. (1996).

The ϵ model is soundly based on physical and physiological processes. The rate of carbon fixation by a plant canopy depends on the radiation absorbed by the canopy and the photosynthetic characteristics of the leaves. The change in the rate of photosynthesis by a single leaf with increasing photon flux density (φ_p) is nonlinear; however, if a canopy is closed so that the foliage intercepts all or most of the incident radiation, and much of the foliage is not photon saturated, photosynthesis by the canopy as a whole is likely to be photon flux limited and the relationship between intercepted energy and net photosynthesis by the canopy tends toward linearity. This tendency is strengthened if a high proportion of incoming radiation is diffuse. Jarvis and Leverenz (1983) demonstrated that the rate of net photosynthesis by a spruce canopy tended toward linearity with φ_p, largely as a consequence of shoot and canopy structure, and an elegant demonstration of the tendency toward linearity, as the period across which integration is carried out increases, was provided by McMurtrie et al. (1992a) using BIOMASS. They showed that, on a daily basis, there was considerable scatter between simulated daily photosynthesis and $\varphi_{p \cdot a}$. However, the variability collapsed to a single, approximately linear, relationship between canopy photosynthesis and $\varphi_{p \cdot a}$ when expressed in annual terms, providing a clear demonstration that the simplified relationship encapsulates the complexities of the photosynthetic process and the interactions of directly illuminated and shaded foliage. (It is commonly found that integrating over time or space scales reduces nonlinearities.)

As we have noted in Chapter 3, radiation absorption by plant canopies has been exhaustively analyzed and modeled in great detail (see, for example, Wang and Jarvis, 1990a; Wang et al., 1992, and references cited therein), but for most purposes, particularly when integrating over time periods longer than a few days, simple models provide adequate descriptions of photon flux interception. The $\varphi_{p \cdot a}$ values needed for the ϵ model can therefore be obtained from the simple Beer's law equation [Eq. (3.8)], converting solar radiation into PAR. (The conversion is generally between 0.45 and 0.5 $\times \varphi_s$.) It is clearly essential, as a basis for using the ϵ model, that we have good estimates of L^* for any stand for which we wish to make productivity estimates (see Chapter 3).

For individual stands, L^* can be estimated by various means (see Chapter 3), including nondestructive methods based on light interception and inversion of the equations describing light interception by canopies, and from relationships between waveband reflectance ratios and L^* (see Chapter 10, Section II and Fig. 10.2). However, if it were possible to estimate $\varphi_{p \cdot a}$ directly there would be no need for L^* values (at least for the purposes of this model). Satellite measurements of the reflectance properties of surfaces also provide an opportunity to estimate $\varphi_{p \cdot a}$ and Sellers *et al.* (1992) have moved in this direction, showing that $\varphi_{p \cdot a}$ is proportional to a simple waveband reflectance ratio (SR = near infrared/ infrared). Hall *et al.* (1990) proposed the use of the second derivative of the reflectance/wavelength function obtained from narrowband spectra to estimate $\varphi_{p \cdot a}$. At the time of writing this book, these relationships were being tested in the field, and we can expect them to become important, particularly in the area of large-scale (regional) and global carbon modeling.

The central problem of the ϵ model is, of course, the value of the coefficient. The model was first proposed in the form given here by Monteith (1977) and has been used and tested with a number of plant communities. Values of ϵ obtained from experimental work vary widely: The upper limit, for intensively managed field crops, appears to be about 2.8 g MJ^{-1} (Russell *et al.*, 1989); young, well-watered tree canopies, with good nutrition, give values of about 1.6 g MJ^{-1} (Cannell *et al.*, 1987, 1988), whereas values obtained for above-ground productivity by plantations and forest canopies are in the range 0.7–1.5 g MJ^{-1} (Linder, 1985; Runyon *et al.*, 1994). The lowest reported ϵ values (≈ 0.2 g MJ^{-1}) appear to be those derived by Saldarriaga and Luxmoore (1991), who investigated the productivity of tropical rain forest stands at 23 sites in Colombia and Venezuela. It is unlikely that environmental variables can account for all the variation in ϵ values; there appears to be a consistent reduction in ϵ with age, illustrated by Landsberg *et al.* (1996) from values in the literature. This has not yet been satisfactorily explained (but see Chapter 8 for a possible explanation).

A development that holds promise for this model is the use of modifiers so that Eq. (9.1) is written

$$\text{NPP} = \epsilon \Sigma \, \varphi_{p \cdot a} \, f_\theta \, f_D \, f_T, \tag{9.2}$$

(see Landsberg, 1986). The modifying factors (f_i) describe, respectively, reductions in the effectiveness of unit $\varphi_{p \cdot a}$ resulting from soil water deficit (Θ), the vapor pressure deficit of the air (D), and temperature (T), all of which affects stomatal conductance. The $\varphi_{p \cdot a}$ value resulting after modification may be regarded as utilizable radiation. A good example of the use of this procedure is provided by McMurtrie *et al.* (1994), who simulated gross primary production (GPP) of pine trees at sites in Australia, New Zealand, the United States (two locations), and Sweden and plotted the results against utilizable $\varphi_{p \cdot a}$ (see Fig. 5.3). They obtained a very strong positive relationship from which they derived $\epsilon = 1.77$ g C MJ^{-1}. There is a good argument for using values of $\varphi_{p \cdot a}$ corrected in this way to derive values of ϵ, although there is a clear need for rigorous research on the calculation of modifier values that are appropriate to the time scales used. The use of unmodified values of $\varphi_{p \cdot a}$ to derive ϵ will continue to result in wide variations because the effects of factors that alter (reduce) the capacity of trees to convert radiation into dry mass are incorporated in the estimates. It is also important to remember that values of ϵ derived from above-ground data will be subject to the unknown variation caused by differences in the amount of carbohydrate allocated below-ground. However, there is now sufficient information about root mass and turnover, and the effects of soil and growing conditions on the allocation of carbon to the roots, to allow useful working estimates to be made.

The ϵ-model implicitly assumes that respiration is not a major cause of variation in the estimates of dry mass production. Prince and Goward (1995) and Waring *et al*. (1995) used the quantum efficiency of leaf photosynthesis, with modified $\varphi_{p \cdot a}$, as the basis for estimating GPP. This solves the problem of ϵ values, but requires that the assumption about respiration holds, which it may well do over relatively long periods.

III. Practical Applications

On the global scale, Ruimy *et al*. (1994) used a biome map of the world and obtained average empirical values of ϵ from the literature, as far as possible, for each biome type, and the ϵ model has been used by Potter *et al*. (1993) to calculate seasonal patterns of net ecosystem productivity across the globe. At the level of local forest management, we contend that a model of this type, based on sound physiological processes (energy absorption and photosynthesis), provides a means of incorporating nutrition and water status into productivity models and is potentially far more useful than statistically based models and growth curves. To realize this potential requires that forest growth rates and yield are either described in terms of conventional volume growth, which can be converted into information about economic timber yields on the basis of stand density and tree height, or that dry matter can be partitioned between stems, branches, and leaves on the basis of allometric ratios.

To describe growth in volume terms simply requires that ϵ be in those units for which values could, presumably, be determined empirically by plotting volume growth increments against $\varphi_{p \cdot a}$. Equation (9.2) would be applied, based on calculated water balances, soil nutrient status, and temperature, with the procedures used to calculate the modifiers being guided by sound ecophysiological principles as discussed in earlier chapters. This has advantages from the practical point of view, but has the major disadvantage that the results cannot be compared to other dry matter production research; the model is reduced to a level of pragmatic empiricism that greatly reduces its value, although useful information can still be obtained.

From the ecological point of view, estimates of biomass production may be the required end point in a modeling exercise, but production forestry requires results in terms of stem volumes. These can be obtained from estimates of growth expressed as dry mass per unit land area by using allometric ratios [Eq. (3.2)] and the procedure described in Chapter 5, Section II,B). Given values for c_i and n_i, the equation can be used to partition total above-ground dry mass $W(t)$ into its component parts:

$$W(t) = w_f + w_{br} + w_{st}, \qquad (9.3)$$

where w_f, w_{br}, and w_{st} denote the mass of foliage, branches, and stems, respectively.

If stem number per unit area in the stand is known, then average stem mass is known and average stem volume is obtained from a value of wood density. More detailed analysis could include the use of standard forestry stem size distribution equations (e.g., the Weibull distribution) and taper equations, which would allow detailed estimates of the value of the product.

The analysis suggested previously clearly depends on values of the coefficients in the allometric equation, but these are readily obtained from analysis of destructive harvest data (see Chapter 5). If the procedure outlined is developed, it will provide the method we need as a substitute for the site index models (see Chapter 10). The calculation of absorbed PAR can be made as simple or as detailed as is felt necessary; research must focus on the characterization of the modifiers f_θ, f_D, and f_T [Eq. (9.2)] as well as a nutritional modifier f_N. The form of these modifying functions will be guided by physiological considerations; f_N may be based on calculations of nutrient uptake rates, particularly in the case of nitrogen, in which a model such as CENTURY can be used. Otherwise, it may be necessary to develop empirical relationships with some measure of fertility such as litterfall nitrogen or nitrogen mineralization. f_θ would be based on soil water content, derived from water balance modeling and (possibly) relationships between integrated predawn water potential (Myers, 1988) and growth; f_D and f_T would be based on physiological considerations (see Runyon *et al.*, 1994; McMurtrie *et al.*, 1994). All the modifiers would be normalized relative to some value at which the factor concerned is considered not to limit growth. From the experimental point of view, it will always be of value to have data from a site or sites at which, as far as possible, all constraints to growth imposed by water and nutrition are removed by fertilization and irrigation (see Landsberg, 1986; Benson *et al.*, 1992). Testing of the model will take the form of seeking linear relationships between utilizable absorbed PAR and the appropriate measure of yield (see McMurtrie *et al.*, 1994). The model should be tested against growth and yield data from a wide range of sites with different growing conditions.

Whether any particular forest management system is sustainable in the long term, or will lead to forest degradation, will always be very difficult to answer experimentally. Models provide the only feasible approach to the problem. The approach used by Dewar and McMurtrie (1996a,b), although conceptual rather than (as yet) practical, illustrates the information that can be obtained from models that deal with soil organic matter and nutrient mineralization (see also Rastetter, 1991), the uptake of nitrogen, and the influence of nitrogen supply rate on carbohydrate partitioning and tree growth patterns. Dewar and McMurtrie analyzed the consequences of fire through its effects on soil nitrogen reserves and, hence, nitrogen supply rates. Developing such approaches so that the models can be parameterized for particular stands, and the consequences of particular management practices (e.g., fertilization, rotation length, residue treatment, etc.) assessed, will provide a tool for evaluating sustainability. However, we must emphasize the (well-recognized) danger of accepting results from models that are elegant and convincing without adequate testing of those models. There is much to be done in this respect in forest production modeling.

Another area in which process-based models can be of particular value for stand-level management is the analysis of the decrease in NPP with stand age. This is a major difference between the empirical yield models and the process models: The yield models are based on age-sequence data, whereas the process models have generally (and necessarily) been developed and parameterized in relation to intensive, short-term field experiments. The optimum point to harvest a wood-producing stand is at the time of maximum mean annual increment (a conventional measure of forest growth). The time when this occurs is strongly dependent on the growth curve of the stand, the shape of which is, in turn, dependent on current growth rates. Few models currently provide much insight into the decline of growth rates with age, but as the appropriate mechanisms are built in, and information on physiology and community dynamics is collected, our ability in this area will improve and the value of the models will increase. Progress in this area has been made by Murty *et al.* (1996) and Gower *et al.* (1996).

IV. Concluding Remarks

A number of points emerge from this brief review and evaluation of some of the current generation of forest ecosystem models. The first question that might be posed is: Do we need so many models? Would it not be possible to amalgamate some of those that have been developed into fewer, soundly based and widely used versions?

There are arguments for and against this. In favor of such amalgamation is the fact that some procedures (photon flux interception and photosynthesis calculation, water balance calculations, and, increasingly, SOM turnover and N availability calculations) are very similar in all models—the principles are established, only the details vary. Therefore, if the algorithms and structure of the models dealing with these processes were available researchers could concentrate on work to improve understanding of the processes and examine the factors causing them to vary. If they find that the algorithms are inadequate to describe the processes, this in itself constitutes progress. It is possible to obtain the code from the authors of most models, parameterize the models for the situations of interest, and run them. Problems arise from the fact that the computer codes, having been written by research scientists, tend to be somewhat idiosyncratic; they are not commercial software and, although the scientists are generally willing to make the code available, they cannot provide support for "debugging" or spend a great deal of time sorting out problems that may arise in parameterizing the models and adapting code to different machines. However, these are technical problems and can be overcome. They will be overcome more rapidly if complex models are developed, as far as possible, in modular form. Landsberg *et al.* (1991, p. 13) commented that "it is time for rational reconstruction of model development, which must include technology for easy visualization and flexible graphics. The challenge is for information technologists, but scientists concerned with biophysics will have to work with them, at least initially."

The main decision to be made in relation to models relates to the purpose of the model and the level of detail needed. Is it a research tool in itself, or a framework for examining hypotheses about particular parts of the system? In the research area, it is likely that the proliferation of models will continue for some time because each scientist or group has particular requirements and tends to have a particular view of problems. It is important that the use of models in a practical sense is not stopped by the fact that there is uncertainty about various aspects of tree and forest growth. There will always be uncertainty; we need to move forward using the knowledge we have while recognizing that it is, and will remain, incomplete and, in some senses, inadequate. Meanwhile, empirical allometric relationships may prove to be surprisingly stable, or it may prove possible to explain variations in terms of clearly identifiable factors or conditions.

Nevertheless, we sugggest that the family of carbon balance models, ranging from MAESTRO through BIOMASS and FOREST-BGC to G'DAY, if made widely available would obviate the need for development of much new software in the radiation interception, photosynthesis, respiration, and water balance areas. It seems unlikely that we need a great deal more detailed research on stomatal functioning and the development of stomatal response models for tree and stand models; there are much larger uncertainties about other parts of the system. Continued research will be needed in relation to carbon allocation (to foliage, branches, stems, and roots—including fine root and mycorrhizal turnover), respiration, and even factors such as the influence of changes in foliage photosynthetic capacity with tree aging. Description of tree growth in terms of carbon allocation is crucial if models are to be used in a dynamic sense over long periods of time. If we assume that systems are in steady state, then we can make the carbon balance calculations on the assumption that, over periods of the order of seasons, foliage mass does not change. However, if we are dealing with young plantations, or regenerating forests, carbon allocation patterns determine the behavior of the system.

Models such as PnET warrant increased attention. In this respect, we note that the procedure of using nested models of different levels of complexity should be widely adopted. Landsberg *et al.* (1991, p. 9) pointed out that "detailed process descriptions . . . may have considerable value as a means of testing simpler, stand-level models." The procedure has been used by McMurtrie *et al.* (1992a), who used a strategy of developing a series of nested models of widely differing levels of resolution. They said

> "models high in the hierarchy (more detailed) aid in the understanding of biological processes underlying ecosystem functioning, and also aid the development of simpler models which are more likely to be useful in ecosystem management. . . . Data and models at each level of the hierarchy are employed in parameterisation and validation at the next level down."

McMurtrie *et al.* (1992a) and McMurtrie and Wang (1993) also compared MAESTRO and BIOMASS and found the outputs over periods of a day to be very similar.

Given that the more detailed and complex models are well founded in terms of the processes simulated and, as far as possible, well tested, this provides a basis for confidence in stand and regional scale models. Regional scale models, such as PnET and the ϵ model, can be tested in relation to detailed models by the "mosaic" method. Forests can be described in spatially organized data bases (GIS; see Chapter 10) that can hold the state variable and model parameter values for each land unit deemed to be homogenous. Remote sensing provides a means of examining large areas and describing their properties in terms of variables that can be used in models (see Sellers *et al.*, 1992). If a landscape consists of a series of units (elements), each of which can be described in enough detail to provide the information needed to apply one or more detailed models, then the procedure would be to calculate the output of interest (e.g., net CO_2 assimilation over a period) for each element for the time interval of interest. Area-weighted summation (the output from each element is weighted by its area as a fraction of the total landscape area) over the landscape gives the total output. The same output is calculated for the whole landscape using the lumped-parameter, simple model. If the lumped parameters of the simpler model(s) are biophysically realistic averages of the corresponding parameters in the detailed model(s), the results ought to match. In the case of PnET, it would be necessary to use several detailed models to simulate the carbon and water balance results. This test exercise could be carried out with simulated landscapes and stands—it would not be essential to use real, measured values.

The "gap," or community dynamics models, should receive continued attention; the obvious move in this area is to combine this type of model with the physiologically based carbon balance models. A start on this has been made by Friend *et al.* (1993) and the approach used in LINKAGES is certainly worth developing. A modification that might be suggested would be to calculate the (net) carbon balance of a stand, using a carbon balance model, and allocate the carbon on an annual or seasonal basis to the component population classes. Growth would then be calculated by converting carbon to wood volume. In other words, LINKAGES would become more process based.

When we turn to management questions, the need for simple, easily parameterized models becomes clearer. FOREST-BGC has been deliberately designed—at least in its later versions—to be driven through L^* and, as noted earlier, the values of the state and ecophysiological variables can be readily adjusted to those applicable to particular forest ecosystems. It appears, therefore, to have considerable potential as a management tool. This also applies to the ϵ model: We are of the view that there should be strong focus on research into variations in ϵ, particularly into ϵ values derived using estimates of utilizable radiation (see McMurtrie *et al.*, 1994). This provides a clearly defined framework for investi-

IV. Concluding Remarks

gating the effects of environmental factors on dry mass production. The importance of improved knowledge about carbon allocation, to provide the basis for estimating tree growth, as opposed to stand carbon balance has already been discussed.

Up to this point, we have focused largely on the use of process-based models to simulate forest growth and carbon allocation. However, many of the models reviewed previously have been used for other purposes, including the examination of the effects of different vegetation cover types on the water and nutrient budgets of forest catchments, determining the effects of different forest harvest intensities and rotation lengths on soil organic matter turnover and nutrient budgets, and exploring the effects of different disturbance regimes on species diversity. Such uses provide another reason why process-based models are likely to replace empirical growth models, a trend likely to be strengthened by the increasing emphasis now being placed on managing forests on a sustainable basis (Lubchenko *et al.*, 1991). We provide two examples of the use of process-based models to examine the effects of different land use practices on the long-term sustainability of forest ecosystems.

As discussed in Chapter 7, there is great concern about the effects of different forest harvesting intensities (e.g., stem only versus whole tree) and rotation lengths on long-term productivity of sites (Hook *et al.*, 1982). Empirical soil fertility and growth data for forests that have been subjected to different tree harvesting intensities and frequencies for several rotations simply do not exist, and even if the data existed they would be site and nutrient specific (see Chapter 7, Table 7.7). Aber *et al.* (1979a, b) developed one of the first process-based models to examine the effects of two different forest rotation lengths on soil organic matter and nitrogen budgets for northern hardwood forests and found that decreasing the rotation length from 90 to 45 years resulted in a dramatic reduction in soil nitrogen availability. Dewar and McMurtrie (1996a) used the G'DAY model and a graphical analysis approach to determine whether different forest utilization intensities were sustainable based on whether soil organic matter and nitrogen poools were in equilibrium. The use of graphical analysis is particularly appealing and has been shown to be useful in identifying ecophysiological constraints on forest growth in reponse to climate change (McMurtrie and Comins, 1996) and forest aging (Murty *et al.*, 1996).

In some regions of the world, forest catchments are managed for water yield, i.e., the primary objective is to maximize water yield, not timber. Process-based models that incorporate canopy interception and evaporation subroutines, such as those discussed in Chapter 4, have been shown to simulate water yields accurately (Harding *et al.*, 1992; Vertessy *et al.*,

1993). Other important issues related to forest hydrology concern the effects of land use practices (Bruijnzeel, 1990; Fahey and Rowe, 1992) and climate change on water yield. Our understanding of the hydrologic cycle is good enough to allow the use of process-based models to examine the effects of management practices without actually implementing those practices and waiting to study their reponse (Bormann and Likens, 1979).

Recommended Reading

Aber, J. D., and Federer, C. A. (1992). A generalized, lumped-parameter model of photosynthesis, evapotranspiration and net primary production in temperate and boreal forest ecosystems. *Oecologia* **92**, 463–474.

Dale, V. H., Doyle, T. W., and Shugart, H. H. (1985). A comparison of tree growth models. *Ecol. Model.* **29**, 145–169.

Landsberg, J. J., Kaufmann, M. R., Binkley, D., Isebrands, J., and Jarvis, P. G. (1991). Evaluating progress toward closed forest models based on fluxes of carbon, water and nutrients. *Tree Physiol.* **9**, 1–15.

McMurtrie, R. E., Rook, D. A., and Kelliher, F. M. (1990). Modelling the yield of *Pinus radiata* on a site limited by water and nitrogen. *For. Ecol. Manage.* **30**, 381–413.

Pastor, J., and Post, W. M. (1986). Influence of climate, soil moisture, and succession on forest carbon and nitrogen cycles. *Biogeochem.* **2**, 3–27.

Running, S. W., and Coughlan, J. C. (1988). A general model of forest ecosystem processes for regional applications. I Hydrologic balance, canopy gas exchange and primary production processes. *Ecol. Model.* **42**, 125–154.

Running, S. W., and Gower, S. T. (1991). FOREST-BGC, a general model of forest ecosystem processes for regional applications. II Dynamic carbon allocation and nitrogen budgets. *Tree Physiol.* **9**, 147–160.

10

Applications of Modern Technology and Ecophysiology to Forest Management

We discussed in Chapter 1 the various levels and types of forest management and noted some of the decisions that have to be made by managers. Decisions are generally guided by policy, and management tries to achieve specified objectives, which may relate to wood production, or multiple resource use, or the preservation of the forests in some relatively undisturbed form. Whatever the policy and the objectives of management, it is essential that the policymakers and managers have accurate information about the forests: what is in them, their condition, and the historical changes and impacts of events such as drought, fire, disease, or management practices. This chapter is concerned with some of the important tools available for obtaining, storing, and manipulating that information. They are Geographical Information Systems (GIS) and remote sensing technology. Models provide the means of manipulating information and making the predictive calculations so essential for decision making; in the penultimate section of the chapter, we work through a scenario in which GIS is used as a basis for forest modeling in a way that would be of value to those concerned with the management of forests for timber production, although examples of other management aims could have been chosen.

GIS is a spatial data management system with an organizational structure that enables collecting, storing, retrieving, transforming, combining, analyzing, and displaying spatial data representing geographic properties or ecological processes. It is, obviously, computer based, and there are a number of commercial packages available. We provide some information about GIS and what it can do but have made no attempt to describe in detail how to use it. We have also made no attempt to review, in any detail, the literature on remote sensing and its use in relation to forests; our intention is simply to outline what can be expected of this technology and how it can be used in practical, management contexts. Some relevant studies are cited to provide examples and illustrations of the points made. A useful introduction to GIS is provided by Burrough (1989) and to remote sensing by Lillesand and Kieffer (1994).

Although each can be used alone, there is a strong synergy in using GIS, remote sensing, and models together. GIS is valuable as a mapping tool and for information storage, but its full potential is realized when it is used to manipulate data and apply models to each element (e.g., the atmosphere, the soil, etc.) of the system. Remote sensing provides information about land surfaces and their attributes, which can supply the input data for ecophysiological models such as those discussed in Chapter 9. These all require as inputs information about the stand to be modeled, the soil characteristics, and weather conditions. Stands vary spatially and temporally in their characteristics; therefore, to explore the effects of particular weather patterns or treatments or events across a forest consisting of a number of stands, models have to be run with a range of starting and environmental conditions, corresponding to the areas of interest. Satellite remote sensing provides the opportunity to obtain sequential measurements that allow evaluation of changes over time.

I. Geographical Information Systems

The fundamental task of a GIS is to record the location of elements of geographical data such as survey sites, roads, rivers, areas of particular forest types, and so on. To be useful, these locations must be defined in terms of a common coordinate system. Usually, it is a geographical coordinate system, such as latitude and longitude or UTM/AMG grid references, but any x,y frame of reference can be used: distance north and east from a survey point or the row and column position in a matrix, for example.

Most geographical features can be represented in a GIS as a point, line, or area. As well as a location, each feature will have one or many "attributes" associated with it—nonlocational characteristics and values that describe what the feature represents and what information is associated with it.

The representation of spatial data in a GIS will depend on the "data model" it uses, the two most common being vector and raster. In the vector model, the location of points, lines, and areas is described by single or multiple coordinate pairs representing the ends of line segments. To describe areas, the lines are closed to form a polygon by giving the first and last points the same location. In the raster model, the data plane is divided up into a rectangular matrix of cells. Points are represented as a single cell, lines by adjoining cells with a maximum width of one cell, and areas by adjoining cells with a width greater than one cell. In this model, location is implied by the row and column position in the matrix and discrete geographical features are defined by assigning the same identifier attribute to cells representing part of the same thing. Originally, the choice of which model to use was a decision made by GIS developers in an effort to balance computational efficiency with the anticipated needs of the user. The most efficient model for an application will depend on the type of data to be dealt with, its spatial variability, the spatial accuracy required, the quality of output required, and the type of operations to be performed on the data (Aronoff, 1989).

A typical example of a "trade-off" between raster and vector systems involves the information content of a scene and how it is split up between recording locations and attributes. In the vector model, locations take more memory to store because they are specified explicitly, but the attribute of even the largest feature is only specified once. When a scene is made up of a small number of features (such as a road and river network and the boundaries of large areas of common forest type and soil), the vector model can store the information in the scene more efficiently. In a raster system, the attribute of a feature may be recorded many times (possibly once for each cell it covers), but the locational information for each cell making up a feature is implied by its relative position in memory so that it comes "free." If there are many small features to account for, the locational information content is high, and the need to specify the position of each feature explicitly becomes a burden. In this case, the raster model can represent the scene more efficiently. A good example of raster efficiency is in the storage of remotely sensed images. Each pixel in a remotely sensed image has the potential to create a separate feature requiring description, especially when the landscape has a high degree of spatial variability, and the pixel brightness (i.e., the attribute) can take a large range of values. In this case, raster-based representation of the data is usually more efficient because vector representation might mean defining the corners of several million spatially discrete elements.

As a result of the decisions made during their early development, most GISs are either dominantly vector or dominantly raster based. In practice, however, there are a number of conversion and information storage algorithms that can be used to get around the disadvantages of both models, allowing modern GISs to deal efficiently with both types of data.

Data, which will usually (but not always) be in digital form in a GIS, are not necessarily the same as information. The dictionary definition of data—a plural word—is "known facts or things, used as a basis for inference or reckoning." The most useful data in a GIS are primary data, which may include things such as biomass per unit area, soil physical characteristics (e.g., saturated hydraulic conductivity), leaf area index, tree height, average rainfall or temperature for a given period (month, season, etc.), and slope or aspect of the land. All these can be expressed as numbers that have units associated with them. Data may also be expressed in words, e.g., soil color or texture and vegetation type (dominant species, etc.). Primary data may be site based (so-called "point"), line based (e.g., rivers, roads, etc.), or polygon based (for example, in the case of landform patterns, or forest boundaries primarily determined by natural features).

Information, or information products, is derived from data; therefore, information may also be data. Comparative information may not be data, unless expressed in quantitative terms; e.g., "polygon a is wetter than polygon b" is an item of information, but is not data. Information about forest type (e.g., temperate deciduous) is not data, although it can be derived from data, for example, about climate and L^* values for each month of the year. Temperature and precipitation data may indicate that a region is climatically temperate, whereas data on L^* for forests in the area, which show that it falls to zero in the winter months and reaches a peak in late summer, indicate that the forests are deciduous. Information products, such as maps, graphs, or tables, may be derived from a particular data layer or by manipulation of several data sets.

The integration of spatial and attribute data in a GIS is often thought of in terms of map layers—overlays having a common theme such as "soils," "topography," "hydrology," "infrastructure," "forest type," and so on, with each layer stored as a separate file. This composite map approach is a natural format for raster-based GISs with each cell in each layer having a single attribute. It is conceptually easy to understand and parallels the data assembly process in which a digital elevation model may be acquired from one source, a soil map from another, and a remotely sensed image from another. A recent approach common to vector-based systems is the georelational model of data integration. In this model, every feature is given a unique identity and the links between attributes and locational information, and among sets of attributes, are established in a relational database (Shepherd, 1991).

In practice, the distinction between vector- and raster-based systems is becoming less and less clear as the technology advances. The capability of dealing with all types of data efficiently is built in, and data base management has become an integral part of the modern GIS as well as a science-based discipline in its own right. Advances in the development of 3D GISs have also added the surface to the list of basic geographical features that can be represented.

An important concept in relation to GISs and geographical data is scale, a term that is often confused and misused in the move from mapped data to digitally stored data. Data stored in a GIS are essentially scaleless. The scale (in the map sense) is continuously variable and depends only on the zoom factor applied when viewing the data on the screen or plotting them out on paper. The analogy is to make a photocopy enlargement of a map, which changes the scale of the original map by the enlargement factor. A more useful spatial measure than scale is resolution, the size of the smallest discernible feature. In aerial photography, this is related to the grain of the film, in satellite imagery to the pixel size on the ground, and in mapping to the minimum mapping unit decided by the cartographer during mapmaking. In the case of maps, this depends on the amount of detail to be shown that in turn is related to the scale of the map and its expected use. In a raster application, the resolution will be the cell size of the matrix (which is decided arbitrarily by the user and may be smaller than the resolution of many of the data sources). In a vector-based application, it will be the length of the smallest line segment used to describe part of a feature, which will depend not only on the scale of digitized maps but also on the method of digitizing (e.g., the resolution of the scanner or the skill of the pen operator). Other considerations are the size of the display monitor (in pixels) and the resolution of the printer (in dots per inch) that ultimately define the look of the data as an output product. It is important to remember that source data digitized at different scales will have different inherent resolutions. This should be considered (among other things) when defining the precision and describing the accuracy of spatial locations in a GIS data base.

The question of scale in relation to GIS—or indeed any mapping system—highlights the problem of scale and variability in relation to forest modeling. The problem applies as much to conventional (site index)-type modeling as to the use of process-based models. The central question relates to homogeneity: What variation can be accepted within an area considered to be a single (homogeneous) unit? The answer depends, to a large extent, on the purpose of the exercise. If we are concerned with yield estimation across a series of forest blocks, where differences of, e.g., 10–15% in annual growth rate would be considered significant, we may work at stand level, where the variation between the site variables is "acceptable"—by some criterion that would have to be decided. If we are concerned with landscape-scale analysis, the criterion for homogeneity will certainly be less rigourous and we are likely to accept composite parameter values that represent the landscape as a whole.

GISs are particularly well suited to the quantitative analysis of landscape structure. Applying the concepts and principles of landscape ecology to assessing the consequences of disturbance is useful for designing and managing landscapes for conserving biological diversity (Noss, 1987; Turner *et al.*, 1989) and increasing our understanding of the impacts of practices such as forest harvesting on landscape-scale ecological processes

(Franklin and Foreman, 1987; Ripple *et al.*, 1991; Mladenoff *et al.*, 1993). Approaches to the analysis of historical changes in land use range from overlaying multidate digital maps (Mladenoff *et al.*, 1993) to using remotely sensed data in a GIS to illustrate changes in forest cover and fragmentation over time (Hall *et al.*, 1991). Studies of this type invariably show increasing rates of forest fragmentation and nonuniform disturbance patterns; other studies have shown that forest fragmentation affects wildlife, especially birds, and needs to be considered if biodiversity, which is a component of sustainability, is to be maintained.

II. Remote Sensing

Remote sensing includes measurements made by instruments mounted on aircraft or satellites. Strictly speaking, aerial photography should be included in the general category of remote sensing because it is a means of obtaining data and information about a region. Photographs can be precisely referenced to objects on the ground (georeferenced), and aerial photography and photogrammetry, involving the use of stereoscopic analysis, have been widely used in forestry as mapping tools. Because vegetation reflects infrared radiation strongly, infrared photography can provide a considerable amount of information about vegetated surfaces. The information obtained from photographs can be digitized and entered in a GIS; the procedure is relatively laborious but the combination of high-quality aerial photography, to provide the baseline maps, and satellite remote sensing can provide a remarkable amount of information about an area.

Aircraft have a great advantage over satellites because they can be flown, and pictures taken or measurements made, at times when the weather is suitable and the targets are clear. As a research tool, the use of multichannel spectroradiometers on light aircraft has considerable potential; they can provide small spatial resolution and high spectral resolution (Waring *et al.*, 1995). It is unlikely that such equipment will become a management tool in the foreseeable future, although there is no reason why enlightened and forward-looking managers, who can command the resources, should not consider them. As an example, a multispectral scanner was used to help monitor and track wildfires in New South Wales in 1993.

Instruments on satellites have the advantage that they can provide data over long periods of time at intervals depending on the satellite's flight path and return time. Disadvantages include the need to correct the data for view angles and atmospheric turbidity and the losses of data caused by cloud. Corrections for atmospheric turbidity are often not made if

II. Remote Sensing

there are no reference data available. To overcome some of these difficulties, images are often composited—the strongest and best quality signals for different parts of an area, obtained on different occasions, are combined to produce a composite image considered to provide the maximum amount of information. Compositing over periods of 2 to 4 weeks is common. Satellite data are now readily available and relatively widely used. Training is required for their processing, which must include care in the georeferencing of data, i.e., ensuring that there is accurate correspondence between ground features and the pixels containing the information obtained from the areas of interest on the ground.

The instruments (usually passive sensors) mounted on satellites measure the flux of electromagnetic radiation emitted or reflected from surfaces. The earth receives energy from, and is heated by, the sun (see Fig. 3.5). The various surfaces of the earth absorb, reflect, and scatter the (shortwave) radiant energy in characteristic ways (see Chapter 3). Absorption in the atmosphere depends on clouds, water vapor, gases, dust, and other particles; absorption and reflection from terrestrial surfaces depend on the properties of those surfaces. All surfaces emit longwave radiation at rates that depend on their temperature (see Chapter 3). Surface radiance does not provide direct information about many of the parameters of interest to foresters, but many of them are correlated with it so instruments that measure surface radiance, in different waveband intervals, can provide us with a great deal of information about those surfaces. Examples are discussed in the following section.

There are currently a number of satellite-mounted instruments that provide data from which information about land surfaces may be obtained. These include the National Aeronautics and Space Administration (NASA) Landsat series of satellites, the French SPOT satellite, and the Advanced Very High Resolution Radiometer (AVHRR) on National Oceanic and Atmospheric Administration (NOAA) series satellites. There are also Japanese, European, Indian, Russian, and Chinese satellites. The widely used Landsat series satellites have carried two instruments: the Multi-Spectral Scanner (MSS) with four wavebands (0.5–1.1 µm), which supplies data at 79-m ground resolution (i.e., pixel size is 79 × 79 m), and the Thematic Mapper (TM), with four wavebands between 0.45 and 0.9 µm, one at 1.55–1.75 µm and one at 2.08–2.35 µm, plus a thermal band at 10.4–12.6 µm. The first six bands provide data at 30-m ground resolution but the resolution of the thermal band is 120 m. The MSS was the only sensor from 1972 to 1983; TM, with better technology, was added in 1983 and MSS was retained to provide a continuous long-term record with the same instrument. Landsat sensors provide a path width of 185 km on the ground. SPOT has three wavebands in the 0.5- to 0.89-µm range and a panchromatic 0.51- to 0.73-µm band. It has the advantage, for some purposes, of 10-m resolution. The AVHRR, currently the preferred instrument for global-scale land surface studies, has five wavebands, a path width of 2700 km, and a resolution of 1.1 km. The first Landsat satellite was launched in 1972, and the first of the NOAA satellites in 1978; therefore, there is considerable potential to use archived data for retrospective studies, which can be of particular value when long-term series of data relating to particular areas are available.

A new generation of satellite-mounted sensors, with much greater waveband discrimination, is expected to be launched by NASA in the late 1990s and early part of the 21st century.

A. Information Obtainable from Remote Sensing

The information contained in measurements of radiant fluxes from natural surfaces has to be extracted. This is done either by the use of inversion models that yield information about the properties of the surfaces or, more commonly, on the basis of some form of correlative calibration between pixel brightness or texture and some physical property of the forest landscape. The information required for forest management purposes, which may be obtained from remote sensing, includes forest type (deciduous, evergreen, broad-leaved, coniferous, etc.), condition (healthy, diseased, drought affected, etc.), leaf area index, standing biomass, structure, and radiant energy absorption.

The simplest form of analysis is land cover classification that takes advantage of the fact that areas of similar land cover will tend to have similar spectral "signatures." It can be used periodically to keep track of forest inventory or as the basis for a more detailed assessment of stand characteristics. Land cover classification can be performed in one of two ways: supervised and unsupervised. In supervised classification, the characteristics of the land cover classes are decided in advance. Sample areas with homogeneous land covers of each type are identified from aerial photographs or field experience, and the multivariate spectral statistics for the corresponding areas in the data are calculated. These "training sites" are used as the basis for judging the probability that each pixel in the rest of the image belongs to one or the other of the chosen classes. In unsupervised classification, only the number of classes is chosen a priori. A clustering algorithm is used to determine the natural groupings of a large sample of pixels in spectral space, and the multivariate statistical properties of these groupings are used as the basis for determining the appropriate classification of each pixel. The land cover type of each class is then determined by analysis after the fact. Supervised classification is generally more accurate, with unsupervised classification used mainly when ground truth information is limited. Straightforward in principle, the success of multispectral land cover classification depends heavily on the type and number of classes chosen, the choice of which spectral bands and auxiliary data to use, the treatment of pixels with mixed land cover, the confounding effects of topography, the characteristics of the clustering and classifying algorithms, the choice of training sites, and, most of all, the analyst's knowledge of the spectral and spatial characteristics of the survey area. A detailed and readable account of multispectral classification strategies is given by Jensen (1986).

Running *et al.* (1995) have suggested that a vegetation classification (unsupervised) based entirely on observable, remotely sensed vegetation properties would yield, at best, six vegetation types: evergreen needle-leaved forests, evergreen broad-leaved forests, deciduous needle-leaved forests, deciduous broad-leaved forests, broad-leaved annuals, and grasses. This information is not available from a single pass, but would require a number of data sets obtained at different times. Running *et al.* tested the approach using AVHRR data and vegetation maps of the continental United States. The results were encouraging. Cohen *et al.* (1995), in a detailed study of the age and structure of forests in the Pacific Northwest of the United States, used a single Landsat TM image and a variety of reference data including infrared aerial photographs, land surveys, and digital elevation models. They were able to distinguish several land cover types— open, semi-open, closed mixed, and closed conifer canopies—and were also able to distinguish between two or three canopy closure classes. Bolstad and Lillesand (1992) used an artificial intelligence program to guide unsupervised land cover classification of a portion of the Nicolet National Forest in Wisconsin and were able to distinguish betweeen upland and lowland conifers and upland hardwoods.

For many management objectives, it is not enough to be able to distinguish forest cover type to the level of canopy cover classes. For example, the ecology of northern hardwoods and aspen is quite different; therefore, the management practices necessary to achieve particular objectives will also differ considerably. More information is needed and can be obtained in some cases as Wolter *et al.* (1995) showed when they used composite Landsat TM imagery from different times of the year, and differences in phenology of the species, to distinguish forest cover types down to species level.

Remotely sensed data also have potential for providing quite detailed information about structural properties of canopies. A number of bidirectional models have been used to characterize structural attributes of vegetation cover: All are based on simple physical principles and attempt to model the reflectance of the canopy as a function of the position of the illuminating source and viewing instrument (Wu and Strahler, 1994). The models are of three main types: two stream, radiative transfer, and geometric optical. From the practical point of view, the geometric-optical type is much simpler and more robust in open-canopy forests. The model treats the forest canopy as a collection of discrete objects on a plane; their reflectance is modeled as a function of the pattern of plants, shadows, and soil from a specified viewpoint. Cohen and Spies (1992) used Landsat TM and SPOT data to characterize stem diameter, crown diameter, tree height, and stand density (trees ha^{-1}) for 41 closed-canopy evergreen conifer stands for which the measurements of canopy characteristics were available. They established strong correlations (most with r^2 values of order 0.8) between stand attributes and spectral image variables obtained from the satellite data. Wu and Strahler (1994) obtained good agreement between observed values of crown size, stand density, and total standing biomass for 9 conifer forests in Oregon and modeled estimates derived from the geometric-optical model with remotely sensed input data.

Because of its importance as the ecophysiological variable that determines the amount of radiation a canopy can intercept and because of transpiration, leaf area index is arguably the most important canopy characteristic, and there have been many studies aimed at estimating it from remotely sensed measurements. These are based on the fact that when incident solar radiation interacts with green leaves there is high absorption in the visible (photosynthetically active) region (0.4–0.7 μm) and corresponding low values for reflectance and transmittance. In the 0.7 to 1.3-μm (infrared) region, there is low absorption, with high reflectance and transmittance (Tucker and Sellers, 1986). These differences lead to strong radiance signals from vegetation, from which indices related to the amount of foliage in the canopy can be derived. The most widely used of these is the normalized difference vegetation index (NDVI), calculated from the reflectance of visible red (Vis) and near-infrared radiation:

$$\text{NDVI} = (\text{near infrared} - \text{Vis}) / (\text{near infrared} + \text{Vis}). \quad (10.1)$$

Detailed treatment of the basis for this relationship is given by Tucker and Sellers (1986).

The correlation between NDVI and L^* has been extensively investigated. Nemani and Running (1989) established a strong log-linear relationship between L^* measured at 17 sites and NDVI. Spanner *et al.* (1990a) found a strong relationship between summer maximum values of NDVI, calculated from AVHRR data, and L^* in forest stands in Oregon, Montana, Washington, and California. However, in another study (Spanner *et al.*, 1990b, p. 109), they noted that "the relationships between the L^* of coniferous forests and (Landsat TM) spectral bands . . . were affected by canopy closure, understorey vegetation and background reflectance . . .

differences in forest stand canopy closure permitted understorey vegetation and the background to contribute to the integrated spectral radiance of the forest stands." Despite this, Curran et al. (1992) used NDVI calculated from Landsat TM to evaluate seasonal variation in L^* in 16 slash pine plots in Florida, which had received different fertilizer treatments. This rigorous test of the method produced results good enough to allow the authors to conclude that the use of remotely sensed data has potential for the study of the seasonal dynamics of a single species in forest canopies. Goward et al. (1994, p. 199) made a thorough study of the spectral reflectance properties of a number of the important landscape constituents across western Oregon and evaluated the implications of their results for remote sensing. They concluded that the simple "green foliage" description provided by remotely sensed NDVI is inadequate, and wrote, "In the extreme it appears possible to record an NDVI of less than 0.55 for a complete canopy of juniper and sagebrush and an NDVI greater than 0.4 from litter and lichen covered landscapes without green foliage."

It appears, from the previous outline of results, that there is no unique relationship between NDVI and L^*, but there seems to be no reason why useful local or regional relationships cannot be developed. Fassnacht et al. (1996) examined a wide variety of vegetation indices calculated from Landsat TM and found that both NDVI and the simple ratio (SR = near infrared/Vis) provided good estimates of L^* for needle-leaved evergreen conifers in the important range up to $L^* \approx 4$ (see Fig. 10.1). There were no data for broad-leaved deciduous forests with $L^*<4$, but for $L^*>4$ the relationship between L^* and the SR appeared consistent with the trend for the northern Wisconsin conifers. The relationship between L^* and the SR for needle-leaved evergreen conifer forests in Oregon, obtained by Spanner et al. (1994), appeared to differ from that for broad-leaved deciduous forests obtained by Fassnacht et al. It was much closer in the case of NDVI but the lack of L^* values less than 4 for broad-leaved forests reduced the value of the comparison. These data confirm the view that good local or regional relationships between vegetation indices and L^* can be developed.

If intercepted radiation could be estimated directly, there would be no need for leaf area index data as the precursor to the calculation of forest productivity (see Chapter 9). There is progress in this direction: Based on analysis using a model, Goward and Huemmrich (1992, p. 136) concluded that "Under typical observing conditions NDVI is near-linearly related to both instantaneous and daily total APAR." They note that the potential influence of landscape variables, including leaf angle distribution, leaf spectral optics, and background spectral reflectance on the APAR/NDVI relationship, is significant and suggest alternative remote

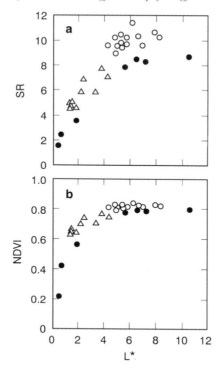

Figure 10.1 The relationship between (a) the simple ratio (SR) and leaf area index (L^*) and (b) normalized difference vegetation index (NDVI) and L^*. Data for evergreen needle-leaved conifer and broad-leaved deciduous hardwood stands in northern Wisconsin are represented by open triangles and circles, respectively. Closed circles represent data for evergreen needle-leaved conifer stands in Oregon (see Spanner *et al.*, 1994) (from Fassnacht, 1996).

sensing approaches to resolve the structural and spectral variations within vegetated landscapes. However, they also recommend that "before either approach is seriously considered, an aggressive effort should be undertaken to demonstrate a need for such added complexity."

One of the great values of remote sensing, as noted earlier, is the opportunity to repeat measurements or observations. Management requires not only predictions but also constant feedback. Were the decisions made correct? What is the new state of the system (forest)? From the point of view of managers, model predictions do not have to be correct to within a few percentage points; they need to be directionally (increase or decrease; system improving or degrading, etc.) and relatively correct (section *a* has a higher growth potential than section *b*) and sufficiently accurate quantitatively to avoid disastrous mistakes. Obviously, the more accurate predictions and estimates are, the better. However, the capacity to monitor change and make decisions about resource management promptly is often as important as accuracy. Remote sensing, particularly satellite remote sensing, offers this facility: The same measurements can be made, of the same system, time after time. This allows forests to be monitored over seasons or years if the information is considered of sufficient value.

III. The Use of GIS, Remote Sensing, and Models as Management Tools

In this section, we work through a hypothetical example in which we assume that remote sensing has been used to determine some of the attributes of a forested area. The information, together with other data pertaining to the area, is held in a GIS that has associated with it a dynamic ecophysiological model. The model we will use as illustration is the ε model, although in fact any appropriate model could be used—the same principles apply. It would also be possible to specify different parameters and variables for each cell—a number of possible schemes could be designed, but as a means of illustrating the points made in the first part of the chapter we work through the one outlined here, commenting on each data layer. Note that different time scales may be used for the variables that change with time: Weather data may be averaged over a month so that the derived variables calculated from them will be calculated for the same interval, but net primary production (NPP) may be summed over a year.

We include comments on the likely origins and availability of the data and the work that may be needed to obtain them. The comments are cross-referenced to earlier chapters wherever pertinent. We make no comment on the computer technology that may be involved in linking GIS files containing information of varying types (site factors, forest state, meterological data, etc.). There should be no difficulty in this area for appropriately trained personnel.

Figure 10.2 is central to this section. It depicts a unit of land, which may be of any size, consisting of a number of subunits each of which may be reasonably homogeneous in terms of, for example, topography and forest type, but within which we may expect to find some variation in forest attributes—in other words, a typical forest management unit. We assume that the remotely sensed images are available for the whole landscape. The section shown with cells (pixels) is a polygon identified as homogeneous in terms of its attributes so that the signals from each pixel within the polygon will not be significantly different. If the area is large,

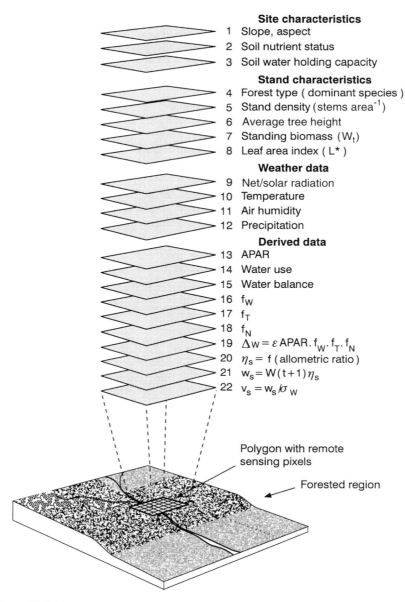

Figure 10.2 Schematic outline of the use of GIS, with remote sensing and other observations and measurements, as a modeling and monitoring tool. The model outlined here is the ε model. It is assumed that a value of ε is available (see text). Notes on the data layers: 1, used with 9 to calculate incident radiation on the surface; 2, may be conventional measure of soil nutrient status and/or may be input from CENTURY-type model of N supply rate; 3, requires an estimate of rooting depth; 4–6, from ground observations. Initial esti-

so each polygon is many square kilometers, the criteria for significant differences will be different from those that would be relevant to forest stands within a forest estate. We assume, as shown, that there are 22 data layers associated with each polygon. This number may vary widely, depending on the information required by, and available to, the land management agency. The procedures outlined below—or some version of them—would be applied to every polygon of the land unit. For the type of analysis suggested, the computing requirements could become prohibitive if we considered the system pixel by pixel; in any case, the information required would not be available at pixel scales.

A. The Data Layers

The following commentary relates to the data layers depicted in Fig. 10.2

1. Site Characteristics

a. Layer 1: Topographic Characteristics (Slope and Aspect) These will be average values for each polygon used to calculate incident radiation (see Chapter 3 and Section III,A,3,a). If a digital elevation map is not available, they can be digitized from standard ordinance survey maps.

b. Layer 2: Soil Nutrient Status These data may be conventional measures of soil chemical properties (e.g., macro- and micronutrients) obtained from digitized soil survey maps, or the input to this layer may come from a submodel of the CENTURY type, in which case the baseline information would consist of soil organic matter, and the submodel would call on information held in other layers to calculate nutrient mineralization rates.

c. Layer 3: Soil Water-Holding Capacity This will be a single number applying to the whole rooting zone or calculated for each soil horizon from the soil physical properties and water-holding characteristics and root depth. It may be estimated from known relationships between soil type, as described in conventional maps, and soil physical characteristics (see Chapter 4) or those estimates may be improved by sampling across the areas of interest. (Note: The data in both layers 3 and 4 will come from

---

mates of 7 on the basis of 4–6; 8, from NDVI and ground observations; 9–11, from regional meteorological station. Interpolate local values on the basis of topography. Local values of precipitation (12) are important; 13, derived from 1,8, and 9; calculate 14 from 9–11 (see Chapters 3 and 4); calculate 15 from 3, 12, and 14; 16, based on average monthly water balance (see Fig. 10.3); 17, based on average monthly temperature (see Fig. 10.4) and frost occurrence; 18, based on 2; 19, as indicated; 20, partitioning coefficient based on allometric ratios (see Chapter 5, Section IV,A); 21, as indicated; 22, wood volume (v_s) = w_s/σ_w, where σ_w is wood density.

independent survey work; for any particular polygon, the quality of the data will depend on the quality of the surveys in terms of sampling densities, the techniques used to interpolate values, and the variation across each polygon.)

2. Stand Characteristics

a. Layer 4: Forest Type Forests may be defined as general types, e.g., as one of the categories used in this book, such as temperate broad-leaved deciduous, temperate needle-leaved evergreen, and so on. This may be relevant in the case of a large area, where homogeneity of forest type may be an adequate criterion for polygon homogeneity. If the polygons are stands, or parts of stands, such criteria are likely to be too general and species-level (composition) information may be required.

b. Layer 5: Stand Density This is measured by stems per unit area. The conventional method of obtaining stand density data is ground inventory, but it would be possible to obtain an indication of stand density from aerial photographs or from direct observation from light aircraft. Calibration against ground observations would be necessary to ensure the accuracy of such estimates.

c. Layer 6: Average Tree Height Average tree height is a conventional forestry variable estimated from ground survey.

d. Layer 7: Standing Biomass [W(t)] An estimate of this should be available from standard forestry yield tables, if they exist for the area, or can be made on the basis of layers 4, 5 and 6, and knowledge of the forest type in question. Every opportunity should be taken to increase the amount of these data available, including determinations made on harvested trees. It is ironic that the end result of commercial forestry is harvested trees but it is not common practice to record the exact location of a few harvested trees, together with their mass, and the component parts of that mass, from which allometric equations could be established. The accumulation of such data over a period of years would provide an invaluable resource that could be used to check and refine model calculations.

e. Layer 8: Leaf Area Index L^* is an important parameter for modeling and the calculation of transpiration rates and hence water balance. It is derived from NDVI or other remotely sensed indexes and calibrated for the forest types under consideration (see Section II,A).

3. Weather Data

a. Layer 9: Solar (φ_s) and Net Radiation (φ_n) Despite its central importance as the variable that drives all plant growth, solar radiation is seldom routinely measured by meteorological networks and virtually never measured by organizations concerned with forest management. The reason for this is, presumably, that there has been little appreciation of the value of the data and, lacking operational ecophysiological models, no use for it. Regional measures of solar radiation would, in most cases, suffice. Daily variation is always high, and spatial variation, over short periods, is also high. However, because solar radiation at the outside of the atmosphere is a function of latitude and season only, for any areas with similar weather patterns a measure of daily solar radiation made within kilometers will provide good monthly average data. It is also possible to estimate solar radiation using the Total Ozone Mapping Spectrometer (TOMS) using a procedure developed by Eck and Dye (1991) and tested by Goward *et al.* (1994), who obtained excellent correspondence between photosynthetically active radiation (PAR) measured at the surface and monthly estimates of PAR obtained from TOMS. Net radiation can be calculated as described in Ch.3 (see also Moore *et al.* (1993) for a procedure appropriate to hilly terrain).

b. Layers 10 and 11: Temperature and Air Humidity Regional temperature and air humidity data can usually be obtained from standard meteorological stations. Extrapolation to areas that may have different exposure and topography can be done using climatological principles; a good example of the application of such principles is the work of Running *et al.* (1987), who produced a computer model called MT-CLIM, which is used to simulate climatological data for areas in the mountainous regions of the western United States. This model can be run on a standard personal computer. The importance of such interpolative techniques will vary depending on topography, distance from the location of the reference measurents, and the averaging time to be used. It is not safe to use MT-CLIM outside the range for which it was developed, although the general principles and ideas will be useful. Simulations should be compared to actual weather station data wherever this is feasible.

c. Layer 12: Precipitation Precipitation, as noted in Chapter 3, tends to be extremely variable spatially as well as temporally. Extrapolation to areas any distance away from reference measurement sites will always carry the danger of large errors. Because the water balance of a forest depends on the temporal distribution of precipitation as well as on the amount, a dense network of precipitation gauges is always desirable. Fully automated precipitation gauges, as well as complete weather stations, are now available and can be set up so that the data are downloaded remotely through a modem. Analysis of the rainfall patterns in any region, in terms of storm size and intensity distributions, will provide useful information and improve the confidence with which amounts of rain can be extrapolated to areas away from the point of measurement.

4. Derived Data These data require calculations that, in most cases, use data stored in layers 1–12. The calculations would be done using subroutines.

a. Layer 13: Absorbed Photosynthetically Active Radiation ($\varphi_{p \cdot a}$) This has been extensively discussed in Chapters 3 and 9. For homogeneous canopies for which L^* is reasonably well known, good estimates of φ_p can be obtained using Beers law [Eq. (3.8)]. The use of remote sensing to provide estimates of $\varphi_{p \cdot a}$ over polygons judged to be uniform in terms of canopy characteristics will always be an advantage, particularly if the estimates are checked against other methods such as calculations based on Beers' law. Clearly, estimates of $\varphi_{p \cdot a}$ are only as good as the values of solar radiation (φ_s) from which they are normally derived—the error in $\varphi_{p \cdot a}$ calculations cannot be less than the errors in φ_s, although it does not follow that the errors are additive.

b. Layer 14: Water Use Water use calculations will normally be made using the Penman–Monteith equation (see Chapter 3), assuming there are sufficient meteorological data, with whatever approximations are necessary, e.g., because of the relatively small effects of the boundary layer conductance term it can be taken as constant in most cases, or a general relationship with average wind speed, based on long-term data for a region, and canopy structure will provide useful values. Current knowledge of canopy conductances was reviewed in Chapter 3 and the effects of soil water content on stomatal conductance in Chapter 4 (see also layer 16). Water use rate may also be estimated from φ_n and Bowen ratio values, where β is assumed to vary with soil moisture status (see Chapter 3, Section III,c).

c. Layer 15: Water Balance Water balance calculations are dealt with in detail in Chapter 4.

d. Layer 16: Water Constraint on Growth (f_w) This is the water balance modifier in Eq. (9.6), which contributes to the calculation of utilizable radiation. We will split this modifier into two parts—a component attributable to ambient vapor pressure deficit [$f_w(D)$] and a component attributable to soil water content [$f_w(\theta_s)$]. This approach has been followed by Runyon et al. (1994), who calculated f_w on the basis of vapor pressure deficits (D) and predawn water potential (ψ_{pd}). They reduced the value of APAR by half (i.e., utilizable radiation = 0.5 APAR) when D was between 1.5 and 2.5 kPa and by 100% when D was greater than 2.5 kPa. They made similar adjustments for ψ_{pd}. Considerably more research is needed to determine the best functional relationships to use for the calculation of f_w; although the form of the relationship may be similar for different species, it is likely that the coefficients will differ. McMurtrie et al. (1994) used a linear relationship with D and daily soil water content. We support this approach in relation to D; in fact, inspection of the results reviewed by Leuning (1995) suggests that the equation given by Thorpe et al. (1980) [Eq. (5.12) in this book] will serve well as a general relationship. Omitting the radiation term in that equation, it becomes

$$g_s = g_{ref}(1 - a'(D - 1.2)). \qquad (10.2)$$

If $D < 1.2$ kPa, $g_s = g_{ref}$, the values of which can be obtained from Kelliher et al. (1993) if an estimate of g_s is required. g_s/g_{ref} is to be set to unity to obtain $f_w(D)$. If $D > 1.2$, the equation applies, with $a' = 0.55$. The accuracy of this equation applied to, for example, monthly mean values of D, has not been established, but it provides a good starting point. We suggest that, because ψ_{pd} is a function of soil water potential in the root zone, and in view of Myers' (1988) result showing that growth is inversely proportional to the integral of ψ_{pd} in terms of time (see Chapter 4), $f_w(\theta_s)$ should be derived as the ratio of actual (average) soil moisture content to that value of θ_s (from layer 15) above which it is considered to have no effect (see Fig. 10.3). This "no effect" point may vary with soil type and rooting depth but as a first estimate might be set to 75% of "available water" (see Chapter 4) The value of f_w will be the lower of $f_w(D)$ and $f_w(\theta)$.

e. *Layer 17: Temperature Constraint on Growth* (f_T) McMurtrie et al. (1994) used a formula for f_T that included frost, spring, and autumn temperatures. Frost on day i reduces utilizable radiation on that day to zero; f_{spring} depends on day degree sums and f_{autumn} is related to day length. McMurtrie et al. do not provide details. We suggest that the effects of frost be incorporated as suggested, but that other temperatures

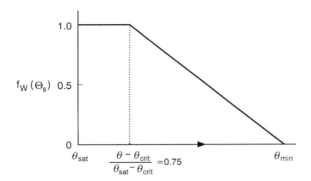

Figure 10.3 Relationship of the type that may be used to calculate the water balance modifier for utilizable APAR. θ denotes volumetric water content. The graph indicates that the ratio $(\theta - \theta_{crit})/(\theta_{sat} - \theta_{crit})$, which applies to the whole root zone, is assumed to be nonlimiting until it reaches a value of 0.75.

(layer 10) be related to an optimum growing temperature for the species in question, e.g., 30°C (see Fig. 10.4). In the case of deciduous forests, in climates where foliage is shed in the autumn (fall) it will be necessary to introduce a declining efficiency factor that reduces from one to zero, starting from the time when foliage begins to change color and ending at leaf fall.

f. Layer 18: Nutritional Constraint on Growth (f_N) Nutritional modifiers have not been tested in research studies. We suggest that f_N be calculated as the ratio of the soil nutritional status—amount of "available" nitrogen, phosporus, potassium, and major trace elements, as indicated by conventional chemical analysis—using the best soils in the region to define $f_N = 1$. The ratios would, presumably, be based on the total nutrients available in the root zone (layer 3). Alternatively, nitrogen availability would be calculated from a model such as CENTURY, which would provide an index of the rate of nitrogen mineralization. The temptation to scale nutrition on the basis of tree growth should be resisted; this would be analogous to the procedures used in deriving site indices, which we have criticized as being circular. Because there are clear indications (see Chapter 5) that the proportion of the carbon available for growth that is allocated to roots is higher in infertile than in fertile soils, it is likely that the allometric ratios of trees on sites of different fertility will vary. This, in common with all other aspects of nutritional modifiers, is an area that requires considerable research.

g. Layer 19: Biomass Increment (ΔW) This will be calculated over the selected interval (month, season, etc.) using an appropriate value of ε with APAR (layer 13) and f_W, f_T, and f_N. Based on the discussion in Chapter 9,

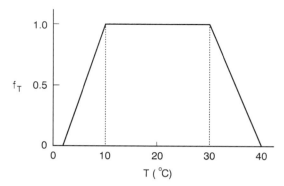

Fig. 10.4 Functional relationship between average air temperature and the temperature modifier for utilizable APAR. Various curves could also be used; the value of more realistic—and complex—forms must be weighed against the accuracy of information about temperature and biological responses.

a starting value of 1.5 g MJ^{-1} for above-ground NPP would be reasonable. Alternatively, ΔW could be calculated using a more complex model, such as BIOMASS or FOREST-BGC, which could be expected to provide better estimates. However, use of one of these models implies more complex calculations, because an equation for photosynthesis is required, and respiration must be calculated. The new value of standing biomass (layer 7) is $W(t) + \Delta W$.

h. Layer 20: Partitioning Coefficients [η_i, Eq. (5.22)] We noted in Chapter 5 that empirical partitioning coefficients can be obtained from biomass and allometric data by differentiating Eq. (3.2) to give $dw_i/d_b = c_i n_i d_b^{n-1}$. It is for this reason that the accumulation of biomass and allometric data, from as wide a range of situations and species as possible, is important. At a more mechanistic level, the approach used by Running and Gower (1991) could be applied. They placed the emphasis on partitioning to foliage growth, which is based on leaf water status and nitrogen availability. Leaf/root partitioning ratios are determined by soil water and nitrogen status and allocation to stems is a residual. This is also, essentially, an empirical scheme, and should be tested and refined by field data wherever possible.

i. Layer 21: Mass of Tree Components These are obtained by applying the partitioning coefficients to standing biomass.

j. Layer 22: Stem Volume If stem mass per unit ground area is w_s, stem density is m ha^{-1}, then individual stem mass is w_s/m, and wood density is σ_w kg m^{-3}, and average stem volume is $w_s/m\sigma_w$. Refinements in terms of stem size distributions can be introduced (see Chapter 3). Combining this information with standard equations for stem taper leads to estimates of tree height, which can be checked against observations recorded in layer 6.

IV. Concluding Remarks

The outline previously discussed could apply to any scale, and the procedures suggested, and data said to be required, will vary with particular applications. Such a scheme offers managers concerned with production forestry the opportunity to update their estimates of standing biomass at any time and use those estimates as the basis for productivity and wood flow calculations. The approach outlined here, and many of the tools described, could be used to manage a watershed for water yield or for wildlife production or to assess the impact of management practices such as clear-cutting or burning (deliberate or inadvertent) on particular areas. It is also possible to link a succession model to a simple biomass production model so that the impact of forest management in each land unit can be evaluated for both forest productivity and biodiversity. When a subunit is logged, or lost by fire, it is "re-set" in terms of its stand characteristics. Run over many cycles, the procedures will allow assessment of the sustainability of the system, which may be reflected in terms of changes in soil organic matter (Fig. 10.2, layer 2), or possibly changes in forest type and stand characteristics. These may be suspected, as a result of particular treatments or events, or observed: Evaluation of the consequences would be speculative, but some such procedure would provide the only means of assessing the consequences of the scenario.

The point has been made several times that there should be continuous observation and measurement to calibrate and check model predictions. Where there are major discrepancies, analysis to identify the reason(s) will lead to improvements in future calculations. Periodic satellite measurements provide the opportunity to update and upgrade information about forest structural properties, L^*, and φ_p. Ground observations made at any time should be entered in the GIS to provide additional information that can be used in interpretation and explanation or analysis.

The philosophy underlying the approach outlined is that management and scientific research should not be separate activities, with scientists providing management with information considered relevant to the problems faced by managers and managers often impatient of the impracticality of the information provided.

The system, as we see it, should work as depicted in Fig. 10.5: Problems

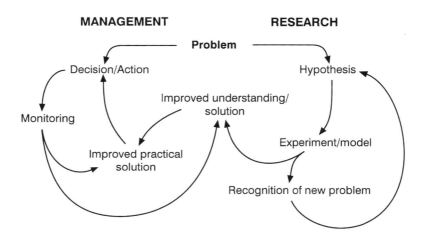

Fig. 10.5 Diagram depicting a good interactive relationship between research and management. A "problem" may be simply a situation in which there are alternative possible decisions and actions based on currently available knowledge. Monitoring the consequences of the action(s) taken will lead to improved knowledge. Research treats alternatives as hypotheses and tests them by experiment or using models. Research may reveal previously unrecognized problems.

with a scientific or technological basis may be identified by managers or scientists; these may be major, multifaceted problems, such as prediction of growth, yield, and sustainability, or single-issue problems such as how to control a pest. (Note: There is no guarantee that single-issue ecological problems will be simple; for example, research aimed at control of a pest will usually involve a great deal of investigation relating to the life cycle of the pest and the circumstances in which it is likely to become serious.) Managers have to make decisions and take action on the basis of the best currently available knowledge; they will be more successful if they have a sound understanding of the processes at work in the system, hence our argument that managers need to understand ecophysiological processes. Scientists undertake research on the problem, in the course of which they increase the knowledge available about it and are able to offer improved advice about solutions. It is also likely that the research will reveal areas in which knowledge is inadequate, indicating the need for further research at basic levels. The improved advice leads to a better basis for further action. Management monitoring should lead to assessment of the consequences of the actions taken or the accuracy of the predictions made; this information will be of value to the scientists, the managers, and presumably to society.

It would be easy to illustrate the previous argument with examples from almost any part of this book but there seems no need to labor the point. If we have to spell it out, we have probably failed in our objective of bridging, at least to some extent, the gap between those who manage forests, and make decisions about their management, and those who study forests, and make pronouncements about the way they function.

V. Peroration

To conclude this book, we note that there is widespread failure to practice sustainable forest management. Long-term societal priorities, and the need to maintain ecological balance and integrity, are ignored or overridden by individuals, groups, or even nations (Mooney and Sala, 1993). The reasons for this range from ecological ignorance to so-called rational economics and the anthropocentric view that the short-term needs of current human societies take precedence over ecological balance and sustainability. The short-sightedness of this attitude is obvious to ecologists, and its dangers are becoming apparent even to some in politics and various fields of human activity unrelated to ecology. It was not one of our objectives to address broad societal problems and attitudes, but we have tried to demonstrate that basic ecophysiological science and understanding are essential as underpinning for the profession of forestry and the management of forests for long-term sustainability. We are in good company in putting forward this argument (Mooney and Sala, 1993; Hornbeck and Swank, 1992; Pitelka and Pitelka, 1993).

Ecologists are often hesitant in presenting their findings to land managers because they are dealing with highly variable systems that are constantly changing and evolving. Information therefore tends to be uncertain and the statistical variability of data is usually high. However, as Ehrlich and Daily (1993) have pointed out, science can never provide certainty, but it can help to make a safe bet. The methods of science, rigorously applied, and knowledge based on science will tilt the balance of probable outcomes from decisions, actions, or predictions in the direction indicated by science. Scientific methods also provide the means of evaluating objectively the consequences of decisions and actions; subjective experience and assertions based on emotional values or "politically correct" attitudes are no substitute for objective criteria and rigorous testing. Moral and emotional values can be applied when the facts—as far as science can establish them—are known. The approach that we advocate in this chapter, and the example we have developed (coupling a process-based model to landscape attributes using GIS and remote sensing) makes it possible to use scientific methods to look back in time and see the results of management and environmental conditions up to the present and project forward in time to predict the effects of forest management practices on long-term sustainability. The use of linked ecosystem-level process-based models opens the possibility of extending the analyses of forest blocks to adjacent watersheds, aquatic ecosystems, and the atmosphere.

Recommended Reading

Aronoff, S. A. (1989). "Geographic Information Systems: A Management Perspective," pp. 294. WDL Publications, Ottawa, Canada.

Burrough, P.A. (1989). "Principles of Geographical Information Systems for Land Resource Assessment." Oxford Univ. Press, New York.

Crist, E. P., and Kauth, R. J. (1986). The tasselled cap de-mystified. *Photogrammetric Engi. Remote Sensing* **52**, 81–86.

Jensen, J. R. (1986). "Introductory Digital Image Processing," pp. 379. Prentice Hall, Englewood Cliffs, NJ.

Maguire, D. J., Goodchild, M. F., and Rhind, D. W. (Eds.) (1991). "Geographical Information Systems. Vol.1: Principles, pp. 647; "Vol.2: Applications," pp. 447. Longman, Harlow, Essex, UK.

Tucker, C. J., and Sellers, P. J. (1986). Satellite remote sensing of primary production. *Int. J. Remote Sensing* **7**, 1395–1416.

Symbols and Definitions

The symbols used in this book are as consistent as possible with those that have become established in the literature. Because of the range of material covered in the book it has been difficult to avoid some duplication, but for the most part this has been overcome by the use of appropriate subscripts. We avoided, as much as possible, the inelegant use of acronyms in place of symbols—a practice that is increasing with the widespread use of computer models. Only the more important symbols are defined here. Symbols used only once, or "minor" constants or coefficients, are not listed. All symbols are defined in the text where they first appear. Standard International (SI) units are used throughout, although in the text and in practical usage, magnitudes may be changed for convenience or for consistency with published data.

A, A_c = CO_2 assimilation rate, canopy assimilation rate (mol m^{-2} s^{-1})

APAR = absorbed photosynthetically active radiation (MJ m^{-2} or mol^{-1} m^{-2})

a_f = foliage density (m^2 foliage surface area m^{-3} canopy volume)

c_a, c_i = CO_2 concentration, or partial pressure of CO_2, at a reference level in the air, and intercellular (mol^{-1} or Pa; also commonly expressed as volume fraction in air, i.e., parts per million)

c_p = specific heat of air at constant pressure (=1.01 × 10^3 J kg^{-1} K^{-1})

D = vapor pressure deficit (kPa)

d (in the wind profile equation) = zero plane displacement

d_B = stem diameter (m) at breast height (1.4 m)

E = mass flux of water vapor (kg m^{-2} s^{-1}

g_a, g_{aM}, g_s, g_c = boundary layer conductance, aerodynamic conductance for momentum, stomatal conductance, canopy conductance (m s^{-1})

H = latent heat flux (W m^{-2})

Symbols and Definitions

h_c = mean canopy top height (m)
I = water intercepted by canopies (mm)
J_{max} = potential electron transport rate (μmol m^{-2} s^{-1})
J_s = volume flux of water across potential gradient (soil water)
k (in the wind profile equation) = van Karman's constant = 0.4
K_c, K_o = Michaelis constants for CO_2 and O_2
K_s = soil hydraulic conductivity
L* = leaf area index (dimensionless)
NDVI = normalized difference vegetation index
NEP = net ecosystem production
NPP = net primary production
P = precipitation (mm)
p_i, p_s = fraction of rain reaching the ground beneath a canopy as throughfall or stemflow, respectively
p = stem population per unit land area
q, q_{sat} = specific humidity, saturated specific humidity (kg m^{-3})
q_R, q_D = (in the hydrologic equation) = runoff, drainage
R_a = autotrophic respiration
$R_{a \cdot m}$, $R_{a \cdot g}$, $R_{a \cdot u}$ = maintenance, growth, and ion uptake respiration
r_a, r_c, r_i = boundary layer resistance ($= 1/g_a$), canopy resistance ($= 1/g_c$), climatological resistance (m s^{-1})
R_s, R_r, R_x, R_f = hydraulic resistances to flow through soil to root, root to xylem, through the xylem, from xylem to evaporating surfaces in leaves
PAR = photosynthetically active radiation (MJ m^{-2} or mol^{-1} m^{-2})
S = water stored on canopy and lost by evaporation (mm)
T = temperature (°C)
u = wind speed (m s^{-1})
u* = friction velocity (m s^{-1})
V_c = carboxylation rate
W, W_{max} = total and maximum tree mass

w_f, w_r, w_s = foliage, root, and stem mass (kg or tons ha^{-1})

z = height (m) or depth in the soil

z_0 = zero plane displacement (m)

α = albedo, or reflection coefficient

β = Bowen ratio (=H/λE)

γ = psychrometric constant (0.066 kPa °C^{-1})

ε = radiation utilization coefficient (g dry mass produced MJ^{-1} photosynthetically active radiation)

ε_s = dimensionless rate of change of q_{sat} with T (=2.2 at 20°C)

λ = latent heat of vaporization of water (J kg^{-1})

$\varphi_s, \varphi_n, \varphi_p, \varphi_{p \cdot a}, \varphi_L\uparrow, \varphi\downarrow$ = short-wave radiant energy (W m^{-2}), net radiation (W m^{-2}), photosynthetically active radiation (W m^{-2} or mol m^{-2} s^{-1}), photosynthetically active radiation absorbed by a canopy, long-wave upward flux (W m^{-2}), long-wave downward flux (W m^{-2})

$\psi_s, \psi_f, \psi_{pd}$ = soil water potential, foliage water potential, predawn (foliage) water potential (MPa)

ρ = air density ($\approx$ 1.3 kg m^{-3} at 20°C)

ρ_s = soil bulk density (kg m^{-3})

σ_f = specific leaf area (m^{2}kg^{-1})

τ = shear stress or momentum flux (N m^{-2})

θ = volumetric soil moisture content

References

Aber, J. A. (1992). Nitrogen cycling and nitrogen saturation in temperate forest ecosystems. *Trends Ecol. Evol.* **7**, 220–224.

Aber, J. A., Nadlehoffer, K. J., Steudler, P., and Mellilo, J. M. (1989). Nitrogen saturation in northern forest ecosystems. *BioScience* **39**, 378–386.

Aber, J. D., and Federer, C. A. (1992). A generalized, lumped-parameter model of photosynthesis, evapotranspiration and net primary production in temperate and boreal forest ecosystems. *Oecologia* **92**, 463–474.

Aber, J. D., and Melillo, J. M. (1991). "Terrestrial Ecosystems." Saunders College Publishing, Philadelphia.

Aber, J. D., Botkin, D. B., and Melillo, J. M. (1979a). Predicting the effects of different harvesting regimes on productivity and yield in northern hardwoods. *Can. J. For. Res.* **9**, 10–14.

Aber, J. D., Melillo, J. M., and Federer, C. A. (1979b). Predicting the effects of rotation length, harvest intensity, and fertilization on fiber yield from northern hardwood forests in New England. *For. Sci.* **28**, 31–45.

Aber, J. D., Melillo, J. M., Nadelhoffer, K. J., McClaugherty, C. A., and Pastor, J. A. (1985). Fine root turnover in forest ecosystems in relation to quantity and form of nitrogen availability: A comparison of two methods. *Oecologia* **66**, 317–321.

Aber, J. D., Magill, A., Boone, R., Melillo, J. M., Steudler, P., and Bowden, R. (1992). Plant and soil responses to three years of chronic nitrogen additions at the Harvard Forest, Petersham, MA. *Ecol. Appl.* **3**, 156–166.

Abrams, M. D., and Mostoller, S. A. (1995). Gas exchange, leaf structure and nitrogen in contrasting successional tree species growing in open and understory sites during drought. *Tree Physiol* **15**, 361–370.

Ågren, G. (1983). Nitrogen productivity of some conifers. *Can. J. For. Res.* **13**, 494–500.

Ågren, G. I., and Axelsson, B. (1980). Population respiration: A theoretical approach. *Ecol. Model.* **11**, 39–54.

Ågren, G. I. and Bosatta, E. (1988). Nitrogen saturation of terrestrial ecosystems. *Environ. Pollut.* **54**, 185–197.

Ågren, G. I., and Ingestad (1987). Root/shoot ratios as a balance between nitrogen productivity and photosynthesis. *Plant Cell Environ.* **10**, 579–586.

Ågren, G. I, McMurtrie, R. E, Parton, W. J, Pastor, J., and Shugart, H. H. (1991). State-of-the art models of production–decomposition linkages in conifer and grassland ecosystems. *Ecol. Appl.* **1**, 118–138.

Alban, D. H., Perala, D. A., and Schlaegel, B. E.(1982). Biomass and nutrient distribution in aspen, pine, and spruce stands on the same soil type in Minnesota. *Can. J. For. Res.* **8**, 290–299.

Alexander, M. (1977). "Introduction to Soil Microbiology," 2nd ed. Wiley, New York.

Allen, M. F. (1991). " The Ecology of Mycorrhizae." Cambridge Univ. Press, New York.

Amthor, J. S. (1989). "Respiration and Crop Productivity." Springer-Verlag, New York.

Amundson, R. G., Chadwick, O. A., and Sowers, J. M. (1990). A comparison of soil climate and biological activity along an elevational gradient in the eastern Mojave Desert. *Oecologia* **80**, 395–400.

Anderson, J. P. E., and Domsch, K. H. (1980). Quantities of plant nutrients in the microbial biomass of selected soils. *Soil Sci.* **130**, 211–216.

Anderson, M. C. (1964). Light relations of terrestrial plant communities and their measurement. *Biol. Rev.* **39**, 425–486.

Anderson, M. C. (1966). Stand structure and light penetration: II A theoretical analysis. *J. Appl. Ecol.* **3,** 41–54.
Anderson, M. C. (1981). The geometry of leaf distribution in some south-eastern Australian forests. *Agric. Meteorol.* **25,** 195–205.
Anderson, T. H., and Domsch, K. H. (1989). Ratios of microbial biomass carbon to total organic carbon in arable soils. *Soil Biol. Biochem.* **21,** 471–479.
Antweiler, R. C., and Drever, J. I. (1983). The weathering of late Tertiary volcanic ash: Importance of organic solutes. *Geochim. Cosmochim. Acta* **47,** 623–629.
Aronoff, S. A. (1989). "Geographic Information Systems: A Management Perspective." WDL, Ottawa, Canada.
Art, H. W., Bormann, F. H., Voigt, G. K., and Woodwell, G. M. (1974). Barrier Island forest ecosystems: role of meteorologic nutrient inputs. *Science* **184,** 60–62.
Attiwill, P. M. (1979). Nutrient cycling in a *Eucalyptus obliqua* (L'Hérit.) forest. III Growth, biomass, and net primary production. *Aust. J. Bot.* **27,** 439–458.
Attiwill, P. M., and Adams, M. A. (1993). Nutrient cycling in forests. *New Phytol.* **124,** 561–582.
Auclair, A. N. D., and Carter, T. B. (1993). Forest wildfires as a recent source of CO_2 at northern latitudes. *Can. J. For. Res.* **23,** 1528–1536.
Axelrod, D. I. (1966). Origin of deciduous and evergreen habits in temperate forests. *Evolution (Lawerence, KS)* **20,** 1–15.
Baldocchi, D. D., and Collineau, S. (1994). The physical nature of solar radiation in heterogeneous canopies: Spatial and temporal attributes. *In* "Exploitation of Environmental Heterogeneity by Plants" (R. Pearcy and M. M. Caldwell, Eds.), pp. 21–71. Academic Press, San Diego.
Baldocchi, D. D., and Vogel, C. (1996). A comparative study of water vapor, energy and CO_2 flux densities above and below a temperate broadleaf and a boreal pine forest. *Tree Physiol.* **16,** 5–16.
Baldocchi, D. D., Vogel, C. A., and Hall, B. (1996). Seasonal variation of carbon dioxide echange rates above and below a boreal jack pine forest.
Baldwin *et al.* (1983)
Ball, J. T., Woodrow, I. E., and Berry, J. A. (1987). A model predicting stomatal conductance and its contribution to the control of photosynthesis under different environmental conditions. *In* "Progress in Photosynthesis Research" (J. Biggens, Ed.), pp. 221–224. Nijhoff, Dordrecht.
Beadle, C. L., Turner, N. C., and Jarvis, P. G. (1978). Critical water potential for stomatal closure in Sitka spruce. *Physiol. Plant.* **43,** 160–165.
Beadle, C. L., Talbot, H., and Jarvis, P. G. (1982). Canopy structure and leaf area index in a mature Scots pine forest. *Forestry* **55,** 105–123.
Beadle, C. L., Jarvis, P. G., Talbot, H., and Neilson, R. E. (1985a). Stomatal conductance and photosynthesis in a mature Scots pine forest. II. Dependence on environmental variables of single shoots. *J. Appl. Ecol.* **22,** 573–586.
Beadle, C. L., Talbot, H., Neilson, R. E., and Jarvis, P. G. (1985b). Stomatal conductance and photosynthesis in a mature Scots pine forest. III. Variation in canopy conductance and canopy photosynthesis. *J. Appl. Ecol.* **22,** 587–595.
Beare, M. H., Neely, C. L., Coleman, D. C., and Hargrove, W. L. (1990). A substrate-induced respiration (SIR) method for measurement of fungal and bacterial biomass on plant residues. *Soil Biol. Biochem.* **22,** 585–594.
Becwar, M. R., and Burke, M. J. (1982). Winter hardiness limitations and physiography of woody timberline flora. *In* "Plant Cold Hardiness and Freezing Stress: Mechanisms and Crop Implications" (P. H. Li and A. Sakai, Eds.), pp. 307–323. Academic Press, New York.
Beets, P. E., and Pollock, D. S. (1987). Accumulation and partitioning of dry matter in a *Pinus radiata* plantation fertilized with nitrogen and potassium. *N. Z. J. For. Sci.* **17,** 246–271.

Behera, N., Joshi, S. K., and Pati, D. P. (1990). Root contribution to total soil metabolism in a tropical forest soil from Orissa, India. *For. Ecol. Manage.* **36,** 125–134.

Bell, R., and Binkley, D. (1989). Soil nitrogen mineralization and immobilization in response to periodic prescribed fire in a loblolly pine plantation. *Can. J. For. Res.* **19,** 816–820.

Benoit, R. E., and Starkey, R. L. (1968). Inhibition of decomposition of cellulose and some other carbohydrates by tannin. *Soil Sci.* **105,** 291–296.

Benson, M. L., Myers, B. J., and Raison, R. J. (1992). Dynamics of stem growth of *Pinus radiata* as affected by water and nutrient supply. *For. Ecol. Manage.* **52,** 117–137.

Berg, B. (1984). Decomposition of root litter and some factors regulating the process long-term root litter decomposition in a Scots pine forest. *Soil Biol. Biochem.* **16,** 609–617.

Berg, B., and Ågren, G. I. (1984). Decomposition of needle litter and its organic chemical constituents: Theory and field experiments. III. Long-term decomposition in a Scots pine forest. *Can. J. Bot.* **62,** 2880–2888.

Berg, B., Ekbohm, G., and McClaugherty, C. (1984). Lignin and holocellulose relations during long-term decompositon of some forest litters. IV. Long term decomposition in a Scots pine forest. *Can. J. Bot.* **62,** 2540–2550.

Berg, B., Johansson, M. B., and McClaugherty, C. (1993). Litter mass-loss rates in late stages of decomposition at some climatically and nutritionally different pine sites. Long-term decomposition in a Scots pine forest VIII. *Can. J. Bot.* **71,** 680–692.

Berish, C. W. (1982). Root biomass and surface area in three successional tropical forests. *Can. J. For. Res.* **12,** 699–704.

Bevege, D. I., Bowen, G. D., and Skinner, M. F. (1975). Comparative carbohydrate physiology of ecto- and endomycorrhizas. *In* "Endomycorrhizas" (F. E. Sanders, B. Mosse, and P. B. Tinker, Eds.), pp. 149–174. Academic Press, Orlando, FL.

Beven, K. (1989). Changing ideas in hydrology—The case of physically based models. *J. Hydrol.* **105,** 157–172.

Bilzi, A. F., and Ciolkosz (1977). Time as a factor in the genesis of four soils developed in recent alluvium in Pennsylvania. *Soil Sci. Soc. Am. J.* **41,** 122–127.

Binkley, D. (1984). Does forest removal increase rates of decomposition and nitrogen release? *For. Ecol. Manage.* **8,** 229–233.

Binkley, D. (1986). "Forest Nutrition Management." Wiley, New York.

Binkley, D. (1992). Mixtures of N_2-fixing and non-N_2-fixing tree species. *In* "The Ecology of Mixed Species Stands of Trees" (M. G. R. Cannell, D. C. Malcolm, and P. Robertson, Eds.), pp. 99–123. Blackwell, Oxford, UK.

Binkley, D., and Brown, T. C. (1993). Forest practices as nonpoint sources of pollution in North America. *Water Res. Bull.* **29,** 729–740.

Binkley, D., and Richter, D. (1987). Nutrient cycles and H^+ budgets of forest ecosystems. *Adv. Ecol. Res.* **16,** 1–51.

Binkley, D., and Valentine, D. (1991). Fifty-year biogeochemical effects of green ash, white pine, and Norway spruce in a replicated experiment. *For. Ecol. Manage.* **40,** 13–25.

Binkley, D., Lousier, J. D., and Cromack, K., Jr. (1984). Ecosystem effects of sitka alder in a Douglas fir plantation. *For. Sci.* **30,** 26–35.

Binkley, D., Driscoll, C. T., Allen, H. L., Schoenberger, P., and McAvoy, D. (Eds.) (1989). Acidic deposition and forest soils: Context and case studies in the southeastern United States. *In* "Ecological Studies," Vol. 72. Springer-Verlag, New York.

Binkley, D., Richter, D., David, M. B., and Caldwell, B. (1994). Soil chemistry in a loblolly/longleaf pine forest with interval burning. *Ecol. Appl.* **2,** 157–164.

Binkley, D., Smith, F. W., and Son, Y. (1995). Nutrient supply and declines in leaf area and production in lodgepole pine. *Can. J. For. Res.* **25,** 621–628.

Birk, E. M., and Vitousek, P. M. (1986). Nitrogen availability and nitrogen use efficiency in loblolly pine stands. *Ecology* **67,** 69–79.

Birkeland, P. W. (1978). Soil development as an indicator of relative age of Quartenary deposits, Baffin Island, N.W.T., Canada. *Arctic Alpine Res.* **10,** 733–747.

Bockheim, J. G. (1979). Properties and relative age of soils of southwestern Cumberland Pennisula, Baffin Island, N.W.T., Canada. *Arctic Alpine Res.* **11,** 289–306.

Bockheim, J. G. (1980). Solution and use of chronofunctions in studying soil development. *Geoderma* **24,** 71–85.

Bockheim, J. G., Leide, J. E., and Tavella, D. S. (1986). Distribution and cycling of macronutrients in a *Pinus resinosa* plantation fertilized with nitrogen and potassium. *Can. J. For. Res.* **16,** 778–785.

Bohn, H. L., McNeal, B. L., and O'Connor, G. A. (1979). "Soil Chemistry." Wiley, New York.

Bolan, N. S. (1991). A critical review on the role of mycorrhizal fungi in the uptake of phosphorus by plants. *Plant Soil* **134,** 189–207.

Bolan, N. S., Robson, A. D., Barrow, N. J., and Aylmore, L. A. G. (1984). Specific activity of phosphorus in mycorrhizal and non-mycorrhizal plants in relation to the availability of phosphorus to plants. *Soil Biol. Biochem.* **16,** 299–304.

Bolin, B., Crutzen P. J., Vitousek, P. M., Woodmansee, R. G., Goldberg, E. D., and Cook, R. B. (1983). Interactions of biogeochemical cycles. *In* "The Major Biogeochemical Cycles and Their Interactions" (B. Bolin and R. B. Cook, Eds.) pp. 1–39. Scope 21, Wiley, Chichester, UK.

Bolstad, P., and Lillesand, T. M. (1992). Improved classification of forest vegetation in northern Wisconsin through a rule-based combination of soils, terrain, and Landsat Thermatic Mapper data. *For. Sci.* **38,** 5–20.

Bonan, G. B. (1991a). A biophysical surface energy budget analysis of soil temperature in the boreal forests of interior Alaska. *Water Resour. Res.* **27,** 767–781.

Bonan, G. B. (1991b). Atmoshpere–biosphere exchange of carbon dioxide in boreal forests. *J. Geophys. Res.* **96,** 7301–7312.

Bonan, G. B., Pollard, D., and Thompson, S. L. (1993). Effects of boreal forest vegetation on global climate. *Nature* **359,** 716–718.

Bonan, G. B., Shugart, H. H., and Urban, D. L. (1991). The sensitivity of some high-latitude boreal forests to climatic parameters. *Climate Change* **16,** 9–29.

Boone, R. D., Sollins, P., and Cromack, K. (1988). Stand and soil change along a mountain hemlock death and regrowth sequence. *Ecology* **69,** 714–722.

Borchert, R. (1980). Phenology and ecophysiology of tropical trees: *Erythrina poeppigiana* O. F. Cook. *Ecology* **61,** 1065–1074.

Borghetti, M., Raschi, A., and Grace, J. (1989). Ultrasound emission after cycles of water stress in *Picea abies. Tree Physiol.* **5,** 229–237.

Borghetti, M., Edwards, W. R. N., Grace, J., Jarvis, P. J., and Raschi, A. (1991). The refilling of embolized xylem in *Pinus sylvestris* L. *Plant Cell Environ.* **14,** 357–369.

Boring, L. R., Swank, W. T., Waide, J. B., and Henderson, G. S. (1988). Sources, fates, and impacts of nitrogen inputs to terrestrial ecosystems: Review and synthesis. *Biogeochem.* **6,** 119–159.

Bormann, B. T., and Gordon, J. C. (1984). Stand density effects in young red alder plantations: Productivity, photosynthate partitioning, and nitrogen fixation. *Ecology* **65,** 394–402.

Bormann, F. J., and Likens, G. E. (1979). "Pattern and Process in a Forested Ecosystem." Springer-Verlag, Berlin.

Borota, J. (1991). "Tropical forests: Some African and Asian case studies of composition and structure." *In* "Developments in Agricultural and Managed Forest Ecology," pp. 274. Elsevier, Amsterdam.

Bosch, J. M., and Hewlett, J. D. (1982). A review of catchment experiments to determine the effects of vegetation changes on water yield and evapotranspiration. *J. Hydrol.* **55,** 3–23.

Botkin, D. B., Janak, J. F., and Wallis, J. R. (1972). Some ecological consequences of a computer model of forest growth. *J. Ecol.* **60,** 849–873.
Bowden, R. D., Nadelhoffer, K. J., Boone, R. D., Melillo, J. M., and Garrison, J. B. (1993). Contributions of aboveground litter, belowground litter, and root respiration to total soil respiration in a temperate mixed hardwood forest. *Can. J. For. Res.* **23,** 1402–1407.
Boyer, W. D., and Miller, J. H. (1994). Effect of burning and brush treatments on nutrients and soil physical properties in young longleaf pine stands. *For. Ecol. Manage.* **70,** 311–318.
Boyle, J. R., and Voigt, G. K. (1973). Biological weathering of silicate minerals. Implications for tree nutrition and soil genesis. *Plant Soil* **38,** 191–201.
Boyle, J. R., Voigt, G. K., and Sawhney, B. L. (1974). Chemical weathering of biotite by organic acids. *Soil Sci.* **117,** 42–45.
Box, E. O. (1981). "Macroclimate and Plant Forms: An Introduction to Predictive Modeling in Phytogeography." Junk, The Hague.
Brasseur, G. P., and Chatfield, R. B. (1991). The fate of biogenic trace gases in the atmosphere. *In* "Trace Gas Emissions by Plants" (T. D. Sharkey, E. A. Holland, and H. A. Mooney, Eds.), pp. 1–27. Academic Press, San Diego.
Braunberger, P. G., Miller, M. H., and Peterson, R. L. (1991). Effect of phosphorus nutrition on morphological characteristics on vesicular-arbuscular mycorrhizal colonization of maize. *New Phytol.* **119,** 107–113.
Bray, J. R., and Gorham, E. (1964). Litter production in forests of the world. *Adv. Ecol. Res.* **2,** 101–157.
Bréda, N., Granier, A., and Aussenac, G. (1995). Effects of thinning on soil and tree water relations, transpiration and growth in an oak forest (*Quercus petraea* (Matt.) Liebl.). *Tree Physiol.* **15,** 295–306.
Bredemeier, M. J. (1988). Forest canopy transformation of atmospheric deposition. *Water Air Soil Pollut.* **40,** 121–138.
Bridgham, S. D., and Richardson, C. J. (1992). Mechanisms controlling soil respiration (CO_2 and CH_4) in southern peatlands. *Soil Biol. Biochem.* **24,** 1089–1099.
Brinkmann, W. L. F., and Nascimento, J. C. (1973). The effect of slash and burn agriculture on plant nutrients in the Tertiary region of Central Amazonia. *Turrialba* **23,** 284–290.
Brinson, M. M. (1977). Decomposition and nutrient exchange of litter in an alluvial swamp forest. *Ecology* **63,** 670–678.
Brix, H. (1981). Effects of nitrogen fertilization source and application rates on foliar nitrogen concentration, photosynthesis and growth of Douglas-fir. *Can. J. For. Res.* **11,** 775–780.
Brix, H., and Ebell, L. F. (1969). Effects of nitrogen fertilization on growth, leaf area and photosynthesis rate in Douglas-fir. *For. Sci.* **15,** 189–196.
Brix, H., and Mitchell, A. (1983). Thinning and fertilization effects on sapwood development and relationships of foliage quantity to sapwood area and basal area in Douglas-fir. *Can. J. For. Res.* **13,** 384–389.
Bruijnzeel, L. A. (1990). "Hydrology of Moist Tropical Forests and Effects of Conversion: A State of Knowledge Review." Free Univ. of Amsterdam, Amsterdam.
Brunig, E. S. (1983). Vegetation structure and function. *In* "Tropical Rain Forest Ecosystems: Structure and Function." (F. B. Golley, Ed.), Vol. 14A, pp. 49–75. Elsevier, Amsterdam.
Bryant, J. P., Chapin, F. S., III, and Klein, D. R. (1983). Carbon/nutrient balance of boreal plants in relation to vertebrate herbivory. *Oikos* **40,** 357–368.
Burges, A., and Drover, D. P. (1953). The rate of podzol development in sands of the Woy Woy District, N.S.W. *Aust. J. Bot.* **1,** 83–94.
Burke, I. C., Yonker, C. M., Parton, W. J., Cole, C. V., Flach, K., and Schimel, D. S. (1989). Texture, climate, and cultivation effects on soil organic matter content in U.S. grassland soils. *Soil Sci. Soc. Am. J.* **53,** 800–805.

Burrough, P. A. (1989). "Principles of Geographic Information Systems for Land Resource Assessment." Oxford Univ. Press, New York.
Burton, A. J., Pregitzer, K. S., and Reed, D. D. (1991). Leaf area and foliar biomass relationships in northern hardwood forests located along an 800 km acid deposition gradient. *For. Sci.* **37,** 1041–1059.
Calder, I. R. (1978). Transpiration observations from a spruce forest and comparisons with predictions from an evaporation model. *J. Hydrol.* **38,** 33–47.
Calder, I. R. (1986). A stochastic model of rainfall interception. *J. Hydrol.* **89,** 65–71.
Calder, I. R. (1990). "Evaporation in Uplands". Wiley, Chichester, UK.
Calder, I. R., Wright, I. R., and Murdiyaso, D. (1986). A study of evaporation from tropical rainforest—West Java. *J. Hydrol.* **89,** 13–31.
Campbell, C. A., Paul, E. A., Rennie, D. A., and McCallum, K. J. (1967). Applicability of the carbon-dating method of analysis to soil humus studies. *Soil Sci.* **104,** 217–224.
Campbell, G. S. (1974). A simple method for determining unsaturated conductivity from moisture retention data. *Soil Sci.* **117,** 311–314.
Campbell, R. G., Ferguson, I. S., and Opie, J. E. (1979). Simulating growth and yield of mountain ash stands: A deterministic model. *Aust. For. Res.* **9,** 189–202.
Canham, C. D., and Loucks, O. L. (1984). Catastrophic windthrow in the presettlement forests of Wisconsin. *Ecology* **65,** 803–809.
Cannell, M. G. R. (1982). "World Forest Biomass and Primary Production Data" Academic Press, New York.
Cannell, M. G. R. (1985). Dry matter partitioning in tree crops. *In* "Attributes of Trees as Crop Plants" (M. G. R. Cannell and J. E. Jackson, Eds.), pp. 160–193. Institute of Terrestrial Ecology, UK.
Cannell, M. G. R. (1989). Physiological basis of wood production: A review. *Scand. J. For. Res.* **4,** 459–490.
Cannell, M. G. R., and Dewar, R. C. (1994). Carbon allocation in trees: A review of concepts for modelling. *Adv. Ecol. Res.* **25,** 59–104.
Cannell, M. G. R., Sheppard, L. J., and Milne, R. (1988). Light use efficiency and woody biomass production of poplar and willow. *Forestry* **61,** 123–136.
Cannell, M. G. R., Milne, R., Sheppard, L. J., and Unsworth, M. H. (1987). Radiation interception and productivity of willow. *J. Appl. Ecol.* **24,** 261–268.
Carbon, B. A., Bartle, G. A., Murray, A. M., and Macpherson, D. K. (1980). The distribution of root length, and the limits to flow of soil water to roots in a dry sclerophyll forest. *For. Sci.* **26,** 656–664.
Carleton, A. M. (1991). "Satellite remote sensing in Climatology," p. 291. Bellhaven Press/CRC Press, Boca Raton.
Carter, K. N., and Prince, S. D. (1991). Epidemic models used to explain biogeographical distribution limits. *Nature* **293,** 644–645.
Chabot, B. F., and Hicks, D. J. (1982). The ecology of leaf life spans. *Annu. Rev. Ecol. Syst.* **13,** 229–259.
Chalot, M., Stewart, G. R., Brun, A., Martin, F., and Bottom, B. (1991). Ammonium assimilation by spruce—*Hebeloma* sp. ectomycorrhizas. *New Phytol.* **119,** 541–550.
Chandler, R. F. (1942). The time required for podzol profile formation as evidenced by the Mendenhall glacier deposits near Juneau, Alaska. *Soil Sci. Soc. Am. J.* **7,** 454–459.
Chapin, F. S., III (1980). The mineral nutrition of wild plants. *Annu. Rev. Ecol. Syst.* **11,** 233–260.
Chapin, F. S., III, and Kedrowski, R. A. (1983). Seasonal changes in nitrogen and phosphorus fractions and autumn retranslocation in evergreen and deciduous taiga trees. *Ecology* **64,** 376–391.
Chapin, F. S., III, Vitousek, P. M., and Van Cleve, K. (1986). The nature of nutrient limitation in plant communities. *Am. Nat.* **127,** 48–58.

Chapman, J. W., and Gower, S. T. (1991). Aboveground production and canopy dynamics in sugar maple and red oak trees in southwestern Wisconsin. *Can. J. For. Res.* **21,** 1533–1543.

Charney, J. G. (1975). Dynamics of deserts and drought in the Sahel. *Q. J. R. Met. Soc.* **101,** 193–202.

Charney, J. G., Stone, P. H., and Quirk, W. J. (1975). Drought in the Sahara: A biogeophysical feedback mechanism. *Science* **187,** 434–435.

Chen, J., and Black, T. A. (1992). Defining leaf area for non-flat surfaces. *Plant Cell Environ.* **15,** 421–429.

Chen, J., Rich, P. W., Gower, S. T., Norman, J. M., and Plummer, S. (1996). Leaf area index of boreal forests: Theory, techniques and measurements. *J. Geophys. Res.* (in press).

Cheng, W., Coleman, D. C., Carroll, C. R., and Hoffman, C. A. (1993). In situ measurement of root respiration and soluble C concentrations in the rhizosphere. *Soil Biol. Biochem.* **25,** 1189–1196.

Christ, M., Zhang, Y., Likens, G. E., and Driscoll, C. T. (1995). Nitrogen retention capacity of a northern hardwood forest soil under ammonium sulfate additions. *Ecol. Appl.* **5,** 802–812.

Chung, H. H., and Barnes, R. L. (1977). Photosynthate allocation in *Pinus taeda*. I. Substrate requirement for synthesis of shoot biomass. *Can. J. For. Res.* **7,** 106–111.

Clancy, K. M., Wagner, M. R., and Reich, P. B. (1995). Ecophysiology and insect herbivory. In "Ecophysiology of Conifers" (W. K. Smith and T. M. Hinckley, Eds.), pp. 125–180. Academic Press, San Diego.

Clapp, R. B., and Hornberger, G. M. (1978). Empirical equations for some soil hydraulic properties. *Water Resour. Res.* **14,** 601–604.

Clarkson, D. T. (1985). Factors affecting mineral nutrient acquisition by plants. *Annu. Rev. Plant Physiol.* **36,** 77–115.

Clayton, J. L., Megahan, W. F., and Hampton, D. (1979). Soil and bedrock properties: Weathering and alteration products and processes in the Idaho batholith. USDA Forest Service Intermountain Forest and Range Experiment Station, Ogden, UT.

Cleaves, E. T., Fisher, D. W., and Bricker, O. P. (1974). Chemical weathering of serpentine in the eastern Piedmont of Maryland. *Geol. Soc. Am. Bull.* **85,** 437–444.

Cleaves, E. T., Godfrey, A. E., and Bricker, O. P. (1970). Geochemical balance of a small watershed and its geomorphic implications. *Geol. Soc. Am. Bull.* **81,** 3015–3032.

Cohen, W. B., and Spies, T. A. (1992). Estimating structural attributes of Douglas-fir/western hemlock forest stands from Landsat and SPOT imagery. *Remote Sensing Environ.* **41,** 1–17.

Cohen, W. B., Spies, T. A., and Fiorella, M. (1995). Estimating the age and structure of forests in a multi-ownership landscape of western Oregon, USA. *Int. J. Remote Sensing* **16,** 721–746.

Cole, D. W., and Gessel, S. P. (1965). Movement of elements through a forest soil as influenced by tree removal and fertilizer additions. In "Forest–Soil Relationships in North America, Proceedings of the Second North American Forest Soils Conference (C. T. Youngberg, Ed.), pp. 95–104. Oregon State Univ. Press, Corvallis.

Cole, D. W., and Rapp, M. (1981). Elemental cycling in forest ecosystems. In "Dynamic Properties of Forest Ecosystems" (D. E. Reichle, Ed.), pp. 341–409. Cambridge Univ. Press, London.

Coley, P. D. (1988). Effects of plant growth rate and leaf lifetime on the amount and type of antiherbivore defense. *Oecologia* **74,** 531–536.

Coley P. D., Bryant, J. P., and Chapin, F. S. (1985). Resource availability and plant antiherbivore defense. *Science* **230,** 895–899.

Collatz, G. J., Ball, T., Grivet, C., and Berry, J. A. (1991). Physiological and environmental regulation of stomatal conductance, photosynthesis and transpiration: A model that includes a laminar boundary layer. *Agric. For. Meteorol.* **54,** 107–136.

Comeau, P. G., and Kimmins, J. P. (1989). Above and below-ground biomass and production of lodgepole pine on sites with differing soil moisture. *Can. J. For. Res.* **19**, 447–454.

Comins, H. N., and McMurtrie, R. E. (1993). Long-term response of nutrient-limited forests to CO_2 enrichment: equilibrium behaviour of plant–soil models. *Ecol. Appl.* **3**, 666–681.

Connell, H. H. (1978). Diversity in tropical rain forests and coral reefs. *Science* **199**, 1302–1309.

Cook, E. R., Johnson, A. H., and Blasing, T. J. (1987). Forest decline: Modeling the effect of climate in tree rings. *Tree Physiol.* **3**, 27–40.

Cooper, A. W. (1981) Aboveground biomass accumulation and net primary production during the first 70 years of succession in *Populus grandidentata* stand on poor sites in northern Lower Michigan. *In* "Forest Succession: Concepts and Applications" (D. C. West, H. H. Shugart, and D. B. Botkin, Eds.), pp. 339–360 Springer-Verlag, New York.

Cornish, P. M. (1993). The effects of logging and forest regeneration on water yields in a moist eucalypt forest in New South Wales, Australia. *J. Hydrol.* **150**, 301–322.

Coûteaux, M. M., Bottner, P., and Berg. B. (1995). Litter decomposition, climate and litter quality. *Trends Ecol. Evol.* **10**, 63–66.

Cowan, I. R. (1968). The interception and absorption of radiation in plant stands. *J. Appl. Ecol.* **5**, 367–379.

Cowan, I. R., and Farquhar, G. D. (1977). Stomatal function in relation to leaf metabolism and environment. *Symp. Soc. Exp. Biol.* **31**, 471–505.

Cowling, E. B., and Merrill, W. (1966). Nitrogen in wood and its role in wood deterioration. *Can. J. Bot.* **44**, 1539–1554.

Cranswick, A. M., Rook, D. A., and Zabkiewicz, J. A. (1987). Seasonal changes in carbohydrate concentration and composition of different tissue types of *Pinus radiata* trees. *N. Z. J. For. Sci.* **17**, 229–245.

Cremer, K. (1992). Relations between reproductive growth and vegetative growth of *Pinus radiata*. *For. Ecol. Manage.* **52**, 179–198.

Crocker, R. L., and Dickson, B. A. (1957). Soil development on the recessional moraines of the Herbert and Mendenhall Glaciers, south-eastern Alaska. *J. Ecol.* **45**, 169–185.

Crocker, R. L., and Major, J. (1955). Soil development in relation to vegetation and surface age at Glacier Bay, Alaska. *J. Ecol.* **43**, 427–448.

Cromer, R. N., Cameron, D. M., Rance, S. J., Ryan, P. A., and Brown, M. (1993). Response to nutrients in *Eucalyptus grandis*. 1. Biomass accumulation. *For. Ecol. Manage.* **62**, 211–230.

Cronan, C. S., and Reiners, W. A.(1983). Canopy processing of acid precipitation by coniferous and hardwood forests in New England. *Oecologia* **59**, 216–223.

Cronan, C. S., Reiners, W. A., Reynolds, R. C., Jr., and Lang, G. E. (1978). Forest floor leaching: Contributions from mineral, organic, and carbonic acids in New Hampshire subalpine forests. *Science* **200**, 309–311.

Cuevas, E., and Medina, E. (1986). Nutrient dynamics within amazonian forest ecosystems I. Nutrient flux in fine litter fall and efficiency of nutrient utilization. *Oecologia* **68**, 466–472.

Cuevas, E., and Medina, E. (1988). Nutrient dynamics within amazonian forest ecosystems I. Fine root growth, nutrient availability and leaf litter production. *Oecologia* **76**, 222–235.

Curran, P. J., Dungan, J. L., and Gholz, H. L. (1992). Seasonal LAI in slash pine estimated with Landsat TM. *Remote Sensing Environ.* **39**, 3–13.

Cushon, G. H., and Feller, M. C. (1989). Asymbiotic nitrogen fixation and denitrification in a mature forest in coastal British Columbia. *Can. J. For. Res.* **19**, 1194–1200.

Dahir, S. E. (1994). Tree mortality and gap formation in old-growth hemlock-hardwood forests of the Great Lakes region, pp. 100. MS. thesis, Univ. of Wisconsin—Madison.

Dahlgren, R. A., and Marrett, D. J. (1991). Organic carbon sorption in arctic and subalpine spodosol B horizons. *Soil. Sci. Soc. Am. J.* **55**, 1382–1390.

Dale, V. H., Doyle, T. W., and Shugart, H. H. (1985). A comparison of tree growth models. *Ecol. Model.* **29**, 145–169.

Das, A. K., and Ramakrishnan, P. S. (1987). Aboveground biomass and nutrient contents in an age series of khasi pine *Pinus kesiya*. *For. Ecol. Manage.* **18**, 61–72.

Davidson, E. A., Stark, J. M., and Firestone, M. K. (1990). Microbial production and consumption of nitrate in an annual grassland. *Ecology* **77**, 1968–1975.

Davidson, E. A., Hart, S. C., and Firestone, M. K. (1992). Internal cycling of nitrate in soils of a mature coniferous forest. *Ecology* **73**, 1148–1156.

Davidson, R. L. (1969). Effect of root/leaf temperature differentials on root/shoot ratios in some pasture grasses and clover. *Ann. Bot.* **33**, 471–505.

Davies, R. I., Coulson, C. B., and Lewis, D. A. (1964). Polyphenols in plant, humus, and soil. III. Stabilization of gelatin by polyphenol tanning. *J. Soil Sci.* **15**, 299–309.

Davies, W. J., and Zhang, J. (1991). Root signals and the regulation of growth and development of plants in drying soil. *Annu. Rev. Plant Physiol. Plant. Mol. Biol.* **42**, 55–76.

Davis, M. B. (1981). Quarternary history and the stability of forest communities. *In* "Forest Succession" (D. C. West, H. H. Shugart, and D. B. Botkin, Eds.), pp. 132–154. Springer-Verlag, New York.

Davis, M. B. (1986). Climatic instability, time lags and community disequilibrium. *In* "Community Ecology" (J. Diamond and T. Case, Eds.), pp.269–284. Harper & Row, New York.

Davoren, A. (1986). Land use basin programs: Comparison of hydrological indices from different land use types. Ministry of Works and Development, Report No. WS 1032. Water and Soil Science Centre, Christchurch, NZ.

Dawson, H. J., Ugolini, F. C., Hrutfiord, B. F., and Zuchara, J. (1978). Role of soluble organics in the soil processes of a podzol, Central Cascades, WA. *Soil Sci.* **126**, 290–296.

Day, F. P., Jr. (1982). Litter decomposition rates in the seasonally flooded Great Dismal Swamp. *Ecology* **63**, 670–678.

Deans, J. D. (1979). Flucuations of the soil environment and fine root growth in a young sitka spruce plantation. *Plant Soil* **52**, 195–208.

DeConinck, F. (1980). Major mechanisms in the formation of spodic horizons. *Geoderma* **24**, 101–128.

Delcourt, H. R., and Harris, W. F. (1980). Carbon budget of the southeastern U.S. biota: Analysis of historical change in trend from source to sink. *Science* **210**, 321–323.

Denmead, O. T., and Bradley, E. F. (1985). Flux–gradient relationships in a forest canopy. *In* "The Forest–Atmosphere Interaction" (B. A. Hutchison and B. B. Hicks, Eds.), pp. 421–442. Reidal, Dordrecht.

Denslow, J. S. (1980). Gap partitioning among tropical rainforest trees. *Biotropica* **12**, 47–55.

Denslow, J. S. (1987). Tropical rain forest gaps and tree species diversity. *Annu. Rev. Ecol. Syst.* **18**, 431–451.

Detwiler, R. P. (1986). Land use change and the global carbon cycle: The role of tropical soils. *Biogeochem.* **2**, 67–93.

Dewar, R., and McMurtrie, R. E. (1996a). Analytical model of stemwood growth in relation to nitrogen supply. *Tree Physiol.* **16**, 161–171.

Dewar, R., and McMurtrie, R. E. (1996b). Sustainable stemwood yield in relation to the nitrogen balance of forest plantations: a model analysis. *Tree Physiol.* **16**, 173–182.

Dick, J. M., Jarvis, P. G., and Leakey, R. R. B. (1991). Influence of male and female cones on needle CO_2 exchange rates of field-grown *Pinus contorta* Douglas trees. *Funct. Ecol.* **5**, 422–432.

Dickinson, R. E., and Henderson-Sellers, A. (1988). Modelling tropical deforestation: A study of GCM land–surface parameterizations. *Q. J. R. Met. Soc.* **114**, 439–462.

Dighton, J. (1983). Phosphatase production by mycorrhizal fungi. *Plant Soil* **71**, 455–462.

Dodd, J. C., Burton, C. C., Burns, R. G., and Jeffries, P. (1987). Phosphatase activity associated with the roots and the rhizosphere of plants infected with vesicular–arbuscular mycorrhizal fungi. *New Phytol.* **107**, 163–172.

Doley, D. (1981). Tropical and sub-tropical forests and woodlands. *In* "Woody Plant Communities" (T. T. Kozlowski, Ed.), Vol. VI. pp. 209–323. Academic Press, New York.

Doley, D. (1982). Photosynthetic productivity of forest canopies in relation to solar radiation and nitrogen cycling. *Aust. For. Res.* **12**, 245–261.

Dommergues, Y. R., Diem, H. G., Gauthier, D. L., Dreyfus, B. L., and Cornet, F. (1984). Nitrogen-fixing trees in the tropics: Potentialities and limitations. *In* "Advances in Nitrogen Fixation Research" (C. Veeger and W. E. Newton, Eds.), pp. 7–15. Nijhoff/Junk, The Hague.

Dons, A. (1986). The effect of large-scale afforestation on Tarawera river flows. *J. Hydrol. (New Zealand)* **25**, 61–73.

Dosskey, M. G., Linderman, R. G., and Boersma, L. (1990). Carbon-sink stimulation of photosynthesis in Douglas-fir seedlings by some ectomycorrhizas. *New Phytol.* **115**, 269–274.

Doyle, T. W. (1981). The role of disturbance in the gap dynamics of a montane rain forest: An application of a tropical succession model. *In* "Forest Succession: Concepts and Application" (D. C. West, H. H. Shugart, and D. B. Botkin, Eds.), pp. 56–73. Springer-Verlag, New York.

Drew, A. P., and Ledig, F. T. (1981). Seasonal patterns of CO_2 exchange in the shoot and root of loblolly pine seedlings. *Bot. Gaz.* **142**, 200–205.

Drew, T. J., and Flewelling, J. W. (1977). Some recent Japanese theories of yield-density relationships and their application to Monterey pine plantations. *For. Sci.* **23**, 517–534.

Duke, S. E., Jackson, R. B., and Caldwell, M. M. (1994). Local reduction of mycorrhizal arbuscule frequency in enriched soil microsites. *Can. J. Bot.* **72**, 998–1001.

Dunin, F. X., McIlroy, I. C., and O'Loughlin, E. M. (1985). A lysimeter characterization of evaporation by eucalypt forest and its representativeness for the local environment. *In* "The Forest–Atmosphere Interaction" (B. A. Hutchison and B. B. Hicks, Eds.), pp. 271–291. Reidal, Dordrecht.

Dyrness, C. T., Viereck, L. A., and Van Cleve, K. (1986). Fire in taiga communities of interior Alaska. *In* "Forest Ecosystems in the Alaskan Taiga" (K. Van Cleve, F. S. Chapin, III, P. W. Flanagan, L. A. Viereck, and C. T. Dyrness, Eds.). Springer-Verlag, New York.

Eck, T. F., and Dye, D. G. (1991). Satellite estimation of incident photosynthetically active radiation using ultraviolet reflectance. *Remote Sensing Environ.* **38**, 135–146.

Edmonds, R. L. (1979). Decomposition and nutrient release in Douglas fir needle litter in relation to stand development. *Can. J. For. Res.* **9**, 132–140.

Edmonds, R. L., Thomas, T. B., and Rhodes, J. J. (1991). Canopy and soil modification of precipitation chemistry in a temperate rain forest. *Soil Sci. Soc. Am. J.* **55**, 1685–1693.

Edwards, N. T. (1982). The use of soda lime for measuring respiration rates in terrestrial ecosystems. *Pedobiologia* **23**, 321–330.

Edwards, N. T., and Ross-Todd, B.M. (1983). Soil carbon dynamics in a mixed deciduous forest following clear-cutting with and without residue removal. *Soil Sci. Soc. Am. J.* **47**, 1014–1021.

Edwards, N. T., and Sollins, P. (1973). Continuous measurement of carbon dioxide evolution from partitioned forest floor components. *Ecology* **54**, 406–412.

Edwards, P. J. (1977). Studies of mineral cycling in a montane rain forest in New Guinea. II. The production and disappearance of litter. *J. Ecol.* **65**, 971–992.

Edwards, P. J., and Grubb, P. J. (1982). Studies of mineral cycling in a montane rain forest in New Guinea. IV. Soil characteristics and the division of mineral elements between the vegetation and soil. *J. Ecol.* **70**, 645–666.

Ehleringer, J. R., and Field, C. B. (Eds.) (1993). "Scaling Physiological Processes: Leaf to Globe." Academic Press, San Diego.

Ehrlich, P. R., and Daily, G. C. (1993). Science and the management of natural resources. *Ecol. Appl.* **3,** 558–560.

Ehrlich, P. R., and Ehrlich, A. H. (1981). "Extinction, the Causes and Consequences of the Disappearance of Species." Random House, New York.

Eissenstat, D. M., and Yanai, R. D. (1996). The ecology of root lifespan. *Adv. Ecol. Res.* **27** (in press).

Ek, A. R., and Monserud, R. A. (1974). FOREST: A computer model for growth and reproduction of mixed species forest stands. Univ. Wisconsin, College of Agriculture and Life Sciences Research Report No. R2635, Madison.

Ek, A. R., and Monserud, R. A. (1979). Performance and comparison of stand growth models based on individual tree and diameter-class growth. *Can. J. For. Res.* **9,** 231–244.

Emanuel, W. R., Shugart, H. H., and Stevenson, M. P. (1985). Climatic change and the broad-scale distribution of terrestrial ecosystem complexes. *Climate Change* **7,** 29–43.

Ericsson, A., and Persson, H. (1980). Seasonal changes in starch reserves and growth of fine roots of 20-year-old Scots pine. *Ecol. Bull. (Stockholm)* **32,** 239–250.

Erickson, H. E., Edmonds, R. L., and Peterson, C. E. (1985). Decomposition of logging residue in Douglas-fir, Western hemlock, Pacific silver fir, and ponderosa pine ecosystems. *Can. J. For. Res.* **15,** 914–921.

Evans, L. J., and Cameron, B. H. (1979). A chronosequence of soils developed from granitic morainal material, Baffin Island, N.W.T. *Can. J. Soil Sci.* **59,** 203–210.

Ewel, J., Berish, C., Brown, B., Price, N., and Raich, J. (1981). Slash and burn impacts on a Costa Rican wet forest site. *Ecology* **62,** 816–829.

Ewel, K. C., Cropper, W. L., and Gholz, H. L. (1987). Soil CO_2 evolution in Florida slash pine plantations. I. Changes through time. *Can. J. For. Res.* **17,** 325–329.

Ewel, K. C., Cropper, W. P., Jr., and Gholz, H. L. (1987). Soil CO_2 evolution in Florida slash pine plantations. II. Importance of root respiration. *Can. J. For. Res.* **17,** 330–333.

Fahey, B. D., and Rowe, L. K. (1992). Land-use impacts. *In* "Water of New Zealand," pp. 265–284. New Zealand Hydrological Society.

Fahey, T. J., and Arthur, M. A. (1994). Further studies of root decomposition following harvest of northern hardwood forests. *For. Sci.* **40,** 618–629.

Fahey, T. J., and Young, D. R. (1984). Soil and xylem water potential and soil water content in contrasting *Pinus contorta* ecosystems, southeastern Wyoming, USA. *Oecologia* **61,** 346–351.

Fan, S.-M., Goulden, M. L., Munger, J. W., Daube, B. C., Bakwin, P. S., Wofsy, S. C., Amthor, J. S., Fitzjarrald, D. R., Moore, K. E., and Moore, T. R. (1995). Environmental controls on the photosynthesis and respiration of a boreal lichen woodland: A growing season of whole-ecosystem exchange measurements by eddy correlation. *Oecologia* **102,** 443–452.

Farquhar, G. D., and Sharkey, T. D. (1982). Stomatal conductance and photosynthesis. *Annu. Rev. Plant Physiol.* **33,** 317–345.

Farquhar, G. D., and von Caemmerer, S. (1982). Modelling of photosynthetic response to environmental conditions. *In* "Physiological Plant Ecology II. Water Relations and Carbon Assimilation" (O. L. Lange, P. S. Nobel, C. B. Osmond, and H. Ziegler, Eds.), pp. 549–588. Springer-Verlag, Berlin.

Farquhar, G. D., O'Leary, M. H., and Berry, J. A. (1982). On the relationship between stable carbon isotope discrimination and the intercellular carbon dioxide concentration in leaves. *Aust. J. Plant Physiol.* **9,** 121–137.

Fassnacht, K. S. (1996). Characterization of the structure and function of upland forest ecosystems in north central Wisconsin. Ph.D. Dissertation, Univ. of Wisconsin, Madison, WI.

Fassnacht, K. S., Gower, S. T., Norman, J. M., and McMurtrie, R. E. (1994). A comparison of optical and direct methods for estimating foliage surface area index in forests. *Agric. For. Meteorol.* **71,** 183–207.

Fassnacht, K. S., Gower, S. T., MacKenzie, M. D., Nordheim, E. V., and Lillesand, T. M. (1996). Estimating the leaf area index of north central Wisconsin forests using the Landsat Thermatic Mapper. *Remote Sensing Environ.* (in press).
Federer, C. A. (1968). Spatial variation of net radiation, albedo and surface temperature of forests. *J. Appl. Meteorol.* **7**, 789–795.
Feeny, P. P. (1976). Plant apparency and chemical defense. *Rec. Adv. Phytochem.* **10**, 1–40.
Feller, M. C. (1977). Nutrient movement through western hemlock–western red cedar ecosystems in southwestern British Columbia. *Ecology* **58**, 1269–1283.
Feller, M. C. (1980). Biomass and nutrient distribution in two eucalypt forest ecosystems. *Aust. J. Ecol.* **5**, 309–333.
Feller, M. C. (1981). Water balances in *Eucalyptus regnans, E. obliqua,* and *Pinus radiata* forests in Victoria. *Aust. For.* **44**, 153–161.
Feller, M. C., and Kimmins, J. P. (1979). Chemical characteristics of small streams near Haney in Southwestern British Columbia. *Water Resour. Res.* **15**, 247–258.
Field, C. (1983). Allocating leaf nitrogen for the maximization of carbon gain: Leaf age as a control on the allocation program. *Oecologia* **56**, 341–347.
Field, C., and Mooney, H. A. (1986). The photosynthesis–nitrogen relationship in wild plants. *In* "On the Economy of Plant Form and Function" (T. J. Givnish, Ed.), pp. 25–55. Cambridge Univ. Press, Cambridge, UK.
Fielding, J. M. (1960). Branching and flowering characteristics of Monterey pine, Bulletin No. 37. Forestry and Timber Bureau, Canberra, Australia.
Firestone, M. K., Firestone, R. B., and Tiedje, J. M. (1980). Nitrous oxide from soil denitrification: Factors controlling its biological production. *Science* **208**, 749–751.
Fisher, R. F. (1995). Amelioration of degraded rain forest soils by plantations of native trees. *Soil Sci. Soc. Am. J.* **59**, 544–549.
Fitzjarrald, D. R., and Moore, K. E. (1995). Physical mechanisms of heat and mass exchange between forests and the atmosphere. *In* "Forest Canopies" (M. D. Lowman and N. M. Nadkarni, Eds.), pp. 45–72. Academic Press, San Diego.
Flanagan, P. W., and Van Cleve, K. (1983). Nutrient cycling in relation to decomposition and organic matter quality in taiga ecosystems. *Can. J. For. Res.* **13**, 795–814.
Fog, K. (1988). The effect of added nitrogen on the rate of decomposition of organic matter. *Biol. Rev.* **63**, 433–462.
Fogel, R., and Cromack, K., Jr. (1977). Effect of habitat and substrate quality on Douglas fir (*Pseudotsuga menziesii*) litter decomposition in western Oregon. *Can. J. Bot.* **55**, 1632–1644.
Ford, E. D., and Deans, J. D. (1978). The effects of canopy structure on stemflow, throughfall and interception loss in a young sitka spruce plantation. *J. Appl. Ecol.* **15**, 905–917.
Foster, J. R., and Lang, G. E. (1982). Decomposition of red spruce and balsam fir boles in the White Mountains of New Hampshire. *Can. J. For. Res.* **12**, 617–626.
Foster, N. W., and Morrison, I. K. (1976). Distribution and cycling of nutrients in a natural *Pinus banksiana* ecosystem. *Ecology* **57**, 110–120.
Foster, N. W., Beauchamp, E. G., and Corke, C. T. (1980). Microbial activity in a *Pinus banksiana:* Lamb. forest floor amended with nitrogen and carbon. *Can. J. For. Res.* **60**, 199–209.
Fowkes, N. D., and Landsberg, J. J. (1981). Optimal root systems in terms of water uptake and movement. *In* "Mathematics and Plant Physiology" (D. A. Rose and D. A. Charles-Edwards, Eds.), pp. 109–125. Academic Press, San Diego.
Fox, T. R., Comerford, N. B., and McFee, W. W. (1990). Phosphorus and aluminum release from a spodic horizon mediated by organic acids. *Soil Sci. Soc. Am. J.* **54**, 1763–1767.
Franklin, J. F., and Forman, R. T. T. (1987). Creating landscape patterns by forest cutting: Ecological consequences and principles. *Landscape Ecol.* **1**, 5–18.

Franxmeier, D. P., Whiteside, E. P., and Mortland, M. M. (1963). A chronosequence of podzols in northern Michigan, III. Mineralology, micromorphology, and net changes during soil formation. *Mich. Ag. Exp. Stn. Q. Bull.* **46**, 37–57.

Fredriksen, R. L. (1972). Nutrient budget of a Douglas-fir forest on an experimental watershed in western Oregon. *In* "Proc. Research on Coniferous Forest Ecosystems." (J. F. Franklin, ed.) pp. 115–131. USDA For. Serv. Pacific Northwest Forest Range Exp. Stn. Portland, OR.

Fredriksen, R., Moore, D., and Norris, L. (1975). The impact of timber harvest, fertilization, and herbicide treatment on streamwater quality in western Oregon and Washington. *In* "Forest Soils and Land Management" (B. Bernier and C. Winget, Eds.), pp. 283–313. Laval Univ. Press, Quebec, Canada.

Freedman, B., Meth, F., and Hickman, C. (1992). Temperate forest as a carbon-storage reservoir for carbon dioxide emitted by coal-fired generating stations. A case study for New Brunswick, Canada. *For. Ecol. Manage.* **55**, 15–29.

Frelich, L. E., and Lorimer, C. G. (1991a). Natural disturbance regimes in the hemlock–hardwood forests of the upper Great Lakes region. *Ecol. Monogr.* **61**, 145–164.

Frelich, L. E., and Lorimer, C. G. (1991b). A simulation of landscape-level stand dynamics in the northern hardwood region. *J. Ecol.* **79**, 223–233.

Friend, A. D., Shugart, H. H., and Running, S. W. (1993). A physiology-based gap model of forest dynamics. *Ecology* **74**, 792–797.

Fuchs, E., and Livingston, N. (1994). Hydraulic control of stomatal conductance in Douglas-fir (*Pseudotsuga menziesii*) and alder (*Alnus oregona*) in drying soils. *Plant Physiol.* **105** (Suppl. 14). [Abstract]

Fujimori, T., Kawanabe, S., Saito, H., Grier, C. C., and Shidei, T. (1976). Biomass and net primary production in forests in three major vegetation zones of the northwestern United States. *J. Jpn. For. Soc.* **58**, 360–373.

Fung, I. Y., Tucker, C. J., and Prentice, K. C. (1987). Application of advanced very high resolution radiometer vegetation index to study atmosphere–biosphere exchange of CO_2. *J. Geophys. Res.* **92**, 2999–3015.

Gale, M. R., and Grigal, D. F. (1987). Vertical root distributions of northern tree species in relation to successional status. *Can. J. For. Res.* **17**, 829–834.

Gallardo, A., and Schlesinger, W. H. (1994). Factors limiting microbial biomass in the mineral soil and forest floor of a warm-temperate forest. *Soil Biol. Biochem.* **26**, 1409–1415.

Gardner, W. R., Hillel, D., and Beyaminini, Y. (1970). Post-irrigation movement of soil water: Redistribution. *Water Resour. Res* **6**, 851–886.

Gash, J. H. C. (1979). An analytical model of rainfall interception by forests. *Q. J. R. Met. Soc.* **105**, 43–55.

Gash, J. H. C., and Shuttleworth, W. J. (1991). Tropical deforestation: Albedo and the surface energy balance. *Climatic Change* **19**, 123–133.

Gentry, J. B., and Whitford, W. G. (1982). The relationship between wood litter in fall and relative abundance and feeding activity of subterranean termites Reticulitermes spp. in three southeastern coastal plain habitats. *Oecologia* **54**, 63–67.

Gerwitz, A., and Page, E. R. (1974). An empirical mathematical model to describe plant root systems. *J. Appl. Ecol.* **11**, 773–782.

Gholz, H. L. (1982). Environmental limits on above-ground net primary production, leaf area, and biomass in vegetation zones of the Pacific Northwest. *Ecology* **63**, 469–481.

Gholz, H. L., and Cropper, W. P. (1991). Carbohydrate dynamics in mature *Pinus elliottii* var *elliottii* trees. *Can. J. For. Res.* **21**, 1742–1747.

Gholz, H. L., and Fisher, R. F. (1982). Organic matter production and distribution in slash pine (*Pinus elliotti*) plantations. *Ecology* **63**, 1827–1839.

Gholz, H. L., Grier, C. C., Campbell, A. G., and Brown, A. T. (1979). Equations and their use for estimating biomass and leaf area of Pacific Northwest plants. Research Paper No. 41, Forest Research Laboratory, Oregon State Univ., Corvallis.

Gholz, H. L., Fisher, R. F., and Pritchett, W. L. (1985a). Nutrient dynamics in slash pine plantation ecosystems. *Ecology* **66**, 647–659.

Gholz, H. L., Perry, C. S., Cropper, W. P., Jr. and Hendry, L. C. (1985b). Litterfall, decomposition and nitrogen and phosphorus dynamics in a chronosequence of slash pine (*Pinus elliottii*) plantations. *For. Sci.* **31**, 463–478.

Gholz, H. L., Hendry, L. C., and Cropper, W. P., Jr. (1986). Organic matter dynamics of fine roots in plantations of slash pine (*Pinus elliottii*) in north Florida. *Can. J. For. Res.* **16**, 529–538.

Gholz, H. L., Vogel, S. A., Cropper, W. P., McKelvey, K., Ewel, K. C., Teskey, R. O., and Curran, P. J. (1991). Dynamics of canopy structure and light interception in *Pinus elliottii* stands, North Florida. *Ecol. Monogr.* **61**, 33–51.

Gifford, R. M. (1979). Carbon dioxide and plant growth under water and light stress: implications for balancing the global carbon budget. *Search* **10**, 316–318.

Gifford, R. M. (1994). The global carbon cycle: a viewpoint on the missing sink. *Aust. J. Plant Physiol.* **21**, 1–15.

Gleeson, S. K. (1993). Optimization of tissue nitrogen and root–shoot allocation. *Ann. Bot.* **71**, 23–31.

Glerum, C. (1980). Food sinks and food reserves of trees in temperate climates. *N. Z. J. For. Sci.* **10**, 176–185.

Gollan, T., Passioura, J. B., and Munns, R. (1986). Soil water status affects the stomatal conductance of fully turgid wheat and sunflower leaves. *Aust. J. Plant Physiol.* **13**, 459–464.

Goodburn, J. M. (1996). Comparison of forest habitat structure and composition in managed and old-growth northern hardwoods. M.S. thesis, University of Wisconsin, Madison, WI.

Gore, J. A., and Patterson, W. A., III (1986). Mass of downed wood in northern hardwood forests in New Hampshire: potential effects of forest management. *Can. J. For. Res.* **16**, 335–339.

Gosz, J. R. (1981). Nitrogen cycling in coniferous ecosystems. *Ecol. Bull. (Stockholm)* **33**, 405–426.

Gosz, J. R. (1986). Biogeochemistry research needs: observations from the ecosystem studies program of the National Science Foundation. *Biogeochem.* **2**, 101–102.

Goulden, M. L., Munger, J. W., Fan, S.-M., Daube, B. C., and Wofsy, S. C. (1996). Measurements of carbon storage by long-term eddy correlation: Methods and a critical evaluation of accuracy. *Global Change Biol.* (in press).

Goward, S. N., and Huemmrich, K. F. (1992). Vegetation canopy PAR absorptance and the normalized difference vegetation index: An assessment using the SAIL model. *Remote Sensing Environ.* **39**, 119–140.

Goward, S. N., Huemmrich, K. F., and Waring, R. H. (1994). Visible-near infrared spectral reflectance of landscape components in Western Oregon. *Remote Sensing Environ.* **47**, 190–203.

Gower, S. T. (1987). Relations between mineral nutrient availability and fine root biomass in two Costa Rican tropical wet forest. *Biotropica* **19**, 252–257.

Gower, S. T., and Richards, J. H. (1990). Larches: Deciduous conifers in an evergreen world. *BioScience* **40**, 818–826.

Gower, S. T., and Son, Y. (1992). Differences in soil and leaf litterfall nitrogen dynamics for five forest plantations. *Soil Sci. Soc. Am. J.* **56**, 1959–1966.

Gower, S. T., and Vitousek, P. M. (1989). Effect of nutrient amendments on fine root biomass of a primary successional tropical forest in Hawaii. *Oecologia* **81**, 566–568.

Gower, S. T., Vogt, K. A., and Grier, C. C. (1992). Carbon dynamics of Rocky Mountain Douglas-fir: Influence of water and nutrient availability. *Ecol. Monogr.* **62**, 43–65.

Gower, S. T., Reich, P. B., and Son, Y. (1993b). Canopy dynamics and aboveground production of five tree species with different leaf longevities. *Tree Physiol.* **12,** 327–345.

Gower, S. T., Haynes, B. E., Fassnacht, K. S., Running, S. W., and Hunt, E. R., Jr. (1993a). Influence of fertilization on the allometric relations for two pines in contrasting environments. *Can. J. For. Res.* **23,** 1704–1711.

Gower, S. T., Gholz, H. L., Nakane, K., and Baldwin, C. V. (1994). Production and carbon allocation patterns of pine forests. *Ecol. Bull. (Copenhagen)* **43,** 115–135.

Gower, S. T., Kloeppel, B. D., and Reich, P. B. (1995). Carbon, nitrogen, and water use by larches and co-occurring evergreen conifers. *In* "Ecology and Management of Larix Forests: A Look Ahead," pp. 110–117. USDA For. Serv. Gen. Tech. Report No. GTR-INT-319. USDA Forest Service, Ogden, UT.

Gower, S. T., Running, S. W., Gholz, H. L., Haynes, B. E., Hunt, E. R., Ryan, M. G., Waring, R. H., and Cropper, W. P., Jr. (1996a). Influence of climate and nutrition on carbon allocation and net primary production of four conifer forests. *Tree Physiol.* **16** (in press).

Gower, S. T., McMurtrie, R. E., and Murty, D. (1996b). Aboveground net primary production decline with stand age, potential causes. *Trends Ecol. Evol.* **11,** 378–382.

Gower, S. T., Pongracic, S., and Landsberg, J. J. (1996c). A global trend in below-ground carbon allocation: Can we use the relationship at smaller scales? *Ecology* **77,** 1750–1755.

Gower, S. T., Vogel, J., Stow, T. K., Norman, J. M., Steele, S. J., and Kucharik, C. J. (1996). Carbon distribution and aboveground net primary production of upland and lowland boreal forests in Saskatchewan and Manitoba, Canada. *J. Geophys. Res.* (submitted for publication).

Grace, J., Lloyd, J., McIntyre, J., Miranda, A., Meir, P., Miranda, H., Moncrieff, J., Massheder, J., Wright, I., and Gash, J. (1995). Fluxes of carbon dioxide and water vapour over an undisturbed tropical forest in south-west Amazonia. *Global Change Biol.* **1,** 1–12.

Grace, J. C. (1987). Theoretical ratio between 'one-sided' and total surface area for pine needles. *N. Z. J. For. Sci.* **17,** 292–295.

Graham, R. L., Wright, L. L., and Turhollow, A. F. (1992). The potential for short-rotation woody crops to reduce U.S. CO_2 emissions. *Climate Change* **22,** 223–238.

Granhall, U. (1981). Biological nitrogen fixation in relation to environmental factors and functioning of natural ecosystems. *Ecol. Bull. (Stockholm),* **33,** 131–144.

Gray, J. T. (1983). Nutrient use by evergreen and deciduous shrubs in southern California I. Community nutrient cycling and nutrient-use efficiency. *J. Ecol.* **71,** 21–41.

Grier, C. C. (1975). Wildfire effects on nutrient distribution and leaching in a coniferous forest ecosystem. *Can. J. For. Res.* **5,** 599–607.

Grier, C. C. (1978). A *Tsuga heterophylla–Picea sitchensis* ecosystem of coastal Oregon: Decomposition and nutrient balances of fallen logs. *Can. J. For. Res.* **8,** 198–206.

Grier, C. C., and Logan, R. S. (1977). Old-growth *Pseudotsuga menziesii* communities of a western Oregon watershed: biomass distribution and production budgets. *Ecol. Monogr.* **47,** 373–400.

Grier, C. C., and Running, S. W. (1977). Leaf area of mature northwestern coniferous forests: Relation to site water balance. *Ecology* **58,** 893–899.

Grier, C. C., and Waring, R. H. (1974). Conifer foliage mass related to sapwood area. *For. Sci.* **20,** 205–206.

Grier, C. C., Lee, K. M., and Archibald, R. M. (1985). Effect of urea fertilization on allometric relations in young Douglas-fir trees. *Can. J. For. Res.* **15,** 900–904.

Grier, C. C., Vogt, K. A., Keyes, M. R., and Edmonds, R. L. (1981). Biomass distribution and above- and belowground production in young and mature *Abies amabilis* zone ecosystems on the Washington Cascades. *Can. J. For. Res.* **11,** 155–167.

Grierson, P. F., and Attiwill, P.M. (1989). Chemical characteristics of the proteoid root mat of *Banksia integrifolia* L. f. *Aust. J. Bot.* **37,** 137–143.

Gries, J. F. (1995). Biomass and net primary production for a northern hardwood stand development sequence in the Upper Peninsula, Michigan. M.S. Thesis. University of Wisconsin, Madison, WI.

Groffman, P., and Tiedje, J. M. (1989). Denitrification in north temperate forest soils: Relationships between denitrification and environmental factors at the landscape scale. *Soil Biol. Biochem.* **21**, 621–626.

Gunderson, C. A., and Wullschleger, S. D. (1994). Photosynthetic acclimation in trees to rising atmospheric CO_2: A broader perspective. *Photosynth. Res.* **39**, 369–388.

Gupta, S. C., and Larson, W. E. (1979). Estimating soil water retention characteristics from particle size distribution, organic matter content and bulk density. *Water Resour. Res.* **15**, 1633–1635.

Gutschick, V. P. (1981). Evolved strategies in nitrogen acquisition by plants. *Am. Nat.* **118**, 607–637.

Hall, F. G., Huemmrich, K. F., and Goward, S. N. (1990). Use of narrow-band spectra to estimate the fraction of absorbed photosynthetically active radiation. *Remote Sensing Environ.* **32**, 47–54.

Hall, F. G., Botkin, D. B., Strebel, D. E., Woods, K. D., and Goetz, S. J. (1991). Large-scale patterns of forest succession as determined by remote sensing. *Ecology* **72**, 628–640.

Halldin, S. (1985). Leaf and bark area distribution in a pine forest. *In* "The Forest–Atmosphere Interaction" (B. A. Hutchison and B. B. Hicks, Eds.), pp. 39–58. Reidel, Dordrecht.

Handley, W. R. C. (1961). Further evidence for the importance of residual leaf protein complexes in litter decomposition and the supply of nitrogen for plant growth. *Plant Soil* **15**, 37–73.

Hansen, E. A., and Baker, J. B. (1979). Biomass and nutrient removal in short rotation intensively cultured plantations. *In* Proc. "Impact of Intensive Harvesting on Forest Nutrient Cycling." College of Environmental Science and Forestry, State University of New York, Syracuse, New York.

Harding, R. J., Hall, R. L., Swaminath, M. H., and Srinivasa Murthy, K. V. (1992). The soil moisture regimes beneath forest and an agricultural crop in southern India—measurements and modelling. *In* "Growth and Water Use of Forest Plantations" (I. R. Calder, R. L. Hall, and P. G. Adlard, Eds.), pp. 244–269. Wiley, Chichester, UK.

Harmon, M. E., Franklin, J. F., Swanson, F. J., Sollins, P., Gregory, S. V., Lattin, J. D., Anderson, N. H., Cline, S. P., Aumen, N. G., Sedell, J. R., Lienkaemper, G. W., Cromack, K., Jr., and Cummins, K. W. (1986). Ecology of coarse woody debris in temperate ecosystems. *Adv. Ecol. Res.* **15**, 133–302.

Harris, S. A. (1971). Podzol development on volcanic ash deposits in the Talamanea Range, Costa Rica (D. H. Yaalon, Ed.), pp. 191–209. *In* "Paleopedology: Origin, Nature and Dating of Paleosols" International Society of Soil Science, Jerusalem.

Hart, S. C., Nason, G. E., Myrold, D. D., and Perry, D. A. (1994). Dynamics of gross nitrogen transformations in an old-growth forest: the carbon connection. *Ecology* **75**, 880–891.

Hartshorn, G. S. (1989). Gap-phase dynamics and tropical tree species richness. *In* "Tropical Forests: Botanical Dynamics, Speciation and Diversity" (L. B. Holm-Nielsen, I. C. Nielsen, and H. Balslev, Eds.), pp. 65–73. Academic Press, London.

Hase, H. and Foelster, H. (1983). Impact of plantation forestry with teak (*Tectona grandis*) on the nutrient status of young alluvial soils in west Venezuela. *For. Ecol. Manage.* **6**, 33–57.

Hasse, W. D., and Ek, A. R. (1981). A simulated comparison of yields for even- versus uneven-aged management of northern hardwood stands. *J. Environ. Manage.* **12**, 235–246.

Havranek, W. M., and Tranquillini, W. (1995). Physiological processes during winter dormancy and their ecological significance. *In* "Ecophysiology of Coniferous Forests" (W. K. Smith and T. M. Hinckley, Eds.), pp. 99–124. Academic Press, San Diego.

Haynes, B. E., and Gower, S. T. (1995). Belowground carbon allocation in unfertilized and fertilized red pine plantations in northern Wisconsin. *Tree Physiol.* **15**, 317–325.

Hellkvist, J., Richards, G. P., and Jarvis, P. G. (1974). Vertical gradients of water potential and tissue water relations in Sitka spruce trees measured with the pressure chamber. *J. Appl. Ecol.* **11**, 637–668.

Henderson, G. S., Swank, W. T., Waide, J. B., and Grier, C. C. (1977). Nutrient budgets of Appalachian and Cascade Region watersheds: A comparison. *For. Sci.* **24**, 385–397.

Henderson-Sellers, A., and Gornitz, V. (1984). Possible climatic impacts of land cover transformations, with particular emphasis on tropical deforestation. *Climatic Change* **6**, 231–257.

Hendrick, R. L., and Pregitzer, K. S. (1993). Patterns of fine root mortality in two sugar maple forests. *Nature (London)* **361**, 59–61.

Hendrick, R. L., and Pregitzer, K. S. (1992). The demography of fine roots in a northern hardwood forest. *Ecology* **73**, 1094–1104.

Hendricks, J. J., Nadelhoffer, K. J., and Aber, J. D. (1993). Assessing the role of fine roots in carbon and nutrient cycling. *Trends Ecol. Evol.* **8**, 174–178.

Henry, J. D., and Swan, J. M. A. (1974). Reconstructing forest history from live and dead plant material—an approach to the study of forest succession in southwest New Hampshire. *Ecology* **55**, 772–783.

Herbohn, J. L., and Congdon, R. A. (1993). Ecosystem dynamics at disturbed and undisturbed sites in north Queensland wet tropical rain forest. II. Litterfall. *J. Trop. Ecol.* **9**, 365–380.

Hilbert, D. W., Larigauderie, A., and Reynolds, J. F. (1991). The influence of carbon dioxide and daily photo-flux density on optimal leaf nitrogen concentration and root/shoot ratio. *Ann. Bot.* **68**, 365–376.

Hillel, D. (1980). "Applications of Soil Physics." Academic Press, New York.

Hinckley, T. M., Lassoie, J. P., and Running, S. W. (1978). "Temporal and Spatial Variations in the Water Status of Forest Trees," Monograph 20, pp. 72. Society of American Foresters, Washington, DC.

Hingston, F. J., and Gailitis, V. (1976). The geographic variation of salt precipitated over western Australia. *Aust. J. Soil Res.* **14**, 319–335.

Hingston, F. J., Galbraith, J. H., and Dimmock, G. M. (1995). "Evaluating the Effects of Soils and Climate on the Productivity of *Eucalyptus globulus* Plantations on Contrasting Sites in South-West of Western Australia, CSF-41A, pp. 67.

Hobbie, S. E. (1992). Effects of plant species on nutrient cycling. *Trends Ecol. Evol.* **7**, 336–339.

Hodges, J., and Lorio, P. (1975). Moisture stress and xylem oleoresin in loblolly pine. *For. Sci.* **21**, 283–290.

Holbrook, N. M. (1995). Stem water storage. *In* "Plant Stems: Physiology and Functional Morphology" (B. L. Gartner, ed.), pp. 151–174. Academic Press, San Diego, CA.

Holbrook, N. M., and Lund, C. P. (1995). Photosynthesis in forest canopies. *In* "Forest Canopies" (M. D. Lowman and N. M. Nadkarni, Eds.), pp. 411–430. Academic Press, London.

Holdridge, L. R. (1947). Determination of world plant formations from simple climatic data. *Science* **105**, 367–368.

Holdridge, L. R. (1967). "Life Zone Ecology," Rev. ed. Tropical Science Center, San Jose, Costa Rica.

Hollinger, D. Y. (1989). Canopy organization and foliage photosynthetic capacity in a broad-leaved evergreen montane forest. *Funct. Ecol.* **3**, 53–62.

Hollinger, D. Y., Kelliher, F. M., Byers, J. N., Hunt, J. E., McSeveny, T. M., and Weir, P. L. (1994). Carbon dioxide exchange between an undisturbed old-growth temperate forest and the atmosphere. *Ecology* **75**, 134–150.

Hook, W. R., Livingston, N. J., Sun, Z. J., and Hook, P. B. (1992). Remote diode shorting improves measurement of soil water by time domain reflectometry. *Soil Sci. Soc. Am. J.* **56,** 1384–1391.

Horn, R., Schulze, E.D., and Hantschel, R. (1989). Nutrient balance and element cycling in healthy and declining Norway spruce stands. *In* "Forest Decline and Air Pollution: A Study of Spruce (*Picea abies*) on Acid Soils" (E. D. Schulze, O. L. Lange, and R. Oren, Eds.), pp. 444–455. Ecological Studies. Springer-Verlag, Berlin.

Hornbeck, J. W., and Swank, W. T. (1992). Watershed ecosystem analysis as a basis for multiple-use management of eastern forests. *Ecol. Appl.* **2,** 238–247.

Hornbeck, J. W., Adams, M. B., Corbett, E. S., Verry, E. S., and Lynch, J. A. (1993). Long-term impacts of forest treatments on water yield: A summary for northeastern USA. *J. Hydrol.* **150,** 323–344.

Horner, J. D. (1987). The effects of manipulation of nitrogen and water availability on the polyphenol content of Douglas-fir foliage: Implications for ecosystem theory. Ph.D. dissertation, Univ. of New Mexico, Albuquerque.

Horner, J. D., Gosz, J. R., and Cates, R. G. (1988). The role of carbon-based plant secondary metabolites in terrestrial ecosystems. *Am. Nat.* **132,** 869–883.

Horton, R. E. (1933). The role of infiltration in the hydrologic cycle. *Trans. Am. Geophys. Union* **14,** 446–460.

Houghton, J. T., Jenkins, G. J., and Ephraumus J. J. (Eds.) (1991). "Climate Change: The IPCC Scientific Assessment." Cambridge Univ. Press, New York.

Houghton, R. A., Hobbie, J. E., Melillo, J. M., Moore, B., Peterson, B. J., Shaver, G. R., and Woodwell, G. M. (1983). Changes in the carbon content of terrestrial biota and soils between 1860 and 1980: A net release of CO_2 to the atmosphere. *Ecol. Monogr.* **53,** 235–262.

Howard, D. M., and Howard, P. J. A. (1980). Effect of species, source of litter, type of soil and climate on litter decomposition. *Oikos* **34,** 115–124.

Humboldt, A. V. (1807). "Ideen zu einer Geographie der Pflanzen nebst einem Naturgemalde der Tropenlander." Tubingen.

Humboldt, A. V., and Bonpland, A. (1805). "Essai sur la Geographie des Plantes: Accompagne dun Tableau Physique des Regions Equinoxiales." Paris.

Humphries, F. R., and Craig, F. G. (1981). Effects of fire on soil chemical, structural and hydrological properties. *In* "Fire and the Australian Biota" (A. M. Gill, R. H. Groves, and I. R. Noble, Eds.), pp. 177–200. Australian Academy of Science, Canberra.

Hunt, E. R., Martin, F. C., and Running, S. W. (1991). Simulating the effects of climatic variation on stem carbon accumulation of a ponderosa pine stand: comparison with average growth increment data. *Tree Physiol.* **9,** 161–171.

Huntington, T. G. (1995). Carbon sequestration in an aggrading forest ecosystem in the southeastern USA. *Soil Sci. Soc. Am. J.* **59,** 1459–1467.

Huston, M. (1979). A general hypothesis of species diversity. *Am. Nat.* **113,** 81–101.

Huston, M. A., and Smith, T. M. (1987). Plant succession: Life history and competition. *Am. Nat.* **130,** 168–198.

Hutchison, B. A., Matt, D. R., McMillen, R. T., Gross, L. J., Tajchman, S. J., and Norman, J. M. (1986). The architecture of a deciduous forest canopy in eastern Tennessee, U.S.A. *J. Ecol.* **74,** 635–646.

Ino, Y., and Monsi, M. (1969). An experimental approach to the calculation of CO_2 amount evolved from several soils. *Jpn. J. Bot.* **20,** 153–188.

IPCC (1995) Intergovernmental Panel on Climate Change: Synthesis Report, p. 27. World Meteorological Organisation, Geneva, Switzerland.

Iwasa, Y., Sato, K., Kakita, M., and Kubo, T. (1993). Modelling biodiversity: Latitudinal gradient of forest species diversity. *In* "Biodiversity and Ecosystem Function, Ecological Studies 99" (E.-D. Schulze and H. A. Mooney, Eds.), pp. 433–451. Springer-Verlag, New York.

Jackson, P. S. (1981). On the displacement height in the logarithmic velocity profile. *J. Fluid Mech.* **111**, 15–25.
Jackson, R. B., and Caldwell, M. M. (1993). The scale of nutrient heterogeneity around individual plants and its quantification with geostatistics. *Ecology* **74**, 612–614.
Jardine, P. M., Weber, N. L., and McCarthy, J. F. (1989). Mechanisms of dissolved organic carbon adsorption on soil. *Soil Sci. Soc. Am. J.* **53**, 1378–1385.
Jarvis, P. G. (1976). The interpretation of the variations in leaf water potential and stomatal conductance found in canopies in the field. *Phil. Trans. R. Soc. London B.* **273**, 593–610.
Jarvis, P. G. (1987). Water and carbon fluxes in ecosystems. In "Potentials and Limitations of Ecosystem Analysis, Ecological Studies 61" (E.-D. Schulze and H. Zwolfer, Eds.), pp. 50–67. Springer-Verlag, New York.
Jarvis, P. G., and Leverenz, J. W. (1983). Productivity of temperate, deciduous and evergreen forests. In "Encyclopedia of Plant Physiology, 12D Physiological Plant Ecology. IV. Ecosystem Processes: Mineral Cycling, Productivity and Man's Influence" (O. L. Lange, P. S. Nobel, C. B. Osmond, and H. Ziegler, Eds.), pp. 234–280. Springer-Verlag, Berlin.
Jarvis, P. G. and McNaughton, K. G. (1986). Stomatal control of transpiration. *Adv. Ecol. Res.* **15**, 1–49.
Jarvis, P. G., James, G. B., and Landsberg, J. J. (1976). Coniferous forest. In "Vegetation and the Atmosphere" (J. L. Monteith, Ed.), Vol. 2, pp. 171–239. Academic Press, London.
Jayasuriya, M. D. A., Dunn, G., Benyon, R., and O'Shaughnessy, P. J. (1993). Some factors affecting water yield from mountain ash (*Eucalyptus regnans*) dominated forests in southeast Australia. *J. Hydrol.* **150**, 345–367.
Jenkinson, D. S. (1963). The turnover of organic matter in soil. In "The Use of Isotopes in Soil Organic Matter Studies," pp. 187–198. FAO/IAEA Technical Meeting Report Volkenrode, Pergamon, Elmsford, NY.
Jenkinson, D. S., and Rayner, J. H. (1977). The turnover of soil organic matter in some of the Rothamsted classical experiments. *Soil Sci.* **123**, 298–305.
Jenkinson, D. S., Adams, D. E., and Wild, A. (1991). Model estimates of CO_2 emissions from soil in response to global warming. *Nature* **351**, 304–306.
Jenkinson, D. S., Harkness, D. D., Vance, E. D., Adams, D. E., and Harrison, A. F. (1992). Calculating net primary production and annual input of organic matter to soil from the amount and radiocarbon content of soil organic matter. *Soil Biol. Biochem.* **24**, 295–308.
Jenny, H. (1980). "The Soil Resource: Origin and Behavior." Springer-Verlag, New York.
Jensen, J. R. (1986). "Introductory Digital Image Processing." Prentice Hall, New York.
Johnson, A. H., Cook, E. R., and Siccama, T. G. (1988). Climate and red spruce growth and decline in the northern Appalachians. *Proc. Natl. Acad. Sci. USA.* **85**, 5369–5373.
Johnson, D. W. (1984). Sulfur cycling in forests. *Biogeochem.* **1**, 29–43.
Johnson, D. W. (1992a). Nitrogen retention in forest soils. *J. Envir. Qual.* **21**, 1–21.
Johnson, D. W. (1992b). Effects of forest management on soil carbon storage. *Water Air Soil Poll.* **64**, 83–120.
Johnson, D. W., and Cole, D. W. (1980). Anion mobility in soils: Relevance to nutrient transport from forest ecosystems. *Environ. Int.* **3**, 79–90.
Johnson, D.W., and Lindberg, S.E. (Eds.) (1992). "Atmospheric Deposition and Forest Nutrient Cycling, Ecological Studies 91." Springer-Verlag, New York.
Johnson, D. W., and Todd, D. E. (1987). Nutrient export by leaching and whole-tree harvesting in a loblolly pine and mixed oak forest. *Plant Soil* **102**, 99–109.
Johnson, D. W., Cole, D. W., Gessel, S. P., Singer, M. J., and Minden, R. V. (1977). Carbonic acid leaching in a tropical, temperate, subalpine, and northern forest soil. *Arctic Alpine Res.* **9**, 329–343.
Johnson, D. W., Cole, D. W., Van Miegroet, H., and Horng, F. W. (1986). Factors affecting anion movement and retention in four forest soils. *Soil Sci. Soc. Am. J.* **50**, 776–783.

Johnson, I. R., and Thornley, J. H. M. (1987). A model of shoot/root partitioning with optimal growth. *Ann. Bot.* **60,** 133–143.

Johnson, K. H., Vogt, K. A., Clark, H. J., Schmitz, O. J., and Vogt, D. J. (1996). Biodiversity and the productivity and stability of ecosystems. *Trends. Ecol. Evol.* **11,** 372–377.

Jones, H. G. (1992). "Plants and Microclimate," 2nd ed. Cambridge Univ. Press, Cambridge, UK.

Jones, J. M., and Richards, B. N. (1977). Effect of reforestation on turnover of 15 N-labelled nitrate and ammonia in relation to changes in soil microflora. *Soil Biol. Biochem.* **9,** 383–392.

Jordan, C. F. (1983). Productivity of tropical rain forest ecosystems and the implications for their use as future wood and energy systems. *In* "Tropical Rainforest Ecosystems: Structure and Function" (F. B. Golley, Ed.), Vol. 14A, pp. 117–136. Elsevier, Amsterdam.

Jordan, C. F., Golley, F., Hall, J., and Hall, J. (1980). Nutrient scavenging of rainfall by the canopy of an Amazonian rain forest. *Biotropica* **12,** 61–66.

Jorgenson, J. R., and Wells, C. G. (1980). Apparent nitrogen fixation in soils influenced by prescribed burning. *Soil Sci. Soc. Am. J.* **35,** 806–810.

Junge, C. E., and Werby, R. T. (1958). The concentrations of chloride, sodium, potassium, calcium, and sulfate in rainwater over the United States. *J. Meteorol.* **15,** 417–425.

Kabeda, O. (1991). Aboveground biomass production and nutrient accumulation in an age sequence of *Pinus caribaea* stands. *For. Ecol. Manage.* **41,** 237–248.

Kaimal, J. C., and Finnigan, J. J. (1994). "Atmospheric Boundary Layer Flows: Their Structure and Measurement." Oxford Univ. Press, New York.

Kalma, J. P., and Fuchs, M. (1976). Citrus orchards. *In* "Vegetation and the Atmosphere" (J. L. Monteith, Ed.), pp. 309–329. Academic Press, London.

Karizumi, N. (1974). The mechanisms and function of tree root in the process of forest production. II. Root biomass and distribution in stands. *Bull. Gov. For. Exp. Stn.* **267,** 1–88.

Karizumi, N. (1976). The mechanisms and function of tree root in the process of forest production. III. Root density and absorptive structure. *Bull. Gov. For. Exp. Stn.* **285,** 43–149.

Kaufmann, M. R., and Troendle, C. A. (1981). The relationship of leaf area and foliage biomass to sapwood conducting area in four subalpine forest tree species. *For. Sci.* **27,** 472–482.

Kazimirov, N. I., and Morozova, R. M. (1973). Biological cycling of matter in spruce forest of Karelia, Nauka, Leningrad Branch, Academy of Sciences.

Keeves, A. (1966). Some evidence of a loss of productivity with successive plantations of *Pinus radiata* in the south-east of South Australia. *Aust. For.* **30,** 51–63.

Kelliher, F. M., Leuning, R., Raupach, M. R., and Schulze, E.-D. (1995). Maximum conductances for evaporation from global vegetation types. *Agric. For. Meteorol.* **73,** 1–16.

Kelliher, F. M., Leuning, R., and Schulze, E.-D. (1993). Evaporation and canopy characteristics of coniferous forests and grasslands. *Oecologia* **95,** 153–163.

Kelliher, F. M., Whitehead, D., and Pollock, D. S. (1992). Rainfall interception by trees and slash in a young *Pinus radiata* D. Don stand. *J. Hydrol.* **131,** 187–204.

Kerruish, C. M., and Rawlins, W. H. M., (eds.) (1991b). "The young eucalypt report: some management options for Australia's forests." Anonymous, p. 272. CSIRO, Melbourne.

Keyes, M. R., and Grier, C. C. (1981). Above- and below-ground net production in 40-year-old Douglas-fir stands on low and high productivity sites. *Can. J. For. Res.* **11,** 599–605.

Kicklighter, D. W., Melillo, J. M., Peterjohn, W. T., Rastetter, E. B., McGuire, A. D., Steudler, P. A., and Aber, J. D. (1994). Aspects of spatial and temporal aggregation in estimating regional carbon dioxide fluxes from temperate forest soils. *J. Geophys. Res.* **99,** 1303–1315.

Kikuzawa, K. (1991). A cost–benefit analysis of leaf habit and leaf longevity of trees and their geographical pattern. *Am. Nat.* **138,** 1250–1263.

Kilgore, B. M., and Taylor, D. (1979). Fire history of a sequoia–mixed conifer forest. *Ecology* **60**, 129–142.

Kinerson, R., and Fritschen, L. J. (1971). Modeling a coniferous forest canopy. *Agric. Meteorol.* **8**, 439–445.

Kinerson, R. S., Higginbotham, K. O., and Chapman, R. C. (1974). The dynamics of foliage distribution within a forest canopy. *J. Appl. Ecol.* **11**, 347–353.

Kinjo, T., and Pratt, P. F. (1971). Nitrate adsorption. II. In competiton with chloride, sulfate, and phosphate. *Soil Sci. Soc. Am. Proc.* **35**, 725–728.

Kira, T. (1983). "Ecology of Tropical Forests." Jinbun-Shoin, Kyoto, Japan.

Kira, T., and Shidei, T. (1967). Primary production and turnover or organic matter in different forest ecosystems of the western Pacific. *Jpn. J. Ecol.* **13**, 273–283.

Kloeppel, B.D., Gower, S. T., and Reich, P. B. (1995). Net photosynthesis for western larch and sympatric evergreen conifers along a precipitation gradient in western Montana. *In* W. C. Schmidt, and R. C. Shearer, Eds., "1993 Symposium—Ecology and Management of Larix Forests: A Look Ahead" pp. 486–488. USDA Forest Service, Intermountain Research Station.

Knight, D. H., Vose, J. M., Baldwin, V. C., Ewel, K. C., and Grodzinska, K. (1994). Contrasting patterns in pine forest ecosystems. *Ecol. Bull. (Copenhagen)* **43**, 9–19.

Knowles, R. (1982). Denitrification. *Microbiol. Rev.* **46**, 43–70.

Kohyama, T., and Fujita, N. (1981). Studies on the *Abies* population of Mt. Shimagare. I. Survivorship curve. *Bot. Manage.* **94**, 55–68.

Koide, R. T., and Elliott, G. (1989). Cost, benefit and efficiency of the vesicular–arbuscular mycorrhizal symbiosis. *Funct. Ecol.* **3**, 252–255.

Koide, R. T., and Schreiner, R. P. (1992). Regulation of the vesicular–arbuscular mycorrhizal symbiosis. *Annu. Rev. Plant Physiol. Mol. Biol.* **43**, 557–581.

Körner, C. (1994). Leaf diffusive conductances in the major vegetation types of the globe. *In* "Ecophysiology of Photosynthesis" (E.-D. Schulze and M. M. Caldwell, Eds.), Vol. 100, pp. 463–490. Springer, Heidelberg.

Körner, C. H., Scheel, J. A., and Bauer, H. (1979). Maximum leaf diffusive conductance in vascular plants. *Photosynthetica* **13**, 45–82.

Kotar, J., Kovach, J., and Locey, C. (1988). "Field Guide to Forest Habitat Types of Northern Wisconsin." Department of Forestry, University of Wisconsin—Madison and Wisconsin Department of Natural Resources, Madison.

Kozlowski, T. T. (1992). Carbohydrate sources and sinks in woody plants. *Bot. Rev.* **59**, 107–223.

Krankina, O. N. (1992). Contributions of forest fires to the carbon flux of the USSR. *In* "Proceedings of Carbon Cycling in Boreal Forest and Subarctic Ecosystems: Biospheric Responses and Feedbacks to Global Climate Change." T. Kolchugina and T. S. Vinson (Eds.). Oregon State Univ. Press, Corvallis.

Kucharik, C. J., Norman, J. M., Murdock, L. M., and Gower, S. T. (1996). Characterizing canopy non-randomness with a multiband vegetation imager (MVI). *J. Geophys. Res.* (in press).

Kurz, W. A. (1989). Net primary production, production allocation and foliage efficiency in second growth Douglas-fir stands with differing site quality. Ph.D. dissertation, University of British Columbia, Vancouver, British Columbia, Canada.

Lamb, D.(1975). Soil nitrogen mineralization in a secondary rainforest succession. *Oecologia* **47**, 257–263.

Lamb, R. J., and Richards, B. N. (1971). Effect of mycorrhizal fungi on the growth and nutrient status of slash and radiata pine seedlings. *Aust. For.* **35**, 1–7.

Lambers, H., and Poorter, H. (1992). Inherent variation in growth rate between higher plants: A search for physiological causes and ecological consequences. *Adv. Ecol. Res.* **23**, 187–261.

Lambert, R. L., Lang, G. E., and Reiners, W. A. (1980). Loss of mass and chemical change in decaying boles of a subalpine balsam fir forest. *Ecology* **61,** 1460–1473.

Landsberg, J. J. (1986). "Physiological Ecology of Forest Production." Academic Press, Sydney.

Landsberg, J. J., Beadle, C. L., Biscoe, P. V., Butler, D. R., Davidson, B., Incoll, L. D., James, G. B., Jarvis, P. G., Martin, P. J., Neilson, R. E., Powell, D. B. B., Slack, E. M., Thorpe, M. R., Turner, N. C., Warrit, B., and Watts, W. R. (1975). Diurnal energy, water and CO_2 exchanges in an apple (*Malus pumila*) orchard. *J. Appl. Ecol.* **12,** 659–684.

Landsberg, J. J., Blanchard, T. W., and Warrit, B. (1976). Studies on the movement of water through apple trees. *J. Exp. Bot.* **27,** 579–596.

Landsberg, J. J., Kaufmann, M. R., Binkley, D., Isebrands, J., and Jarvis, P. G. (1991). Evaluating progress toward closed forest models based on fluxes of carbon, water and nutrients. *Tree Physiol.* **9,** 1–15.

Landsberg, J. J., Linder, S., and McMurtrie, R. E. (1995). "Effects of Global Change on Managed Forests," GTCE Report No. 4, p. 35. IGBP, IUFRO, Canberra, Vienna.

Landsberg, J. J., Prince, S. D., Jarvis, P. G., McMurtrie, R. E., Luxmoore, R., and Medlyn, B. E. (1996). Energy conversion and use in forests: An analysis of forest production in terms of radiation utilisation efficiency. *In* "The Use of Remote Sensing in the Modeling of Forest Productivity at Scales from the Stand to the Globe" (H. L. Gholz, K. Nakane, and H. Shimoda, Eds.), in press. Kluwer, Dordrecht.

Lang, A. R. G., Yueguin, X., and Norman, J. M. (1985). Crop structure and the penetration of direct sunlight. *Agric. For. Meteorol.* **35,** 83–101.

Lang, G. E., and Formann, R. T. T. (1978). Detrital dynamics in a mature oak forest, Hutcheson Memorial Forest, New Jersey. *Ecology* **59,** 580–595.

Langford, K. J. (1976). Change in yield of water following a bushfire in a forest of *Eucalyptus regnans*. *J. Hydrol.* **29,** 87–114.

Larcher, W., and Bauer, H. (1981). Ecological significance of resistance to low temperature. *In* "Encyclopedia of Plant Physiology" (O. L. Lange, P. S. Nobel, C. B. Osmond, and H. Ziegler, Eds.), Vol. 12A, pp. 403–437. Springer-Verlag, Berlin.

Larsen, J. A. (1982). "The Northern Forest Border in Canada and Alaska: Biotic Communities and Ecological Relationships, Ecological Studies 70." Springer-Verlag, New York.

Lauer, W. (1989). Climate and weather. *In* "Tropical Rainforest Ecosystems: Biogeographical and Ecological Studies" (H. Leith and M. J. A. Werger, Eds.), Vol. 14B, pp. 7–53. Elsevier, Amsterdam.

Law, R. (1985). Evolution in a mutualistic environment. *In* "The Biology of Mutualism, Ecolgy, and Evolution" (D. H. Boucher, Ed.), pp. 145–170. Oxford Univ. Press, New York.

Lawlor, D. (1987). "Photosynthesis: metabolism, control and physiology." Longman, New York.

Lawlor, D. W. (1993). "Photosynthesis: Molecular, Physiological and Environmental Processes," 2nd ed., p. 318. Longmans, Harlow, UK.

Lee, R. (1980). "Forest Hydrology." Columbia Univ. Press, New York.

Leech, J. W. (1985). Modelling for forest management. *In* "Research for Forest Management" (J. J. Landsberg and W. Parsons, Eds.), pp. 229–233. CSIRO, Melbourne.

Leemans, R., and Cramer, W. (1990). "The IIASA Climate Database for Land Area on a Grid with 0.5° Resolution," No. WP-90-41. International Institute for Applied Systems Analysis, Laxenburg, Austria.

Leuning, R. (1989). Leaf energy balances: Developments and applications. *Phil. Trans. R. Soc. London B* **324,** 191–206.

Leuning, R. (1990). Modelling stomatal behaviour and photosynthesis of *Eucalyptus grandis*. *Aust. J. Plant. Physiol.* **17,** 159–175.

Leuning, R. (1995). A critical appraisal of a combined stomatal–photosynthesis model for C_3 plants. *Plant Cell Environ.* **18,** 339–355.

Leuning, R., Kelliher, F. M., De Pury, D. G. G., and Schulze, E.-D. (1995). Leaf nitrogen, photosynthesis, conductance and transpiration: Scaling from leaves to canopies. *Plant Cell Environ.* **18,** 1183–1200.

Likens, G. E., Boxmann, F. H., Pierce, R. S., Eaton, J. S., and Johnson, N. M. (1995). "Biogeochemistry of a Forested Ecosystem, 2nd Ed." Springer-Verlag, New York.

Lillesand, T. M., and Kieffer, R. W. (1994). "Remote Sensing and Image Interpretation," 3rd ed. Wiley, New York.

Linacre, E. (1992). "Climate, Data and Resources: A Reference and Guide." Routledge, London.

Lindberg, S. E. (1992). Atmospheric deposition and canopy interactions of sulfur. *In* "Atmospheric Deposition and Forest Nutrient Cycling, Ecological Studies 91" (D. W. Johnson, and S. E. Lindberg, Eds.), Springer-Verlag, New York.

Lindberg, S. E., Lovett, G. M., Richter, D., and Johnson, D. W. (1986). Atmospheric deposition and canopy interactions of major ions in a forest. *Science* **231,** 141–145.

Linder, S. (1985). Potential and actual production in Australian forest stands. *In* "Research for Forest Management" (J. J. Landsberg and W. Parsons, Eds.), pp. 11–35. CSIRO, Melbourne.

Linder, S., and Axelsson, B. (1982). Changes in carbon uptake and allocation patterns as a result of irrigation and fertilization in a young *Pinus sylvestris* stand. *In* "Carbon Uptake and Allocation: Key to Management of Subalpine Forest Ecosystems, Internal Union Forest Research Organization (IUFRO) Workshop" (R. H. Waring, Ed.), pp. 38–44. Forest Research Laboratory, Oregon State Univ., Corvallis.

Linder, S., and Rook, D. A. (1984). Effects of mineral nutrition on carbon dioxide and partitioning of carbon in trees. *In* "Nutrition of Plantation Forests" (G. D. Bowen and E. K. S. Nambiar, Eds.), pp. 211–236. Academic Press, London.

Linder, S., and Troeng, E. (1981). The seasonal variation in stem and coarse root respiration of 20-year-old Scots pine (*Pinus sylvestris* L.). *Mitt. Forstl. Bundes-Versuchsanst. Wien* **142,** 125–139.

Linder, S. L., Benson, M. L., Myers, B. J., and Raison, R. J. (1987). Canopy dynamics and growth of *Pinus radiata*. I. Effects of irrigation and fertilization during drought. *Can. J. For. Res.* **17,** 1157–1165.

Lindroth, A. (1985). Canopy conductance of coniferous forests related to climate. *Water Resour. Res.* **21,** 297–304.

Lloyd, D. G. (1980). Sexual strategies in plants. I. A hypothesis of serial adjustment of maternal investment during one reproductive season. *New Phytol.* **86,** 69–79.

Lockaby, B. G., Miller, J. H., and Clawson, R. G. (1995). Influences of community composition on biogeochemistry of loblolly pine (*Pinus taeda*) systems. *Am. Midl. Nat.* **134,** 176–184.

Lovett, G. (1994). Atmospheric deposition of nutrient and pollutants in North America: An ecological perspective. *Ecol. Appl.* **4,** 629–650.

Lorimer, C. G. (1992). Causes of the oak regeneration problem. *In* "Oak Regeneration: Serious Problems and Practical Recommendations," pp. 14–39. USDA For. Serv. Gen. Tech. Rpt. SE-84. Southeastern For. Exp. Stn., Asheville, NC.

Lubchenko, J., Olson, A., Brubaker, L. B., Carpenter, S. R., Holland, M. M., Hubell, S. P., Levin, S. A., MacMahon, H. A., Matson, P. A., Melillo, J. M., Mooney, H. A., Peterson, C. H., Pulliam, R., Real, L., Regal, P. J., and Risser, P. G. (1991). The Sustainable Biosphere initiative: An ecological research agenda. *Ecology* **72,** 371–412.

Ludlow, M. M. (1980). Adaptive significance of stomatal response to water stress. *In* "Adaptation of Plants to Water and High Temperature Stress" (N. C. Turner and P. J. Kramer, Eds.), pp. 123–138. Wiley, New York.

Lugo, A. E. (1992). Comparison of tropical tree plantations with secondary forests of similar age. *Ecol. Monogr.* **62,** 1–41.

Lugo, A. E., Sanchez, A. J., and Brown, S. (1986). Land use and organic carbon content of some subtropical soils. *Plant Soil,* **96,** 185–196.

Lusk, C., and Ogden, J. (1992). Age structure and dynamics of a podocarp-broadleaf forest in Tongariro National Park, New Zealand. *J. Ecol.* **80,** 379–393.

Lyr, H., and Hoffmann, G. (1967). Growth rates and growth periodicity of tree roots. *Int. Rev. For. Res.* **2,** 39–64.

MacDonald, L., Smart, A., and Wismar, R. (1991). "Monitoring Guidelines to Evaluate Effects of Forestry Activities on Streams in the Pacific Northwest and Alaska." SEPA/910/9-91-001, Region 10, Seattle, WA.

MacLean, D. A., and Wein, R. W. (1978). Weight loss and nutrient changes in decomposing litter and forest floor material in New Brunswick forest stands. *Can. J. Bot.* **56,** 2730–2749.

Mäkela, A., and Hari, P. (1986). Stand growth model based on carbon uptake and allocation in individual trees. *Ecol. Model.* **33,** 205–229.

Mann, L. K. (1986). Changes in soil carbon storage after cultivation. *Soil Sci.* **142,** 279–288.

Marchland, D. E. (1971). Rates and modes of denudation, White Mountains, eastern California. *Amer. J. Sci.* **270,** 109–135.

Markgraf, V., McGlone, M., and Hope, G. (1995). Neogene paleoenvironmental and paleoclimatic change in southern temperate ecosystems—A southern perspective. *Trends Ecol. Evol.* **10,** 143–147.

Marquis, D. A. (1981). Effect of deer browsing on timber production in Allegheny hardwood forests of northwestern Pennsylvania. USDA For. Serv. Res. Pap. NE-475.

Marshall, J. D., and Waring, R. H. (1986). Comparison of methods of estimating leaf area index in old-growth Douglas-fir. *Ecology* **67,** 975–979.

Marshall, J. D., and Waring, R. H. (1995). Predicting fine root production and turnover by monitoring root starch and soil temperature. *Can. J. For. Res.* **15,** 791–800.

Martin, F., Canet, D., Rolin, D., Marchal, J.P., and Larher, F. (1983). Phosphorus-31 nuclear magnetic resonance study of polyphosphate metabolism in intact mycorrhizal fungi. *Plant Soil* **71,** 469–476.

Matson, P. A., and Boone, R. (1984). Natural disturbance and nitrogen mineralization: Wave form dieback of mountain hemlock in the Oregon Cascades. *Ecology* **65,** 1511–1516.

Matson, P. A., and Waring, R. H. (1984). Effects of nutrient and light limitation on mountain hemlock: Susceptibility to laminated root-rot. *Ecology* **65,** 1517–1524.

Matson, P. A., Vitousek, P. M., Ewel, J. J., Mazzarino, M. J., and Robertson, G. P. (1987). Nitrogen transformations following tropical forest felling and burning on a volcanic soil. *Ecology* **68,** 491–502.

Matson, P. A., Gower, S. T., Volkmann, C., Billow, C., and Grier, C. C. (1992). Soil nitrogen cycling and nitrous oxide flux in a Rocky Mountain Douglas-fir forest: Effects of fertilization, irrigation and carbon addition. *Biogeochemistry* **18,** 101–117.

Matson, P. A., Johnson, L., Billow, C., Miller, J., and Pu, R. (1994). Seasonal changes in canopy chemistry across the Oregon transect: Patterns and spectral measurement with remote sensing. *Ecol. Appl.* **4,** 280–298.

Mattson, K. G., and Smith, H. C. (1993). Detrital organic matter and soil CO_2 efflux in forests regenerating from cutting in West Virginia. *Soil Biol. Biochem.* **25,** 1241–1248.

Mattson, W. J., and Addy, N. D. (1975). Phytophagous insects as regulators of forest primary production. *Science* **190,** 515–521.

Mattson-Djos, E. (1981). The use of pressure bomb and porometer for describing plant water stress in tree seedlings. *In* "Proceedings of the Nordic Symposium on Vitality and Quality of Nursery Stock," (P. Puttonen, Ed.), pp. 45–57. Department of Silviculture, Univ. of Helsinki, Helsinki.

McCarthy, B. C, and Bailey, R. R. (1994). Distribution and abundance of coarse woody debris in a managed forest landscape of the central Appalachians. *Can. J. For. Res.* **24,** 1317–1329.

McCauley, K. J., and Cook, S. A. (1980). *Phellinus weirii* infestation of two mountain hemlock forests in the Oregon Cascades. *For. Sci.* **26**, 23–29.

McClaugherty, C. A., Aber, J. D., and Melillo, J. M. (1982). The role of fine roots in the organic matter and nitrogen budgets of two forested ecosystems. *Ecology* **63**, 1481–1490.

McClaugherty, C. A., Pastor, J., Aber, J. D., and Melillo, J. M. (1985). Forest litter decomposition in relation to soil nitrogen dynamics and litter quality. *Ecology* **66**, 266–275.

McColl, J. G. (1978). Ionic composition of forest soil solutions and effects of clearcutting. *Soil Sci. Soc. Am. J.* **42**, 358–367.

McDowell, W. H., and Likens, G. E. (1988). Origin, composition, and flux of dissolved organic carbon in the Hubbard Brook Valley. *Ecol. Monogr.* **58**, 177–195.

McIntosh, R. P. (1981). Succession and ecological theory *In* "Forest Succession: Concepts and Application" (D. C. West, H. H. Shugart, and D. B. Botkin, Eds.), pp. 227–304. Springer-Verlag, New York.

McKee, W. H. (1982). Changes in soil fertility following prescribed burning on coastal plain pine sites. USDA Forest Service Research Paper No. SE–234.

McKey, D., Waterman, P. G., Mbi, C. N., Gartlan, J. S., and Strusaker, T. T. (1978). Phenolic content of vegetation in two African rain forests: Ecological implications. *Science* **202**, 61–64.

McMurtrie, R. E. (1985). Forest productivity in relation to carbon partitioning and nutrient cycling: A mathematical model. *In* "Attributes of Trees as Crop Plants" (M. G. R. Cannell and J. E. Jackson, Eds.), pp. 194–207. Natural Environment Research Council. U. K.

McMurtrie, R. E., and Comins, H. N. (1996). The temporal response of forest ecosystems to doubled atmospheric CO_2 concentration, *Global Change Biol.* **2**, 49–57.

McMurtrie, R. E., and Landsberg, J. J. (1992). Using a simulation model to evaluate the effects of water and nutrients on the growth and carbon partitioning of *Pinus radiata*. *For. Ecol. Manage.* **52**, 243–260.

McMurtrie, R. E., and Wang, Y.-P. (1993). Mathematical models of the photosynthetic response of plant stands to rising CO_2 levels and temperatures. *Plant Cell Environ.* **16**, 1–13.

McMurtrie, R. E., and Wolf, L. J. (1983). Above- and below-ground growth of forest stands: A carbon budget model. *Ann. Bot.* **52**, 437–448.

McMurtrie, R. E., Linder, S., Benson, M. L., and Wolf, L. (1986). A model of leaf area development for pine stands. *In* "Crown and Canopy Structure in Relation to Productivity" (T. Fujimori and D. Whitehead, Eds.), pp. 284–307. Forestry and Forest Products Research Institute, Japan.

McMurtrie, R. E., Rook, D. A., and Kelliher, F. M. (1990). Modelling the yield of *Pinus radiata* on a site limited by water and nutrition. *For. Ecol. Manage.* **30**, 381–413.

McMurtrie, R. E., Comins, H. N., Kirschbaum, M. U. F., and Wang, Y.-P. (1992a). Modifying exisitng forest growth models to take account of effects of elevated CO_2. *Aust. J. Bot.* **40**, 657–677.

McMurtrie, R. E., Leuning, R., Thompson, W. A., and Wheeler, A. M. (1992b). A model of canopy photosynthesis and water use incorporating a mechanistic formulation of leaf CO_2 exchange. *For. Ecol. Manage.* **52**, 261–278.

McMurtrie, R. E., Gholz, H. L., Linder, S., and Gower, S. T. (1994). Climatic factors controlling the productivity of pine stands: A model-based analysis. *Ecol. Bull. (Copenhagen)* **43**, 173–188.

McMurtrie, R. E., Gower, S. T., Ryan, M. G., and Landsberg, J. J. (1995). Forest productivity: Explaining its decline with stand age. *Bull. Ecol. Soc. Am.* **76**, 152–154.

McNaughton, K. G., and Black, T. A. (1973). A study of evapotranspiration from a Douglas fir forest using the energy balance approach. *Water Resour. Res.* **9**, 1579–1590.

McNaughton, K. G., and Jarvis, P. G. (1983). Predicting effects of vegetation changes on transpiration and evaporation. *In* "Water Deficits and Plant Growth" (T. T. Kozlowski, Ed.), Vol. VII, pp. 1–47. Academic Press, New York.

McNulty, S. G., Aber, J. D., and Boone, R. D.(1991). Spatial changes in forest floor and foliar chemistry of spruce-fir forests across New England. *Biogeochem.* **14,** 13–29.

Meentemeyer, V. (1978). Macroclimate and lignin control of litter decomposition rates. *Ecology* **59,** 465–472.

Meentemeyer, V., and Berg, B. (1986). Regional variation in rate of mass loss of Scots pine needle litter in Swedish pine forests as influenced by climate and litter quality. *Scand. J. For. Res.* **1,** 167–180.

Meier, C. E, Grier, C. C., and Cole, D. W. (1985). Below- and above-ground N and P use by *Abies amabilis* stands. *Ecology* **66,** 1928–1942.

Meinzer, F. C. (1993). Stomatal control of transpiration. *Trends Ecol. Evol.* **8,** 289–294.

Meinzer, F. C., and Grantz, D. G. (1990). Stomatal and hydraulic conductance in growing sugarcane: Stomatal adjustment to water transport capacity. *Plant Cell Environ.* **13,** 383–388.

Melillo, J. M., Aber, J. D., and Muratore, J. F. (1982). Nitrogen and lignin control of hardwood leaf litter decomposition dynamics. *Ecology* **63,** 621–626.

Melillo, J. M., McGuire, A. D., Kicklighter, D. W., Moore, B., III, Vorosmarty, C. J., and Schloss, A. L. (1993). Global climate change and terrestrial net primary production. *Nature* **363,** 234–240.

Mencuccini, M., and Grace, J. (1995). Climate influences the leaf area/sapwood area ratio in Scots pine. *Tree Physiol.* **15,** 1–10.

Menge, J. A., and Grand, L. F. (1978). Effect of fertilization on production of epigeous basdiocarps by mycorrhizal fungi in loblolly pine plantations. *Can. J. Bot.* **56,** 2357–2362.

Menge, J. A., Grand, L. F., and Haines, L. W. (1977). The effect of fertilization on growth and mycorrhizal numbers in an 11-year-old loblolly pine plantation. *For. Sci.* **23,** 37–44.

Mengel, K., and Kirkby, E. A. (1979). "Principles of Plant Nutrition." International Potash Institute, Berne.

Miles, J. (1985). The pedogenic effects of different species and vegetation types and the implications of succession. *J. Soil Sci.* **36,** 571–584.

Miller, H. G. (1984). Dynamics of nutrient cycling in plantation ecosystems. *In* "Nutrition of Forest Plantation" (E. K. S. Nambiar and G. D. Bowen, Eds.), pp. 53–78. Academic Press, London.

Miller, S. L., Durall, D. M., and Rygiewicz, P. T. (1989). Temporal allocation of ^{14}C to extramatrical hyphae of ectomycorrhizae ponderosa pine seedlings. *Tree Physiol.* **5,** 239–249.

Mitchell, N. D., and Williams, J. E. (1996). The consequences for native biota of anthropogenic induced climate change. *In* "Greenhouse 94" (G. Pearman, Ed.). CSIRO, Melbourne.

Mitchell, R. G., Waring, R. H., and Pitman, G. B. (1983). Thinning lodgepole pine increases tree vigor and resistance to mountain pine beetle. *For. Sci.* **29,** 204–211.

Mladenoff, D. J., White, M. A., Pastor, J., and Crow, T. R. (1993). Comparing spatial pattern in unaltered old-growth and disturbed forest landscapes. *Ecol. Appl.* **3,** 294–306.

Moller, C. M., Muller, D., and Nielsen, J. (1954). Graphic presentation of dry matter production in European beech. *Forstl. Forsoegsvaes. Dan.* **21,** 327–335.

Monk, C. D. (1966). An ecological significance of evergreeness. *Ecology* **47,** 504–505.

Monteith, J. L. (1995). A reinterpretation of stomatal responses to humidity. *Plant Cell Environ.* **18,** 357–364.

Monteith, J. L., and Unsworth, M. H. (1990). "Principles of Environmental Physics," 2nd ed. Arnold, London.

Monteith, J. L. (1977). Climate and the efficiency of crop production in Britain. *Phil. Trans. R. Soc. London,* 277–294.

Mooney, H. A. (1972). The carbon balance of plants. *Annu. Rev. Ecol. Syst.* **3,** 315.

Mooney, H. A., and Sala, O. E. (1993). Science and sustainable use. *Ecol. Appl.* **3,** 564–566.

Moore, C. J. (1976). Eddy flux measurements above a pine forest. *Q. J. R. Met. Soc.* **102**, 913–918.

Moore, I. D., Norton, T. W., and Williams, J. E. (1993). Modelling environmental heterogeneity in forested landscapes. *J. Hydrol.* **150**, 717–747.

Morrison, J. I. L., and Gifford, R. M. (1984). Plant growth and water use with limited water supply in high CO_2 concentrations. II. Plant dry weight, partitioning and water use efficiency. *Aust. J. Plant. Physiol.* **11**, 375–384.

Mulholland, P. J., and Kuenzler, E. J. (1979). Organic carbon export from upland and forested wetlands watersheds. *Limnol. Oceanogr.* **24**, 960–966.

Muller, D. (1879). Studier over Skovjord, som bidrag til skovdyrkningens theori. II. Om bogemuld og bogemor paa sand og ler *Tidssk. Skov.* **3**, 1–124.

Muller, D. (1884) Studier over Skovjord, som bidrag til skovdyrkningens theori. II. Om muld og mor I egeskove og paa heder. *Tidssk. Skov.* **7**, 1–232.

Müller, M. J. 1982. "Selected climatic data for a global set of standard stations for vegetation science," p. 306. Junk, The Hague.

Murphy, C. E., Jr., Schubert, J. F., and Dexter, A. H. (1981). The energy and mass exchange characteristics of a loblolly pine plantation. *J. Appl. Ecol.* **18**, 271–281.

Murphy, P. G., and Lugo, A. (1986). Ecology of tropical dry forest. *Annu. Rev. Ecol. Syst.* **17**, 67–88.

Murty, D., McMurtrie, R. E., and Ryan, M. G. (1996). Declining forest productivity in ageing forest stands—A modeling analysis of alternative hypotheses. *Tree Physiol.* **16**, 187–200.

Myers, B. J. (1988). Water stress integral—A link between short term stress and long-term growth. *Tree Physiol.* **4**, 315–323.

Myers, B. J., and Talsma, T. (1992). Site water balance and tree water status in irrigated and fertilised stands of *Pinus radiata*. *For. Ecol. Manage.* **52**, 17–42.

Myers, N. (1996). "The Primary Source: Tropical Forests and Our Future." Norton, New York.

Myrold, D. D., Matson, P. A., and Peterson, D. L. (1989). Relationships between soil microbial properties and above-ground stand characteristics of conifer forests in Oregon. *Biogeochemistry* **8**, 265–281.

Nadelhoffer, K. J., and Raich, J. W. (1992). Fine root production estimates and belowground carbon allocation in forest ecosystems. *Ecology* **73**, 1139–1147.

Nadelhoffer, K. J., Aber, J. D., and Melillo, J. M. (1985). Fine roots, net primary production, and soil nitrogen availability: A new hypothesis. *Ecology* **66**, 1377–1390.

Nadkarni, N. M. (1984). Biomass and mineral capital of epiphytes of an *Acer macrophyllum* community of a temperate moist coniferous forest, Olympic Peninisula, Washington state. *Can. J. Bot.* **62**, 2023–2028.

Nadkarni, N. M., and Matelson, T. J. (1992). Biomass and nutrient dynamics of fine litter of terrestrially rooted material in a neotropical montane forest, Costa Rica. *Biotropica* **24**, 113–120.

Nakane, K., Yamamoto, M., and Tsubota, H. (1983). Estimation of root respiration rate in a mature forest ecosystem. *Jpn. J. Ecol.* **33**, 397–408.

Nambiar, E. K. S., and Fife, D. N. (1991). Nutrient retranslocation in temperate conifers. *Tree Physiol.* **9**, 185–207.

Nemani, R. R., and Running, S. W. (1989). Estimation of regional surface resistance to evapotranspiration from NDVI and Thermal-IR AVHRR data. *J. Appl. Meteorol.* **28**, 276–284.

Nepstad, D. C., de Carvalho, C. R., Davidson, E. A., Jipp, P. H., Lefebvre, P. A., Negreiros, G. H., da Silva, E. D., Stone, T. A., Trumbore, S. E., and Vieira, S. (1994). The role of deep roots in the hydrological and carbon cycles of Amazonian forests and pastures. *Nature* **372**, 666–669.

Newman, E. I. (1969). Resistance to water flow in soil and plant I. Soil resistance in relation to amounts of root: Theoretical estimates. *J. Appl. Ecol.* **6,** 1–12.

Newman, E. I. (1978). Root microorganisms, their significance in the ecosystem. *Biol. Rev.* **53,** 511–554.

Nikolov, N., and Helmisaari, H. (1992). Silvics of the circumpolar boreal forest tree species. *In* "A Systems Analysis of the Global Boreal Forest" (H. H. Shugart, R. Leemans, and G. B. Bonan, Eds.), pp. 13–84. Cambridge Univ. Press, Cambridge, UK.

Nilson, T. (1977). A theory of radiation penetration into non-homogeneous plant canopies. *In* "Penetration of Solar Radiation into Plant Canopies," Academy Sciences Estonian SSR, Tartu (in Russian).

Nilsson, S. I. (Ed.) (1986). "Critical Loads for Nitrogen and Sulfur." Nordic Council of Ministers, Stockholm, Sweden.

Nilsson, S. I., and Duinker, P. (1987). The extent of forest decline in Europe: A synthesis of survey results. *Environment* **29,** 4–9.

Nilsson, S. I., Berden, M., and Popovic, B. (1988). Experimental work related to nitrogen deposition, nitrification, and soil acidification—a case study. *Environ. Pollut.* **54,** 233–248.

Norman, J. M. (1979). Modeling the complete crop canopy. *In* "Modification of the Aerial Environment of Plants" (B. J. Barfield and J. F. Gerber, Eds.), pp. 249–277. American Society of Agricultural Engineering, St. Joseph, MI.

Norman, J. M. (1980). Interfacing leaf and canopy light interception models. *In* "Predicting Photosynthesis for Ecosystem Models" (J. D. Hesketh and J. W. Jones, Eds.), Vol. II, pp. 49–67. CRC Press, Boca Raton, FL.

Norman, J. M. (1982). Simulation of microclimates. *In* "Biometeorology in Integrated Pest Management" (J. L. Hatfield and I. J. Thomason, Eds.), pp. 65–99. Academic Press, New York.

Norman, J. M., and Jarvis, P. G. (1975). Photosynthesis in Sitka Spruce (*Picea sitchensis* (Bong) Carr.) V. Radiation pentration theory and a test case. *J. Appl. Ecol.* **12,** 839–878.

Norman, J. M., Garcia, R., and Verma, S. B. (1992). Soil surface CO_2 fluxes and the carbon budget of a grassland. *J. Geophys. Res.* **97,** 18845–18853.

Noss, R. F. (1987). Protecting natural areas in fragmented landscapes. *Natural Areas J.* **7,** 2–13.

Nowacki, G. J., Abrams, M. D., and Lorimer, C. G. (1990). Composition, structure, and historical development of northern red oak stands along an edaphic gradient in north-central Wisconsin. *For. Sci.* **36,** 276–292.

Nye, P. H., and Greenland, D. J. (1960). "The Soil under Shifting Cultivation," Tech. Commun. No. 51 of the Commonwealth Bureau of Soils, Harpenden. Commonwealth Agriculture Bureau, Farnham, Royal, UK.

O'Brien, B. J. (1984). Soil organic matter fluxes and turnover rates estimated from radiocarbon enrichments. *Soil Biol. Biochem.* **16,** 115–120.

O'Connell, A. M. (1988). Decomposition of leaf litter in karri (*Eucalyptus diversicolor* F. Muell.) forest of varying age. *For. Ecol. Manage.* **24,** 113–125.

O'Connell, A. M., and Menage, P. M. A. (1982). Litter fall and nutrient cycling in karri (*Eucalyptus diversicolor* F. Muell.) forest in relation to stand age. *Aust. J. Ecol.* **7,** 49–62.

Odum, E. (1969). The strategy of ecosystem development. *Science* **164,** 262–270.

Oker-Blom, P. (1986). Photosynthetic radiation regime and canopy structure in modeled forest stands. *Acta. For. Fenn.* **197,** 1–44.

Oker-Blom, P. (1989). Relationships between radiation interception and photosynthesis in forest canopies: Effect of stand structure and latitude. *Ecol. Model.* **49,** 73–87.

Oker-Blom, P., and Kellomäki, S. (1983). Effect of grouping of foliage on the within-stand and within-crown light regime: comparison of random and grouping canopy models. *Agric. Meteorol.* **28,** 143–155.

Oliver, C. D. (1981). Forest development in North America following major disturbances. *For. Ecol. Manage.* **3,** 153–168.

Olson, J. S. (1963). Energy storage and the balance of producers and decomposers in ecological systems. *Ecology* **41,** 322–331.

Olson, R. K., and Reiners, W. A. (1983). Nitrifiication in subalpine balsam fir soils: Tests for inhibitory factors. *Soil Biol. Biochem.* **15,** 413–418.

Oosting, H. J. (1956). "The Study of Plant Communities." Freeman, San Francisco.

Opie, J. E. (1972). STANDSIM—a general model for simulating growth of even-aged stands. *In* "Proceedings of the Institute National de la Recherche Agronomique," Publ No. 72-3, pp. 217–239.

Oppenheimer, M., Epstein, C. B., and Yuhnke, R. E. (1985). Acid deposition, smelter emissions, and the linearity issue in the western United States. *Science* **229,** 859–862.

Ovington, J. D. (1983). Temperate broad-leaved evergreen forests. *In* "Ecosystems of the World" (J. D. Ovington, Ed.). Elsevier, Amsterdam.

Packham, D. R. (1970). Heat transfer above a small ground fire. *Aust. For. Res.* **5,** 19–24.

Pallardy, S. G., Cermack, J., Ewers, F. W., Kaufmann, M. R., Parker, W. C., and Sperry, J. S. (1995). Water transport dynamics in trees and stands. *In* "Resource Physiology of Conifers: Acquisition, Allocation and Utilization" (W. R. Smith and T. M. Hickley, Eds.), pp. 301–389. Academic Press, San Diego.

Parker, G. G. (1983). Throughfall and stemflow in the forest nutrient cycle. *Adv. Ecol. Res.* **13,** 57–133.

Parton, W. J., Schimel, D. S., Cole, C. V., and Ojima, D. S. (1987). Analysis of factors controlling soil organic matter levels in Great Plains grasslands. *Soil. Sci. Soc. Am. J.* **51,** 1173–1179.

Parton, W. J., Stewart, W. B., and Cole, C. V. (1988). Dynamics of C, N, P and S in grassland soils: A model. *Biogeochem.* **5,** 109–131.

Parton, W. J., Scurlock, J. M. O., Ojima, D. S., Gilmanov, T. G., Scholes, R. J., Schimel, D. S., Kirchner, T., Menaut, J., Seastedt, T., Garcia Moya, E., Kamnalrut, A., and Kinyamario, J. I. (1993). Observations and modeling of biomass and soil organic matter dynamics for the grassland biome worldwide. *Global Biogeochem. Cycles* **7,** 785–809.

Passioura, J. B. (1982). Water in the soil–plant–atmosphere continuum. *In* "Encyclopedia of Plant Physiology. Physiological Plant Ecology II" (O. L. Lange, P. S. Nobel, C. B. Osmond, and H. Ziegler, Eds.), Vol.12B, pp. 5–33. Springer-Verlag, Berlin.

Passioura, J. B., and Munns, R. (1984). Hydraulic resistance of plants. II. Effects of rooting medium, and time of day, in barley and lupin. *Aust. J. Plant Physiol.* **11,** 341–350.

Pastor, J., and Post, W. M. (1986). Influence of climate, soil moisture, and succession on forest carbon and nitrogen cycles. *Biogeochemistry* **2,** 3–27.

Pastor, J., and Post, W. M. (1993). Linear regressions do not predict the transient responses of eastern North American forests to CO_2 induced climate change. *Climatic Change* **23,** 111–119.

Pastor, J., Aber, J. D., McClaugherty, C. A., and Melillo, J. M. (1984). Above-ground production and N and P cycling along a nitrogen mineralization gradient on Blackhawk Island, Wisconsin. *Ecology* **65,** 256–268.

Pastor, J., Naiman, R. J., Dewey, B., and McInnes, P. (1988). Moose, microbes, and the boreal forest. *BioScience* **38,** 770–777.

Paul, E., and Clark, F.E. (1989). "Soil Microbiology and Biochemistry." Academic Press, San Diego, CA.

Payette, S. (1992). Fires as a controlling process in the North American boreal forest. *In* "A Systems Analysis of the Global Boreal Forest" (H. H. Shugart, R. Leemans, and G. B. Bonan, Eds.), pp. 13–84. Cambridge Univ. Press, Cambridge, UK.

Pearson, J. A., Fahey, T. J., and Knight, D. H. (1984). Biomass and leaf area in contrasting lodgepole pine forests. *Can. J. For. Res.* **14,** 259–265.

Peet, R. K. (1981). Changes in biomass and production during secondary forest succession. In "Forest Succession: Concepts and Applications" (D. C. West, H. H. Shugart, and D. B. Botkin, Eds.), pp. 324–338. Springer-Verlag, New York.

Penning de Vries, F. W. T. (1975). The cost of maintenance processes in plant cells. Annu. Bot. **39**, 77–92.

Penning de Vries, F. W. T., Brunsting, A. H. M., and van Laar, H. H. (1974). Produts, requirements and efficiency of biosynthesis: a quantitative approach. J. Theor. Biol. **45**, 339–377.

Perala, D. A., and Alban, D. H. (1982). Biomass, nutrient distribution and litterfall in *Populus*, *Pinus* and *Picea* stands on two different soils in Minnesota. Plant Soil **64**, 177– 192.

Pereira, H. C. (1973). "Land Use and Water Resources in Temperate and Tropical Climates." Cambridge Univ. Press, Cambridge.

Persson, H. (1980a). Death and repalcement of fine roots in a mature Scots pine stand. Ecol. Bull. (Stockholm) **32**, 251–260.

Persson, H. (1980b). Spatial distribution of fine root growth, mortality and decomposition in a young Scots pine stand in central Sweden. Oikos **34**, 77–87.

Peterjohn, W. T., Melillo, J. M., Steudler, P. A., Newkirk, K. M., Bowles, F. P., and Aber, J. D. (1994). Responses of trace gas fluxes and N availability to experimentally elevated soil temperatures. Ecol. Appl. **4**, 617–625.

Peterson, D. L., and Waring, R. H. (1994). Overview of the Oregon transect ecosystem research project. Ecol. Appl. **4**, 211–225.

Phillips, C. J., and Watson, A. J. (1994). "Structural Tree Root Research in New Zealand: A Review," Landcare Research Science Series No. 7. Manaaki Whenua Press, Lincoln, New Zealand.

Phillipson, J., Abel, R., Steele, J., and Woodell, S. R. J. (1978). Earthworm numbers, biomass and respiratory metabolism in a beech woodland—Wytham Woods, Oxford. Oecologia **33**, 291–309.

Pinker, R. T., Thompson, O. E., and Eck, T. F. (1980a). The energy balance of a tropical evergreen forest. J. Appl. Meteor **19**, 1341–1350.

Pinker, R. T., Thompson, O. E., and Eck, T. F. (1980b). The albedo of a tropical evergreen forest. Q. J. R. Met. Soc. **106**, 551–558.

Pitelka, L. F., and Pitelka, F. A. (1993). Environmental decision making: Multidimensional delimmas. Ecol. Appl. **3**, 566–568.

Post, W. M., Emanuel, W. R., Zinke, P. J., and Stangenberger, A. G. (1982). Soil carbon pools and world life zones. Nature **298**, 156–159.

Potter, C. S., Randerson, J. T., Field, C. B., Matson, P. A., Vitousek, P. M., Mooney, H. A., and Klooster, S. A. (1993). Terrestrial ecosystem production: A process model based on global satellite and surface data. Global Biogeochem. Cycles **7**, 811–841.

Potter, G. L., Ellsaesser, W., MacCracken, M. C., and Luther, F. M. (1975). Possible climatic impact of tropical deforestation. Nature **258**, 697–698.

Prance, G. T. (1989). American tropical forests. In "Tropical Rainforest Ecosystems: Biogeographical and Ecological Studies" (H. Leith and M. J. A. Werger, Eds.), Vol. 14B, pp. 99–132. Elsevier, Amsterdam.

Prentice, K. C., and Fung, J. Y. (1990). The sensitivity of terrestrial carbon storage to climate change. Nature **346**, 48–50.

Press, W. H., Flannery, B. P., Teukolsky, and Vetterling, W. T. (1986). "Numerical Recipes." Cambridge Univ. Press, New York.

Prince, S. D. (1991). A model of regional primary production for use with coarse resolution satellite data. Int. J. Remote Sensing **12**, 1313–1330.

Prince, S. D., and Goward, S. N. (1995). Global primary production: A remote sensing approach. J. Biogeogr. **22**, 815–835.

Pritchett, W. L. (1979). "Properties and Management of Forest Soils." Wiley, New York.

Proctor, J., and Woodell, S. R. J. (1975). The ecology of serpentine soils. *Adv. Ecol. Res.* **9**, 255–366.
Protz, R., Ross, G. J., Martini, I. P., and Terasmae, J. (1988a). Rate of podzolic soil formation near Hudson Bay, Ontario. *Can. J. Soil Sci.* **64**, 31–49.
Protz, R., Ross, G. J., Shipitalo, M. J., and Terasmae, J. (1988b). Podzolic soil development in the southern James Bay lowlands, Ontario. *Can. J. Soil Sci.* **68**, 287–305.
Pulliam, W. M. (1993). Carbon dioxide and methane exports from a southeastern floodplain swamp. *Ecol. Monogr.* **63**, 29–53.
Puttonen, P. (1986). Carbohydrate reserves in *Pinus sylvestris* seedling needles as an attribute of seedling vigor. *Scand. J. For. Res.* **1**, 181–193.
Pye, J. M., and Vitousek, P. M. (1985). Soil and nutrient removals by erosion and windrowing at a southeastern U.S. Piedmont site. *For. Ecol. Manage.* **11**, 145–155.
Quall and Haines (1991).
Rafter, T. A., and Stout, J. D. (1970). Radiocarbon measurements as an index of the rate of turnover of organic matter in forest and grassland ecosystems in New Zealand. *In* "Proceedings of the 12th Nobel Symposium" (I. U. Olsson, Ed.), pp. 401–417. Wiley, New York.
Raich, J. A. (1980). Fine roots regrow rapidly after forest felling. *Biotropica* **2**, 231–232.
Raich, J. W., and Nadelhoffer, K. J. (1989). Below ground carbon allocation in forest ecosystems: Global trends. *Ecology* **70**, 1346–1354.
Raich, J. W., and Schlesinger, W. A. (1992). The global carbon dioxide flux in soil respiration and its relationship to vegetation and climate. *Tellus* **44b**, 81–99.
Raison, J. K., Chapman, E. A., Wright, I. C., and Jacobs, S. W. L. (1979). Membrane lipid transitions: Their correlation with the climatic distribution of plants. *In* "Low Temperature Stress in Crop Plants: The Role of the Membrane" (J. M. Lyons, D. Graham, and J. K. Raison, Eds.), pp. 177–186. Academic Press, New York.
Raison, R. J. (1979). Modification of the soil environment by vegetation fires, with particular reference to nitrogen transformations: A review. *Plant Soil* **51**, 73–108.
Raison, R. J., Khanna, P., and Woods, P. (1985). Mechanisms of element transfer to the atmosphere during vegetation fires. *Can. J. For. Res.* **15**, 132–140.
Raison, R. J., Myers, B. J., and Benson, M. L. (1992). Dynamics of *Pinus radiata* foliage in relation to water and nitrogen stress: I. Needle production and properties. *For. Ecol. Manage.* **52**, 139–158.
Rasmussen, D. H., and MacKenzie, A. P. (1972). Effect of solute on ice-solution interfacial free energy: Calculation from measured homogeneous nucleation temperatures. *In* "Water Structure at the Water Polymer Interface" (H. H. G. Jellinok, Ed.), pp. 126–145. Plenum, New York.
Rastetter, E. B., Ryan, M. G., Shaver, G. R., Melillo, J. M., Nadelhoffer, K. J., Hobbie, J. E., and Aber, J. D. (1991). A general biogeochemical model describing the responses of the C and N cycles in terrestrial ecosystems to changes in CO_2, climate and N deposition. *Tree Physiol.* **9**, 101–126.
Rauner, J. L. (1976). Deciduous forests. *In* "Vegetation and the Atmosphere" (J. L. Monteith, Ed.), Vol. 2, pp. 241–264. Academic Press, London.
Raupach, M. R. (1989). Stand overstorey processes. *Phil. Trans. R. Soc. B* **324**, 175–190.
Raupach, M. R., and Finnigan, J. J. (1988). Single-layer models of evaporation from plant canopies are incorrect but useful, whereas multilayer models are correct but useless: Discuss. *Aust. J. Plant Physiol.* **15**, 705–716.
Read, D. J. (1991). Mycorrhizas in ecosystems—Nature's response to the "Law of the Minimum." *In* "Frontiers in Mycology" (D. L. Hawksworth, Ed.), pp. 101–130. CAB International, Regensburg, Germany.
Redfield, A. C. (1958). The biological control of chemical factors in the environment. *Am. Sci.* **46**, 206–226.

Reich, P. B., Koike, T., Gower, S. T., and Schoettle, A. W. (1995). Causes and consequences of variation in conifer leaf life-span. *In* "Ecophysiology of Conifers" (W. K. Smith and T. M. Hinckley, Eds.), pp. 225–254. Academic Press, San Diego, CA.

Reich, P. B., Uhl, C., Walters, M. B., and Ellsworth, D. S. (1991). Leaf life-span as a determinant of leaf structure and function among 23 species in Amazonian forest communities. *Oecologia* **86,** 16–24.

Reich, P. B., Walters, M. B., and Ellsworth, D. S. (1992). Leaf life-span in relation to leaf, plant and stand characteristics among diverse ecosystems. *Ecol. Monogr.* **62,** 365–392.

Reid, C. P. P., Kidd, F. H., and Ekwebelam, S. A. (1983). Nitrogen nutrition, photosynthesis and carbon allocation in ectomycorrhizal pine. *Plant Soil* **71,** 415–432.

Reiners, W. (1986). Complimentary models for forest ecosystems. *Am. Nat.* **127,** 59–73.

Reiners, W. A. (1981). Nitrogen cycling in relation to ecosystem succession. *Ecol. Bull. (Stockholm)* **33,** 507–528.

Rennolls, K., Geary, D. N., and Rollinson, T. J. D. (1985). Characterizing diameter distributions by the use of the Weibull distribution. *Forestry* **58,** 58–66.

Reynolds, J. F., and Thornley, J. H. M. (1982). A shoot/root partitioning model. *Ann. Bot.* **49,** 585–597.

Rice, E. L. (1974). "Allelopathy." Academic Press, New York.

Richards, J. H., and Bliss, L. (1986). Winter water relations of a deciduous timberline conifer, *Larix lyallii* Parl. *Oecologia* **69,** 16–24.

Richardson C. J., (Ed.) (1981). "Pocosin Wetlands: An Integrated Analysis of Coastal Plain Freshwater Bogs in North Carolina." Hutchinson Ross, Stroudsburg, PA.

Richter, D. D., Ralston, C. W., and Harms, W. R. (1982). Prescribed fire: Effects on water quality and nutrient cycling. *Science* **215,** 661–663.

Richter, H. (1973). Frictional potential losses and total water potential in plants: A re-evaluation. *J. Exp. Bot.* **27,** 473–479.

Riha, S. J., Campbell, G. S., and Wolfe, J. (1986). A model of competition for ammonium among heterotrophs, nitrifers and roots. *Soil Sci. Soc. Am. J.* **50,** 1463–1466.

Ripple, W. J., Bradshaw, G. A., and Spies, T. A. (1991). Measuring forest landscape patterns in the Cascade range of Oregon, USA. *Biol. Conserv.* **57,** 73–88.

Robertson, G. P., and Tiedje, J. M. (1984). Denitrification and nitrous oxide production in an old-growth and successional Michigan forests. *Soil Sci. Soc. Am. J.* **48,** 383–389.

Robertson, G. P., and Vitousek, P. M. (1981). Nitrification potentials in primary and secondary succession. *Ecology* **62,** 376–386.

Robertson, G. P., Huston, M. A., Evans, F. C., and Tiedje, J. M. (1988). Spatial variability in a successional plant community: Patterns of nitrogen availability. *Ecology* **69,** 1517–1524.

Robertson, G. P., Vitousek, P. M., Matson, P. A., and Tiedje, J. M. (1987). Denitrification in a clearcut loblolly pine (*Pinus taeda* L.) plantation in the southeastern U.S. *Plant Soil* **97,** 119–129.

Rodin, L. E., and Bazilevich, N. I. (1967). "Production and Mineral Cycling in Terrestrial Vegetation." Oliver and Boyd, Edinburgh.

Röhrig, E., and Ulrich, B. (eds.) (1991). "Temperate Deciduous Forests." Ecosystems of the World. (D. W. Goodall, editor), p. 630. Elsevier, Amsterdam, London.

Romme, W. H., Knight, D. H., and Yavitt, J. B. (1986). Mountain pine beetle outbreaks in the Rocky Mountains: Regulators of primary production? *Am. Nat.* **127,** 484–494.

Ross, D. J., Sparling, G. P., Burke, C. M., and Smith, C. T. (1995). Microbial biomass C and N, and mineralizable-N, in litter and mineral soil under *Pinus radiata* on a coastal sand: Influence of stand age and harvest management. *Plant Soil* **175,** 167–177.

Ross, J. (1981). "The Radiation Regime and Architecture of Plant Stands," pp. 391. Junk, The Hague.

Rotty, R. M., and Masters, C. D. (1985). Carbon dioxide from fossil fuel combustion: Trends, resources, and technological implications. *In* "Atmospheric Carbon Dioxide and the Global Carbon Cycle" (J. R. Trabalka, Ed.), DOE/ER-0239, pp. 63–80. U.S. Department of Energy, Washington, DC.

Rovira, A. D. (1969). Plant root exudates. *Bot. Rev.* **35**, 35–37.

Rowe, L. K. (1983). Rainfall interception by an evergreen beech forest, Nelson, New Zealand. *J. Hydrol.* **66**, 143–158.

Ruark, G. A., and Bockheim, J. G. (1988). Biomass, net primary production, and nutrient distribution for an age sequence of *Populus tremuloides* ecosystems. *Can. J. For. Res.* **18**, 435–443.

Rubin, E. S., Cooper, R. N., Frosch, R. A., Lee, T. H., Marland, G., Rosenfeld, A. H., and Stine, D. D. (1992). Realistic mitigation options for global warming. *Science* **257**, 148–266.

Ruess, R. W., Van Cleve, K., Yarie, J., and Viereck, L. A. (1996). Comparative estimates of fine root production in successional taiga forests of interior Alaska. *Can. J. For. Res.* (in press).

Ruimy, A., Saugier, B., and Dedieu, G. (1994). Methodology for the estimation of terrestrial net primary production from remotely sensed data. *J. Geophys. Res.* **99**, 5263–5283.

Runkle, J. R. (1989). Synchrony of regeneration, gaps, and latitudinal differences in tree species diversity. *Ecology* **70**, 546–547.

Running, S. W., and Coughlan, J. C. (1988). A general model of forest ecosystem processes for regional applications. I. Hydrologic balance, canopy gas exchange and primary production processes. *Ecol. Model.* **42**, 125–154.

Running, S. W., and Gower, S. T. (1991). FOREST-BGC, a general model of forest ecosystem processes for regional applications. II Dynamic carbon allocation and nitrogen budgets. *Tree Physiol.* **9**, 147–160.

Running, S. W., Nemani, R. R., and Hungerford, R. D. (1987). Extrapolation of synoptic meteorological data in mountainous terrain and its use for simulating forest evapotranspiration and photosynthesis. *Can. J. For. Res.* **17**, 472–483.

Running, S. W., Nemani, R. R., Peterson, D. L., Band, L. E., Potts, D. F., Pierce, L. L., and Spanner, M. A. (1989). Mapping regional forest evapotranspiration and photosynthesis by coupling satellite data with ecosystem simulation. *Ecology* **70**, 1090–1101.

Running, S. W., Loveland, T. R., and Pierce, L. L. (1994). A vegetation classification logic based on remote sensing for use in global biogeochemical models. *Ambio* **23**, 77–81.

Running, S. W., Loveland, T. R., Pierce, L. L., Nemani, R. R., and Hunt, E. R., Jr. (1995). A remote sensing based vegetation classification logic for global land cover analysis. *Remote Sensing Environ.* **51**, 39–48.

Runyon, J., Waring, R. H., Goward, S. N., and Welles, J. M. (1994). Environmental limits on net primary production and light-use efficiency across the Oregon transect. *Ecol. Appl.* **4**, 226–237.

Ruprecht, J. K., Schofield, N. J., Crombie, D. S., Vertessy, R. A., and Stoneman, G. L. (1991). Early hydrological response to intense forest thinning in southwestern Australia. *J. Hydrol.* **127**, 261–277.

Russell, G., Jarvis, P. G., and Monteith, J. L. (1989). Absorption of radiation by canopies and stand growth. *In* "Plant Canopies: Their Growth, Form and Function" (G. Russell, B. Marshall, and P. G. Jarvis, Eds.), pp. 21–39. Cambridge Univ. Press, Cambridge, UK.

Rutter, A. J. (1975). The hydrological cycle in vegetation. *In* "Vegetation and the Atmosphere" (J. L. Monteith, Ed.), Vol. 1, pp. 111–154. Academic Press, London.

Ryan, M. G. (1991). Effects of climate change on plant respiration. *Ecol. Appl.* **1**, 157–167.

Ryan, M. G., and Waring, R. H. (1992). Maintenance respiration and stand development in a subalpine lodgepole pine forest. *Ecology* **73**, 2100–2108.

Ryan, M. G., Binkley, D., and Fownes, J. H. (1996). Age-related decline in forest productivity: Pattern and process. *Adv. Ecol. Res.* (in press).

Ryan, M. G., Linder, S., Vose, J. M., and Hubbard, R. M. (1994). Respiration of pine forests. *Ecol. Bull. (Copenhagen)* **43**, 50–63.

Ryan, M. G., Gower, S. T., Hubbard, R. M., Waring, R. H., Gholz, H. L., Cropper, W. L., and Running, S. W. (1995). Stem maintenance respiration of conifer stands in contrasting climates. *Oecologia* **101**, 133–140.

Rygiewicz, P. T., and Andersen, C. P. (1994). Mycorrhizae alter quality and quantity of carbon allocated below ground. *Nature* **369**, 58.

Sagan, C., Toon, O. B., and Pollack, J. B. (1979). Anthropogenic albedo changes and the Earth's climate. *Science* **206**, 1363–1368.

St. John, T. V. (1983). Response of tree roots to decomposing organic matter in two lowland Amazonian rain forests. *Can. J. For. Res.* **13**, 346–349.

St. John, T. V., Coleman, D. C., and Reid, C. P. P. (1983). Growth and spatial distribution of nutrient-absorbing organs: Selective exploitation of soil heterogeneity. *Plant Soil* **71**, 487–493.

Sakai, A. (1979). Freezing avoidance mechanism of primordial shoots of conifer buds. *Plant Cell Physiol.* **20**, 1381–1390.

Sakai, A. (1983). Comparative studies on freezing resistance of conifers with special reference to cold adaptation and its evaluative aspects. *Can. J. Bot.* **61**, 2323–2332.

Sakai, A., and Larcher, W. (1987). "Frost Survival of Plants. Responses and Adaptation to Freezing Stress, Ecological Studies 62." Springer-Verlag, Berlin.

Sakai, A., and Weiser, C. J. (1973). Freezing resistances of trees in North America with reference to tree regions. *Ecology* **54**, 118–126.

Sala, A., and Tenhunen, J. D. (1994). Site-specific water relations and stomatal response of *Quercus ilex* in a Mediterranean watershed. *Tree Physiol.* **14**, 601–617.

Saldarriaga, J. G., and Luxmoore, R. J. (1991). Solar energy conversion efficiencies during succession of a tropical rainforest in Amazonia. *J. Trop. Ecol.* **7**, 233–242.

Saliendra, N. Z., Sperry, J. S., and Comstock, J. P. (1995). Influence of leaf water status on stomatal response to humidity, hydraulic conductance, and soil drought in *Betula occidentalis*. *Planta* **196**, 357–366.

Salisbury, F. B., and Ross, C. W. (1992). "Plant Physiology," 4th ed. Wadsworth, Belmont, CA.

Sanchez, P. A. (1976). "Properties and Management of Soils in the Tropics." Wiley, New York.

Sanchez, P. A. (1981). Soils of the humid tropics. *In* "Blowing in the Wind: Deforestation and Long-Range Implications" (Department of Anthropology, Ed.), pp. 347–410. William and Mary College, Williamsburg, VA.

Sanchez, P. A., Bandy, D. E., Villachica, J. H., and Nicholaides, J. J. (1982). Amazon basin soils: Management for continuous crop production. *Science* **216**, 821–827.

Sands, P. J. (1988). Resource modelling: its nature and use. *Mem. Entom. Soc. Can.* **143**, 5–10.

Sanford, R. L., Parton, W. J., Ojima, D. S., and Jean Lodge, D. (1991). Hurricane effects on soil organic matter dynamics and forest production in the Luquillo experimental forest Puerto Rico: Results of simulation modelling. *Biotropica* **23**, 364–372.

Santantonio, D., and Hermann, R. K. (1985). Standing crop, production, and turnover of fine roots on dry, moderate and wet sites of mature Douglas-fir in western Washington. *Ann. Sci.* **42**, 113–142.

Santantonio, D., Hermann, R. K., and Overton, W. S. (1977). Root biomass studies in forest ecosystems. *Pedobiology* **17**, 1–31.

Santos, P. F., and Whitford, W. G. (1981). The effect of microarthropods on litter decomposition in a Chihuahuan desert ecosystem. *Ecology* **62**, 654–663.

Scheu, S. (1990). Changes in microbial nutrient status during secondary succession and its modifications by earthworms. *Oecologia* **84**, 351–358.

Schiffman, P. M., and Johnson, W. C. (1990). Phytomass and detrital storage during forest regrowth in the southeastern United States Piedmont. *Can. J. For. Res.* **19**, 69–78.

Schimper, A. F. W. (1898). "Pflanzengeographie auf physiologischer Grundlage." Jena.

Schlentner, R. L., and Van Cleve, K. (1985). Relationships between CO_2 evolution from soil, substrate temperature, and substrate moisture in four mature forest types in interior Alaska. *Can. J. For. Res.* **15,** 97–106.

Schlesinger, W. H. (1977). Carbon balance in terrestrial detritus. *Annu. Rev. Ecol. Syst.* **8,** 51–81.

Schlesinger, W. H. (1991). "Biogeochemistry: An Analysis of Global Change." Academic Press, New York.

Schlesinger, W. H., DeLucia, E. H., and Billings, W. D. (1989). Nutrient-use efficiency of woody plants on contrasting soils in western Great Basin, Nevada. *Ecology* **70,** 105–113.

Schmaltz, J. (1991). Deciduous forest of southern South America. *In* "Ecosystems of the World 7. Temperate Deciduous Forests" (E. Rohrig and B. Uhlich, Eds.), pp. 557–578. Elsevier, Amsterdam.

Schnitzer, M. (1979). Reactions of humic substances with metals and minerals in soils. *In* "Collogues Internationaux du C.N.R.S. Migrations Organo-Minerales dans les Sols Temperes," No. 303, 229–234.

Schoch, P., and Binkley, D. (1986). Prescribed burning increased nitrogen availability in a mature loblolly pine stand. *For. Ecol. Manage.* **14,** 13–22.

Schoettle, A. W., and Fahey, T. J. (1994). Foliage and fine root longevity of pines. *Ecol. Bull. (Copenhagen)* **43,** 136–153.

Schoettle, A. W., and Smith, W. K. (1991). Interrelation between shoot characteristics and solar irradiance in the crown of *Pinus contorta* ssp. *latifolia*. *Tree Physiol.* **9,** 245–254.

Schouw, J. F. (1823)."Grunzuge einer allgemeinen Pflanzengeographie." Berlin.

Schowalter, T. D. (1981). Insect herbivore relationship to the state of the host plant: biotic regulation of ecosystem nutrient cycling through ecosystem succession. *Oikos* **37,** 126–130.

Schowalter, T. D., and Crossley, D. A. (1984). The influence of arthropods on ecosystems. *BioScience* **34,** 157–161.

Schowalter, T. D., Hargrove, W. W., and Crossley, D. A. (1986). Herbivory in forested ecosystems. *Annu. Rev. Entomol.* **31,** 177–196.

Schuepp, P. H. (1972). Studies of forced-convection heat and mass transfer of fluttering realistic leaf models. *Bound. Layer Meteorol.* **2,** 263–274.

Schuepp, P. H. (1973). Model experiments of heat and mass transfer of leaves and plant elements. *Bound. Layer Meteorol.* **3,** 454–467.

Schulze, E.-D., and Mooney, H. A., (Eds.) (1993). "Biodiversity and Ecosystem Function." Springer-Verlag, New York.

Schulze, E.-D., Fuchs, M., and Fuchs, M. I. (1977a). Spacial distribution of photosynthetic capacity and performance in a mountain spruce forest of Northern Germany III. The significance of the evergreen habit. *Oecologia* **30,** 239–248.

Schulze, E.-D., Fuchs, M. I., and Fuchs, M. (1977b). Spacial distribution of photosynthetic capacity and performance in a mountain spruce forest of northern Germany I. Biomass distribution and daily CO_2 uptake in different crown layers. *Oecologia* **29,** 43–61.

Schulze, E.-D., Kelliher, F. M., Körner, C., Lloyd, J., and Leuning, R. (1994). Relationships among maximum stomatal conductance, ecosystem surface conductance, carbon assimilation rate, and plant nitrogen nutrition: A global ecology scaling exercise *Annu. Rev. Ecol. Syst.* **25,** 629–669.

Schulze, E.-D., Schulze, W., Kelliher, F. M., Vygodskaya, N. N., Ziegler, W., Kobak, K. I., Koch, H., Arneth, A., Kusnetsova, W. A., Sogatchev, A., Issajev, A., Bauer, G., and Hollinger, D. Y. (1995). Aboveground biomass and nitrogen nutrition in a chronosequence of pristine Dahurian *Larix* stands in eastern Siberia. *Can. J. For. Res.* **25,** 943–960.

Schurr, U., Gollan, T., and Schulze, E. D. (1992). Stomatal response tp drying soil in relation to changes in the xylem sap composition of *Helianthus annuus*. II. Stomatal sensitivity to abscisic acid imported from the xylem sap. *Plant Cell Environ.* **15,** 561–567.

Seastedt, T. R., and Crossley, D. A. (1980). Effects of microarthropods on the seasonal dynamics of nutrients in forest litter. *Soil Biol. Biochem.* **12,** 337–342.

Sellers, P. J., Berry, J. A., Collatz, G. J., Field, C. B., and Hall, F. G. (1992). Canopy reflectance, photosynthesis, and transpiration. III. A reanalysis using improved leaf models and a new canopy integration scheme *Remote Sensing Environ.* **42,** 187–216.
Sextone, A. J., Parkin, T. B., and Tiedje, J. M. (1985). Temporal responses of soil denitrification rates to rainfalll and irrigation. *Soil Sci. Soc. Am. J.* **49,** 99–105.
Sharma, M. L., Barron, R. J. W., and Williamson, D. R. (1987). Soil water dynamics of lateritic catchments as affected by forest clearing for pasture. *J. Hydrol.* **94,** 29–46.
Shepherd, I. D. H. (1991). Information integration and GIS. *In* "Geographical Information Systems" (D. J. Maguire, M. F. Goodchild, and D. W. Rhind, Eds.), Vol. 2, pp. 337–360. Longman, Harlow, UK.
Sheridan, R. P. (1989). Nitrogenase activity by *Hapalosiphon flexuosus* associated with *Sphagnum arythrocalyx* in cloud forest on the volcano La Soufriere, French West Indies. *Biotropica* **23,** 134–140.
Sheriff, D. W., Nambiar, E. K. S., and Fife, D. N. (1986). Relationship between nutrient status, carbon assimilation and water use in *Pinus radiata*. *Tree Physiol.* **2,** 73–88.
Sheriff, D. W., Margolis, H. A., Kaufmann, M. R., and Reich, P. B. (1995). Resource use efficiency. *In* "Resource Physiology of Conifers: Acquisition, Allocation, and Utilization" (W. K. Smith and T. M. Hinckley, Eds.), pp. 143–178. Academic Press, San Diego.
Shugart, H. H. (1984). "A Theory of Forest Dynamics: The Ecological Implications of Forest Succession Models," pp. 278. Springer-Verlag, New York.
Shugart, H. H., Jr., and West, D. C. (1977). Development of an Appalachian deciduous forest succession model and its application to assessment of the impact of the chestnut blight. *J. Environ. Manage.* **5,** 161–179.
Shukla, J., Nobre, C., and Sellers, P. J. (1990). Amazonian deforestation and climate change. *Science* **247,** 1322–1325.
Shuttleworth, W. J. (1988). Evaporation from Amazonian rainforest. *Proc. R. Soc. London B* **233,** 321–346.
Shuttleworth, W. J. (1989). Micrometeorology of temperate and tropical forest. *Phil. Trans. R. Soc. London B* **324,** 299–334.
Shuttleworth, W. J., and Calder, I. R. (1979) Has the Priestley–Taylor equation any relevance to forest evaporation? *J. Appl. Meteorol.* **18,** 639–646.
Shuttleworth, W. J., Gash, J. H. C., Lloyd, C. R., Moore, C. J., Roberts, J., De O Marques Filho, A., Fisch, G., De Paula Silva Filho, V., De Nazare Goes Ribeiro, M., Molion, L. C. B., De Abreu Sa, L. D., Nobre, J. C. A., Cabral, O. M. R., Patel, S. R., and De Moraes, J. C. (1984). Eddy correlation measurements of energy partition for Amazonian forest. *Q. J. R. Met. Soc.* **110,** 1143–1162.
Shvidenko, A., and Nilsson, S. (1994). What do we know about the Siberian forests? *Ambio* **23,** 396–404.
Siau, J. F. (1971). "Flow in Wood." Syracuse, NY.
Silvester, W. B. (1989). Molybdenum limitation of asymbiotic nitrogen fixation in forests of Pacific Northwest America. *Soil Biol. Biochem.* **21,** 283–289.
Skole, D., and Tucker, C. (1993). Tropical deforestation and habitat fragmentation in the Amazon: Satellite data from 1978 to 1988. *Science* **260,** 1905–1910.
Slatyer, R. O. (1967). "Plant–Water Relationships," p. 366. Academic Press, London.
Smith, F. W., and Long, J. N. (1992). A comparison of stemwood production in monocultures and mixtures of *Pinus contorta* var. *latifolia* and *Abies lasiocarpa*. *In* "The Ecology of Mixed Species Stands of Trees" (M. G. R. Cannell, D. C. Malcolm, and P. Robertson, Eds.), pp. 87–98. Blackwell, London.
Smith, N. J., and Hann, D. W. (1984). A new analytical model based on the -3/2 power rule of self-thinning. *Can. J. For. Res.* **14,** 605–609.
Smith, S. E., and Smith, F. A. (1990). Structure and function of the interfaces in biotrophic symbioses as they relate to nutrient transport. *New Phytol.* **114,** 1–38.

Smith, T. M., Leemans, R., and Shugart, H. H. (1992). Sensitivity of terrestrial carbon storage to CO_2 induced climate change: Comparison of four scenarios based on general circulation models. *Climate Change* **21**, 367–384.
Smolander, H., and Oker-Blom, P. (1990). The effect of nitrogen content on the photosynthesis of Scots pine needles and shoots. *Ann. Sci. For.* **46**, 473–475.
Sollins, P., and McCorison, F. M. (1981). Nitrogen and carbon solution chemistry of an old growth coniferous forest watershed before and after cutting. *Water Resour. Res.* **17**, 1409–1418.
Sollins, P., Spycher, G., and Topik, C. (1983). Processes of soil organic matter accretion at a mudflow chronosequence, Mt. Shasta, California. *Ecology* **64**, 1273–1282.
Solomon, A. M., and Bartlein, P. J. (1992). Past and future climate change: Response by mixed deciduous–coniferous forest ecosystems in northern Michigan. *Can. J. For. Res.* **22**, 1727–1738.
Solomon, A. M., Tharp, M. L., West, D. C., Taylor, G. E., Webb, J. M., and Trimble, J. L. (1984). "Response of Unmanaged Forests to Carbon Dioxide-Induced Climate Change: Available Information, Initial Tests and Data Requirements," TR-009. US Department of Energy, Washington, DC.
Sommerfeld, R. A., Mosier, A. R., and Musselman, R. C., (1993). CO_2, CH_4 and N_2O flux through a Wyoming snowpack and implications for global budgets. *Nature* **361**, 140–142.
Son, Y., and Gower, S. T. (1991). Aboveground nitrogen and phosphorus use by five plantation-grown tree species with different leaf longevities. *Biogeochem.* **14**, 167–191.
Son, Y., and Gower, S. T. (1992). Nitrogen and phosphorus distribution for five plantation species in southwestern Wisconsin. *For. Ecol. Manage.* **53**, 175–193.
Spanner, M., Johnson, J., Miller, J., McCreighton, R., Freemantle, J., Runyon, J., and Gong, P. (1994). Remote sensing of seasonal leaf area index across the Oregon transect. *Ecol. Appl.* **4**, 258–271.
Spanner, M. A., Pierce, L. L., Running, S. W. and Peterson, D. L. (1990a). The seasonality of AVHRR data of temperate coniferous forests: Relationship with leaf area index. *Remote Sensing Environ.* **33**, 97–112.
Spanner, M. A., Pierce, L. L., Peterson, D. L., and Running, S. W. (1990b). Remote sensing of temperate coniferous forest leaf area index. The influence of canopy closure, understory vegetation and background reflectance. *Int. J. Remote Sensing* **11**, 95–111.
Sperry, J. S., and Sullivan, J. E. M. (1992). Xylem embolism in response to freeze-thaw cycles and water stress in ring-porous, diffuse-porous, and conifer species. *Plant Physiol.* **100**, 605–613.
Sperry, J. S., Alder, N. N., and Eastlack, S. E. (1993). The effect of reduced hydraulic conductance on stomatal conductance and xylem cavitation. *J. Exp. Bot.* **44**, 1075–1082.
Spies, T. A., Franklin, J. F., and Thomas, T. B. (1988). Coarse woody debris in Douglas-fir forest of western Oregon and Washington. *Ecology* **69**, 1689–1702.
Sposito, G. (1984). "The Surface Chemistry of Soils." Oxford Univ. Press, Oxford, UK.
Sprugel, D. G. (1976). Dynamic structure of wave-regenerated *Abies balsamea* forests in northeastern United States. *J. Ecol.* **64**, 889–911.
Sprugel, D. G. (1984). Density, biomass, productivity, and nutrient-cycling changes during stand development in wave-regenerated balsam-fir forests. *Ecol. Monogr.* **54**, 165–186.
Staff, H., and Berg, B. (1982). Accumulation and release of plant nutrients in decomposing Scots pine needle litter II. Long-term decomposition in a Scots pine forest. *Can. J. Bot.* **60**, 1561–1568.
Stanhill, G. (1970). Some results of helicopter measurements of the albedo of different land surfaces. *Solar Energy* **13**, 59–66.
Stapledon, G. (1964). "Human Ecology." Faber and Faber, London.

Stark, N. (1971a). Nutrient cycling. I. Nutrient distribution in some Amazonian soils. *Trop. Ecol.* **12,** 24–50.
Stark, N. (1971a). Nutrient cycling. II. Nutrient distribution in some Amazonian vegetation. *Trop. Ecol.* **12,** 177–201.
Stenberg, P., DeLucia, E. H., Schoettle, A. W., and Smolander, H. (1995). Photosynthetic light capture and processing from the cell to canopy. *In* "Resource Physiology of Conifers: Acquisition, Allocation, and Utilization" (W. K. Smith and T. M. Hinckley, Eds.), pp. 1–38. Academic Press, San Diego.
Stevenson, F. J. (1982). "Humus Chemistry, Genesis, Composition and Reactions." Wiley, New York.
Stewart, J. B. (1977). Evaporation from the wet canopy of a pine forest. *Water Resour. Res.* **13,** 915–921.
Stewart, J. B., and Thom, A. S. (1973). Energy budgets in pine forest. *Q. J. R. Met. Soc* **99,** 154–170.
Stigter, C. J. (1980). Solar radiation as statistically related to sunshine duration: A comment using low-latitude data. *Agric. Meteorol.* **21,** 173–178.
Stoneman, G. L. (1993). Hydrological response to thinning a small jarrah (*Eucalyptus marginata*) forest catchment. *J. Hydrol.* **150,** 393–407.
Strain, B. R., Higginbotham, K. O., and Mulroy, J. C. (1976). Temperature preconditioning and photosynthetic capacity of *Pinus taeda* L. *Photosynthetica* **10,** 47–53.
Strakhov (1967). "Principles of Lithogenesis." Oliver and Boyd, London.
Strong, D. R., Jr. (1977). Epiphyte loads, tree-falls, and perennial forest disruption: A mechanism for maintaining higher species richness in the tropics without animals. *J. Biogeogr.* **4,** 215–218.
Sugai, S. F., and Schimel, J. P. (1994). Decomposition and biomass incorporation of ^{14}C-labeled glucose and phenolics in taiga florest floor: Effect of substrate quality, successional state, and season. *Soil Biol. Biochem.* **25,** 1379–1389.
Swank, W. T., and Douglas, J. E. (1974). Streamflow greatly reduced by converting deciduous hardwood stands to pine. *Science* **185,** 857–859.
Swank, W. T., Waide, J. B., Crossley, D. A., Jr., and Todd, R. L. (1981). Insect defoliation enhances nitrate export from forest ecosystems. *Oecologia* **51,** 297–299.
Swift, M. J., Heal, O. W., and Anderson, J. M. (1979). "Decomposition in Terrestrial Ecosystems." Univ. of California Press, Berkeley.
Syers, J. K., Adams, J. A., and Walker, T. W. (1970). Accumulation of organic matter in a chronosequence of soils developed on wind-blown sand in New Zealand. *J. Soil Sci.* **21,** 146–153.
Tajchman, S. J. (1984). Distribution of the radiative index of dryness and forest site quality in a mountainous watershed. *Can. J. For. Res.* **14,** 717–721.
Takhtajan, A. (1986). "Floristics Regions of the World." Univ. of California Press, Berkeley.
Talsma, T., and Hallam, P. M. (1980). Hydraulic conductivity measurement of forest catchments. *Aust. J. Soil. Res.* **30,** 139–148.
Tan, C. S., Black, T. A., and Nnyamah, J. U. (1978). A simple diffusion model of transpiration applied to a thinned Douglas-fir stand. *Ecology* **59,** 1221–1229.
Tanner, E. V. J. (1981). The decomposition of leaf litter in Jamaican montane rain forests. *J. Ecol.* **69,** 263–275.
Tardieu, F., and Davies, W. J. (1993). Integration of hydraulic and chemical signalling in the control of stomatal conductance and water status of droughted plants. *Plant Cell Environ.* **16,** 341–349.
Tardieu, F., Zhang, J., Katerji, N., Bethenod, O., Palmer, S., and Davies, W. J. (1992). Xylem ABA controls the stomatal conductance of field-grown maize subjected to soil compaction or soil drying. *Plant Cell Environ.* **15,** 193–197.

Tarrant, R. F., Lu, K. C., Bollen, W. B., and Chen, C. S. (1968). "Nutrient Cycling by Throughfall and Stemflow Precipitation in Three Coastal Oregon Forest Types," U.S.D.A. Forest Service Research Paper No. PNW-54. U.S.D.A. Forest Service, Portland, OR.

Tate, K. R., Ross, D. J., O'Brien, B. J., and Kelliher, F. M. (1993). Carbon storage and turnover, and respiratory activity, in the litter and soil of an old-growth southern beech (*Nothofagus*) forest. *Soil Biol. Biochem.* **25**, 1601–1612.

Tate, R. L., III (1987). "Soil Organic Matter: Biological and Ecological Effects." Wiley, New York.

Teklehaimanot, Z., Jarvis, P. G., and Ledger, D. C. (1991). Rainfall interception and boundary layer conductance in relation to tree spacing. *J. Hydrol.* **123**, 261–278.

Teskey, R. O., and Hinckley, T. M. (1981). Influence of temperature and water potential on root growth of white oak. *Physiol. Plant.* **52**, 363–369.

Teskey, R. O., Hinckley, T. M., and Grier, C. C. (1983). Effect of interruption of flow path on stomatal conductance of *Abies amabilis*. *J. Exp. Bot.* **34**, 1251–1259.

Teskey, R. O., Grier, C. C., and Hinckley, T. M. (1984). Changes in photosynthesis and water relations with age and season in *Abies amabilis*. *Can. J. For. Res.* **14**, 77–84.

Teskey, R. O., Whitehead, D., and Linder, S. (1994). Photosynthesis and carbon gain by plants. *Ecol. Bull. (Copenhagen)* **43**, 35–49.

Teskey, R. O., Gholz, H. L., and Cropper, W. P., Jr. (1995a). Influence of climate and fertilization on net photosynthesis of mature slash pine. *Tree Physiol.* **14**, 1215–1227.

Teskey, R. O., Sheriff, D. W., Hollinger, D. Y., and Thomas, R. B. (1995b). External and internal factors regulating photosynthesis. *In* "Resource Physiology of Conifers: Acquisition, Allocation and Utilization" (W. K. Smith and T. M. Hinckley, Eds.), pp. 105–142. Academic Press. New York.

Tezuka, Y. (1961). Development of vegetation in relation to soil formation in the volcanic island of Oshima, Izu, Japan. *Jpn. J. Bot.* **17**, 371–402.

Thom, A. S. (1975). Momentum, mass and heat exchange of plant communities. *In* "Vegetation and the Atmosphere: Principles" (J. L. Montieth, Ed.), Vol. 1, pp. 57–109. Academic Press, London.

Thomas, R. B., and Strain, B. R. (1991). Root restriction as a factor in photosynthetic acclimation of cotton seedlings grown in elevated carbon dioxide. *Plant Physiol.* **96**, 627–634.

Thompson, W. A., and Wheeler, A. M. (1992). Photosynthesis by mature needles of field grown *Pinus radiata*. *For. Ecol. Manage.* **52**, 225–242.

Thornley, J. H. M. (1972a). A model to describe the partitioning of photosynthate during vegetative plant growth. *Ann. Bot.* **36**, 419–430.

Thornley, J. H. M. (1972b). A balanced quantitative model for root/shoot ratios in vegetative plants. *Ann. Bot.* **36**, 431–441.

Thornley, J. H. M., and Johnston, I. R. (1990). "Plant and Crop Modeling." Oxford Science, Oxford, UK.

Thornthwaite, C. W. (1948). An approach toward a rational classification of climate. *Geogr. Rev.* **38**, 55–94.

Thornthwaite, C. W., and Mather, J. R. (1947). Instructions and tables for computing potential evapotranspiration and water balance. *Pubs. Climatol.* **10**, 183–311.

Thorpe, M. R., Warrit, B., and Landsberg, J. J. (1980). Responses of apple leaf stomata: A model for single leaves and a whole tree. *Plant Cell Environ.* **3**, 23–27.

Tiedje, J. M., Sextone, A. J., Parkin, T. B., Revsbech, N. P., and Shelton, D. R. (1984). Anaerobic processes in soil. *Plant Soil* **76**, 197–212.

Tilman, D. (1993). Community diversity and succession: The roles of competition, dispersal, and habitat modification. *In* "Biodiversity and Ecosystem Function, Ecological Studies 99" (E. D. Schulze and H. A. Mooney, Eds.). Springer-Verlag, New York.

Timmer, V. R., and Stone, E. L. (1978). Comparative foliar analysis of young balsam fir fertilized with nitrogen, phosphorus, potassium and lime. *Soil Sci. Soc. Am. J.* **42**, 125–130.

Topp, G. C., Davis, J. L., and Annan, A. P. (1980). Electromagnetic determination of soil water content: measurement in coaxial transmission lines. *Water Resour. Res.* **16,** 574–582.

Torquebiau, E. F. (1988). Photosynthetically active radiation environment, patch dynamics and architecture in a tropical rainforest in Sumatra. *Aust. J. Plant. Physiol.* **15,** 327–342.

Townsend, A. R., Vitousek, P. M., and Trumbore, S. E. (1995). Soil organic matter dynamics along gradients in temperature and land use on the island of Hawaii. *Ecology* **76,** 721–733.

Tranquillini, W. (1979). "Physiological Ecology of the Alpine Timberline, Ecological Studies 31." Springer-Verlag, Berlin.

Troeng, E., and Linder, S. (1982). Gas exchange in a 20-year-old stand of Scots pine. I. Net photosynthesis of current and one-year-old shoots within and between seasons. *Physiol. Plant.* **54,** 7–14.

Trumbore, S. E., Vogel, J. S., and Southon, J. R. (1989). AMS ^{14}C measurements of fractionated soil organic matter: An approach to deciphering the soil carbon cycle. *Radiocarbon* **31,** 644–654.

Tucker, C. J., and Sellers, P. J. (1986). Satellite remote sensing of primary production. *Int. J. Remote Sensing* **7,** 1395–1416.

Tukey, H. B. (1970). The leaching of substances from plants. *Annu. Rev. Ecol. Syst.* **21,** 305–324.

Turner, J. (1975). Nutrient cycling in a Douglas-fir ecosystem with respect to age and nutrient status. Ph.D. Dissertation, University of Washington, Seattle, WA.

Turner, J. (1977). Effects of nitrogen availability on nitrogen cycling in a Douglas-fir stand. *For. Sci.* **23,** 307–316.

Turner, J., and Lambert, M. J. (1986). Effects of forest harvesting nutrient removals on soil nutrient reserves. *Oecologia* **70,** 140–148.

Turner, J., and Long, J. N. (1975). Accumulation of organic matter in a series of Douglas-fir stands. *Can. J. For. Res.* **5,** 681–690.

Turner, M. J., Gardner, R. H., Dale, V. H., and O'Neill, R. V. (1989). Predicting the spread of disturbance in heterogeneous landscapes. *Oikos* **55,** 121–127.

Tyree, M. T., and Jarvis, P. G. (1982). Water in tissues and cells. *In* "Encyclopedia of Plant Physiology. Physiological Plant Ecology II" (O. L. Lange, P. S. Nobel, C. B. Osmond, and H. Ziegler, Eds.), Vol. 12B, pp. 35–77. Springer-Verlag, Berlin.

Tyree, M. T., and Sperry, J. S. (1988). Do woody plants operate near the point of catastrophic xylem dysfunction caused by dynamic water stress? Answers from a model. *Plant Physiol.* **88,** 574–580.

Tyrrell, L. E., and Crow, T. R. (1994). Dynamics of dead wood in old-growth hemlock hardwood forests of northern Wisconsin and northern Michigan. *Can. J. For. Res.* **24,** 1672–1683.

Uehara, G., and Gillman, G. (1981). "The Mineralogy, Chemistry, and Physics of Tropical Soils with Variable Charge Clay." Westview, Boulder, CO.

Ugolini, F. C. (1968). Soil development and alder invasion in a recently deglaciated area of Glacier Bay, Alaska. *In* "Biology of Alder" (J. M. Trappe, J. F. Franklin, R. F. Tarrant, and G. M. Hansen, Eds.), pp. 115–140. U.S. Forest Service Pacific Northwest Forest Range Exp. Stn., Portland, OR.

Ugolini, F. C., and Edmonds, R. L. (1983). Soil Biology. *In* "Pedogenesis and Soil Taxonomy. I. Concepts and Interactions" (L. P. Wilding, N. E. Smeck, and G. F. Hall, Eds.), pp. 193–231. Elsevier, Amsterdam.

Ugolini, F. C., Minden, R., Dawson, H., and Zachara, J. (1977). An example of soil processes in the *Abies amabilis* zone of central Cascades, Washington. *Soil Sci.* **124,** 291–302.

Updegraff, K., Pastor, J., Bridgham, S. D., and Johnston, C. A. (1995). Environmental and substrate controls over carbon and nitrogen mineralization in northern wetlands. *Ecol. Applic.* **5,** 151–163.

USDA Forest Service (1992). National forest fire statistics for the United States: Annual summary statistics. 1916–1990. USDA Forest Service, Washington, DC.

Vance, E. D., and Henderson, G. S. (1984). Soil nitrogen availability following long-term burning in an oak–hickory forest. *Soil Sci. Soc. Am. J.* **48**, 184–190.

Vanclay, J. K. (1994). "Modelling Forest Growth and Yield: Applications to Mixed Tropical Forests," pp. 312. CAB International.

Van Cleve, K., and Alexander, V. (1981). Nitrogen cycling in tundra and boreal ecosystems. *Ecol. Bull (Stockholm)* **33**, 375–404.

Van Cleve, K., and Moore, T. A. (1978). Cumulative effects of nitrogen, phosphorus and potassium fertilizer additions on soil respiration, pH and organic matter content. *Soil Sci. Soc. Am. J.* **42**, 121–124.

Van Cleve, K., and Noonan, L. L. (1975). Litterfall and nutrient cycling in the forest floor of birch and aspen stands in interior Alaska. *Can. J. For. Res.* **5**, 626–639.

Van Cleve, K., Barney, R. J., and Schlentner, R. L. (1981). Evidence of temperature control on production and nutrient cycling in two interior Alaskan black spruce ecosystems. *Can. J. For. Res.* **11**, 258–273.

Van Cleve, K., Oliver, L., Schlentner, R., Viereck, L. A., and Dyrness, C. T. (1983). Productivity and nutrient cycling in taiga forest ecosystems. *Can. J. For. Res.* **13**, 747–767.

Van den Driessche, R. (1973). Different effects of nitrate and ammonium forms of nitrogen on growth and photosynthesis of slash pine seedlings. *Aust. For.* **36**, 125–137.

van der Werf, A., Kooijman, A., and Lambers, H. (1988). Respiratory energy costs for the maintenance of biomass, for growth and for ion uptake in roots of *Carex diandra* and *Carex acutiformis*. *Physiol. Plant.* **72**, 483–491.

Van Hook, R. I., Johnson, D. W., West, D. C., and Mann, L. K. (1982). Environmental effects of harvesting forests for energy. *For. Ecol. Manage.* **4**, 79–94.

Van Lear, D. H., Douglass, J. E., Cox, S. K., and Augspurger, M. K. (1985). Sediment and nutrient export in runoff from burned and harvested pine watersheds in the South Carolina Piedmont. *J. Envir. Qual.* **14**, 169–174.

Van Miegroet, H., and Cole, D. W. (1985). Acidification sources in red alder and Douglas-fir soils—Importance of nitrification. *Soil Sci. Soc. Am. J.* **49**, 1274–1279.

Van Wagner, C. E. (1988). The historical pattern of annual burned area in Canada. *For. Chron.* **64**, 182–185.

Veen, B. W. (1980). Energy cost of ion transport *In* "Genetic Engineering of Osmoregulation: Impact on Plant Productivity for Food, Chemicals, and Energy" (D. W. Rains, R. C. Valentine, and A. Hollaender, Eds.), pp. 187–195. Plenum, New York.

Verma, S. B., Baldocchi, D. D., Anderson, D. E., Matt, D. R., and Clement, R. J. (1986). Eddy fluxes of CO_2, water vapour, and sensible heat over a deciduous forest. *Boundary-Layer Meteorol.* **36**, 71–91.

Vertessy, R. A., Hatton, T. J., O'Shaughnessy, P. J., and Jayasuriya, M. D. A. (1993). Predicting water yield from a mountain ash forest catchment using a terrain analysis based catchment model. *J. Hydrol.* **150**, 665–700.

Vitousek, P. M. (1982). Nutrient cycling and nutrient use efficiency. *Am. Nat.* **119**, 553–572.

Vitousek, P. M. (1984). Litterfall, nutrient cycling, and nutrient limitation in tropical forests. *Ecology* **65**, 285–298.

Vitousek, P. M. (1990). "Reforestation to Offset Carbon Dioxide Emissions," EN-6910 Research Project 3041-1 Final Report. EPRI, Palo Alto, CA.

Vitousek, P. M. (1994). Potential nitrogen fixation during primary succession in Hawaii Volcanoes National Park. *Biotropica* **26**, 234–240.

Vitousek, P. M., and Matson, P. A. (1984). Mechanisms of nitrogen retention in forest ecosystems: A field experiment. *Science* **225**, 51–52.

Vitousek, P. M., and Matson, P. A. (1985). Disturbance, nitrogen availability, and nitrogen losses in an intensively managed loblolly pine plantation. *Ecology* **66**, 1360–1376.

Vitousek, P. M., and Melillo, J. M. (1979). Nitrate losses from disturbed forests: Patterns and mechanisms. *For. Sci.* **25**, 605–619.

Vitousek, P. M., and Reiners, W. A. (1975). Ecosystem succession and nutrient retention: A hypothesis. *BioScience* **27**, 376–381.

Vitousek, P. M., and Sanford, R. L., Jr. (1986). Nutrient cycling in moist tropical forest. *Annu. Rev. Ecol. Syst.* **17**, 137–167.

Vitousek, P. M., and Walker, R. R. (1989). Biological invasion by *Myrica faya* in Hawaii: Plant demography, nitrogen fixation, and ecosystem effects. *Ecol. Monogr.* **59**, 247–265.

Vitousek, P. M., Gosz, J. R., Grier, C. C., Melillo, J. M., and Reiners, W. A. (1982). A comparative analysis of potential nitrification and nitrate mobility in forest ecosystems. *Ecol. Monogr.* **52**, 155–177.

Vitousek, P. M., Van Cleve, K., Balakrishnan, N., and Mueller-Dombois, D. (1983). Soil development and nitrogen turnover in montane rainforest soils in Hawaii. *Biotropica* **15**, 268–274.

Vitousek, P. M., Fahey, T., Johnson, D. W., and Swift, M. J. (1988). Elemental interactions in forest ecosystems: Succession, allometry, and input-output budgets. *Biogeochem.* **5**, 7–34.

Vitousek, P. M., Van Cleve, K., and Matson, P. A. (1989). Nitrogen availability and nitrification: primary, secondary, and old-field seres. *Plant Soil* **115**, 229–239.

Vogt, K. (1991). Carbon budgets of temperate forest ecosystems. *Tree Physiol.* **9**, 69–86.

Vogt, K. A., Edmonds, R. L., Antos, G. C., and Vogt, D. J. (1980). Relationships between CO_2 evolution, ATP concentrations and decomposition in four forest ecosystems in western Washington. *Oikos* **35**, 72–79.

Vogt, K. A., Grier, C. C., Meier, C. E., and Edmonds, R. L. (1982). Mycorrhizal role in net primary production and nutrient cycling in *Abies amabilis* ecosystems in western Washington. *Ecology* **63**, 370–380.

Vogt, K. A., Grier, C. C., Meir, C. E., and Keyes, M. R. (1983). Organic matter and nutrient dynamics in forest floors in young and mature *Abies amabilis* available stands in western Washington, as affected by fine-root input. *Ecol. Monogr.* **53**, 139–157.

Vogt, K. A., Grier, C. C., Gower, S. T., Sprugel, D. G., and Vogt, D. A. (1986a). Overestimation of net root production: A real or imaginary problem. *Ecology* **67**, 577–579.

Vogt, K. A., Grier, C. C., and Vogt, D. J. (1986b). Production, turnover, and nutrient dynamics of above- and belowground detritus of world forests. *Adv. Ecol. Res.* **15**, 303–377.

Vogt, K. A., Vogt, D. J., Gower, S. T., and Grier, C. C. (1990). Carbon and nitrogen interactions for forest ecosystems. *In* "Above and Belowground Interactions in Forest Trees in Acidified Soils. Our Pollution Report Series of the Environmental Research Programme" (H. Persson, Ed.), pp. 203–235. Commission of the European Communities, Belgium.

Vogt, K. A., Vogt, D. J., Moore, E. E., Fatuga, R., Redlin, M. R., and Edmonds, R. L. (1987). Conifer and angiosperm fine root biomass in relation to stand age and site productivity in Douglas-fir forests. *J. Ecol.* **75**, 857–870.

Vose, J. M., and Allen, H. L. (1988). Leaf area, stemwood growth and nutritional relationships in loblolly pine. *For. Sci.* **34**, 547–563.

Waide, J. B., Todd, R. L., Caskey, W. H., Swank, W. T., and Boring, L. R. (1987). Changes in nitrogen pools and transformations following forest clearcutting in the southern Appalachians. *In* "Forest Hydrology and Ecology at Coweeta" (W. T. Swank and D. A. Crossley, Eds.). Springer-Verlag, New York.

Waksman, S. A., and Starkey, R. L. (1931). "The Soil and the Microbe." Wiley, London.

Waldrop, T. A., VanLear, D. H., Lloyd, F. T., and Harms, W. R. (1987). "Long-Term Studies of Prescribed Burning in Loblolly Pine Forests of the Southeastern Coastal Plain, U.S. Forest Service General Technical Report No. SE-45. U.S. Forest Service, Portland, OR.

Walker, J., Raison, R. J., and Khanna, P. K. (1983). Fire. *In* "Australian Soils: The Human Impact" (J. S. Ruissell and R. F. Isbell, eds.), pp. 185–216. University Queensland Press, Australia.

Walter, H. (1979). "Vegetation of the Earth and Ecological Systems of the Geo-Biosphere," 2nd ed. Springer-Verlag, New York.

Walters, M. B., Kruger, E. L., and Reich, P. B. (1993). Growth, biomass distribution and CO_2 exchange of northern hardwood seedlings in high and low light: Relationships with successional status and shade tolerance. *Oecologia* **94,** 7–16.

Wang, D., Lowery, B., Norman, J. M., and McSweeney, K. (1996). Ant burrow effects on water flow and soil hydraulic properties of Sparta sand. *Soil Till. Res.* **37,** 83–93.

Wang, Y.-P., and Jarvis, P. G. (1990a). Description and validation of an array model—MAESTRO. *Agric. For. Meteorol.* **51,** 257–280.

Wang, Y. P., and Jarvis, P. G. (1990b). Effect of incident beam and diffuse radiation on PAR absorption, photosynthesis and transpiration of sitka spruce—A simulation study. *Silva. Carelica* **15,** 167–180.

Wang, Y. P., and Polglase, P. J. (1995). Carbon balance in the tundra, boreal forest and humid tropical forest during climate change: Scaling up from leaf physiology and soil carbon dynamics. *Plant Cell Environ.* **18,** 1226–1244.

Wang, Y.-P., Jarvis, P. G., and Benson, M. L. (1990). Two-dimensional needle-area density distribution within the crowns of *Pinus radiata*. *For. Ecol. Manage.* **32,** 217–237.

Wang, Y.-P., McMurtrie, R. E., and Landsberg, J. J. (1992). Modelling canopy photosynthetic productivity. *In* "Crop Photosynthesis: Spatial and Temporal Determinants." (N. R. Baker and H. Thomas, Eds.), pp. 43–67. Elsevier, New York.

Warcup, J. H. (1981). Effect of fire on the soil micro-flora and other non-vascular plants. *In* "Fire and the Australian Biota" (A. M. Gill, R. H. Groves, and I. R. Noble, Eds.), pp. 203–214. Australia Academy of Science, Canberra.

Ward, R. C. (1984). On the response to precipitation of headwater streams in humid areas. *J. Hydrol.* **74,** 171–189.

Wardle, D. A. (1992). A comparative assessment of factors which influence microbial biomass carbon and nitrogen levels in soil. *Biol. Rev.* **67,** 321–358.

Wardle, J. (1984). "The New Zealand Beeches: Ecology, Utilization and Management." New Zealand Forest Service, Christchurch, NZ.

Waring, R. H. (1987). Characteristics of trees disposed to die. *BioScience* **37,** 569–574.

Waring, R. H., and Franklin, J. F. (1979). Evergreen coniferous forests of the Pacific Northwest. *Science* **204,** 1380–1386.

Waring, R. H., and Pitman, G. B. (1985). Modifying lodgepole pine stands to change susceptibility to mountain pine beetle attack. *Ecology* **66,** 889–897.

Waring, R. H., and Running, S. W. (1978). Sapwood water storage: Its contribution to transpiration and effect upon water conductance through the stems of old-growth Douglas-fir. *Plant Cell Environ.* **1,** 131–140.

Waring, R. H., Schroeder, P. E., and Oren, R. (1982). Application of the pipe model theory to predict canopy leaf area. *Can. J. For. Res.* **12,** 556–560.

Waring, R. H., Law, B. E., Goulden, M. L., Bassow, S. L., McCreight, R. W., Wofsy, S. C., and Bazzaz, F. A. (1995). Scaling gross ecosystem production at Harvard Forest with remote sensing: A comparison of estimates from a constrained quantum-use efficiency model and eddy correlation. *Plant Cell Environ.* **18,** 1201–1213.

Wartinger, A., Heilmeier, H., Hartung, W., and Schultze, E.-D. (1990). Daily and seasonal courses of leaf conductance and abscisic acid in the xylem sap of almond trees (*Prunus dulcis* M.) under desert conditions. *New Phytol.* **116,** 581–587.

Waterloo, M. J. (1994). Water and nutrient dynamics of *Pinus caribaea* plantation forests on former grassland soils in southwest Viti Levu, Fiji. Ph.D. thesis, Free University of Amsterdam.

Watts, W. R. (1977). Field studies of stomatal conductance. *In* "Environmental Effects on Crop Physiology" (J. J. Landsberg and C. V. Cutting, Eds.), pp. 173–189. Academic Press, London.

Webb, L. J., Tracey, J. G., and Williams, W. T. (1972). Regeneration and pattern in the subtropical rainforest. *J. Ecol.* **60,** 675–695.
Webb, L. J., Tracey, J. G., and Williams, W. T. (1991). A floristic framework of Australian rainforests. *Aust. J. Ecol.* **9,** 169–198.
Wedin, D. A., and Tilman, D. (1990). Species effects on nitrogen cycling: A test with perennial grasses. *Oecologia* **84,** 433–441.
Weinstein, D. A., Beloin, R. M., and Yanai, R. D. (1991). Modeling changes in red spruce carbon balance and allocation in response to interacting ozone and nutrient stresses. *Tree Physiol.* **9,** 127–146.
Weller, D. E. (1987). A Reevaluation of the -3/2 power rule of plant self-thinning. *Ecol. Monogr.* **57,** 23–43.
Wells, C. G. (1971). Effects of prescribed burning on soil chemical properties and nutrient availability. *In* "Proceedings of the Prescribed Burning Symposium," pp. 86–99. U.S. Forest Service Southeastern For. Exp. Stn., Asheville, NC.
Wessman, C. A., Aber, J. D., Peterson, D. L., and Melillo, J. M. (1988). Remote sensing of canopy chemistry and nitrogen cycling in temperate forest ecosystems. *Nature* **335,** 154–156.
Weston, C. J., and Attiwill, P. M. (1990). Effects of fire and harvesting on nitrogen transformations and ionic mobility in soils of *Eucalyptus regnans* forests of southeastern Australia. *Oecologia* **83,** 20–26.
White, C. S. (1986). Volatile and water-souble inhibitors of nitrogen mineralization and nitrification in a ponderosa pine ecosystem. *Biol. Fertil. Soils* **2,** 97–104.
White, C. S. (1988). Nitrification inhibition by monoterpenoids: Theoretical model of action based on molecular structures. *Ecology* **69,** 1631–1633.
White, J. (1981). The allometric interpretation of the self-thinning rule. *J. Theor. Biol.* **89,** 475–500.
Whitehead, D. (1978) The estimation of foliage area from basal area in Scots pine. *Forestry* **51,** 137–149.
Whitehead, D., and Jarvis, P. G. (1981). Coniferous forests and plantations. *In* "Water Deficits and Plant Growth: Woody Plant Communities" (T. T. Kozlowski, Ed.), Vol. VI. pp. 49–152. Academic Press, New York.
Whitehead, D., Okali, D. U. U., and Fasehun, F. E. (1981). Stomatal response to environmental variables in two tropical forest species during the wet season in Nigeria. *J. Appl. Ecol.* **18,** 571–587.
Whitehead, D., Livingston, N. J., Kelliher, F. M., Hogan, K. P., Pepin, S., McSeveny, T. M., and Byers, J. N. (1996). Response of transpiration and photosynthesis to a transient change in illuminated foliage area for a *Pinus radiata* D. Don tree. *Plant Cell Environ.* (in press).
Whitford, W. G., Meentemeyer, V., Seastedt, T. R., Cromack, K., Jr., Santos, P., Todd, R. L., and Waide, J. B. (1981). Exceptions to the AET model: Deserts and clear-cut forests. *Ecology* **62,** 275–277.
Whitmore, T. C. (1984). "Tropical Rain Forests of the Far East," 2nd ed. MacMillan, New York.
Whittaker, R. H. (1975). "Communities and Ecosystems." MacMillan, New York.
Whittaker, R. H., and Likens, G. E. (1973). Carbon in the biota. *In* "Carbon and the Biosphere" (G. M. Woodwell and E. V. Pecan, Eds.), CONF 720510, pp. 281–302. National Technical Information Services, Washington, DC.
Whittaker, R. H., and Woodwell, G. M. (1967). Surface area relations of woody plants and forest communities. *Am. J. Bot.* **54,** 931–939.
Whittaker, R. H., Walker, R. B., and Kruckberg, A. R. (1954). The ecology of serpentine soils. *Ecology* **35,** 258–288.
Whittaker, R. H., Likens, G. E., Bormann, F. H., Eaton, J. S., and Siccama, T. G. (1979). The

Hubbard Brook ecosystem study: Forest nutrient cycling, and element behavior. *Ecology* **60,** 203–220.

Williams, J., Prebble, R. E., Williams, W. T., and Hignett, C. T. (1983). The influence of texture, structure and clay mineralogy on the soil moisture characteristic. *Aust. J. Soil. Res.* **21,** 15–32.

Williams, K., Percival, F., Merino, J., and Mooney, H. A. (1987). Estimation of tissue construction cost from heat of combustion and organic nitrogen content. *Plant Cell Environ.* **10,** 725–734.

Williamson, M. (1989). Mathematical models of invasion. *In* "Biological Invasions: A Global Perspective" (J. A. Drake, H. A. Mooney, F. diCastri, R. H. Groves, F. J. Kruger, M. Rejmanek, and M. Williamson, Eds.), pp. 329–360. Wiley, Chichester, UK.

Wilson, E. O. (1992). "The diversity of Life," p. 424. Norton, New York.

Wise, P. K., and Pitman, M. G. (1981). Nutrient removal and replacement associated with short-rotation eucalypt plantations. *Aust. For.* **44,** 142–152.

Wofsy, S. C., Goulden, M. L., Munger, J. W., Fan, S.-M., Bakwin, P. S., Daube, B. C., Bassow, S. L., and Bazzaz, F. A. (1993). Net exchange of CO_2 in a mid-latitude forest. *Science* **260,** 1314–1317.

Wolter, P. T., Mladenoff, D. J., Host, G. E., and Crow, T. R. (1995). Improved forest classification in the northern Lake States using multi-temporal Landsat imagery. *Photo. Eng. Rem. Sensing* **61,** 1129–1143.

Wong, S. C., Cowan, I. R., and Farquhar, G. D. (1979). Stomatal conductance correlates with photosynthetic capacity. *Nature* **282,** 424–426.

Woods, R. V. (1976). Early silviculture for upgrading productivity on marginal *P. radiata* in the south-eastern region of South Australia. Bull. 24. SA. p. 24. Woods and Forests Dept.

Woodmansee, R. G., Vallis, I., and Mott, J. J. (1981). Grassland nitrogen. *Ecol. Bull (Stockholm)* **33,** 443–462.

Woodward, F. I. (1987). "Climate and Plant Distribution." Camb. Univ. Press, New York.

Woodward, F. I. (1995). Ecophysiological controls of conifer distributions. *In* "Ecophysiology of Coniferous Forests" (W. K. Smith and T. M. Hinckley, Eds.), pp. 79–94. Academic Press, San Diego.

Woodwell, G. M., and Whittaker, R. H. (1967). Primary production and the cation budget of the Brookhaven forest. *In* Proc. Symp. "Primary Production and Nutrient Cycling in Natural Ecosystems," pp. 151–166. Univ. Maine Press, Orono, ME.

Wu, Y., and Strahler, A. H. (1994). Remote sensing of crown size, stand density, and biomass on the Oregon transect. *Ecol. Appl.* **4,** 299–312.

Wullschleger, S. D. (1993). Biochemical limitations to carbon assimilation in C_3 plants—A retrospective analysis of the A/C_i curves from 109 species. *J. Exp. Bot.* **44,** 907–920.

Yanai, R. D.(1991). Soil solution phosphorus dynamics in a whole-tree harvested northern hardwood forest. *Soil Sci. Soc. Am. J.* **55,** 1746–1752.

Yarie, J., and Van Cleve, K. (1983) Biomass and productivity of white spruce stands in central Alaska. *Can. J. For. Res.* **13,** 767–772.

Yavitt, J. B., and Fahey, T. J. (1986). Litter decay and leaching from the forest floor in *Pinus contorta (lodgepole pine)* ecosystems. *J. Ecol.* **74,** 525–545.

Yoda, K., Shinosaki, K., Ogawa, H., Hozumi, K., and Kira, T. (1965). Estimation of the total amount of respiration in woody organs of trees and forest communities. *J. Biol. Osaka. City. Univ.* **16,** 15–26.

Yoder, B. J., and Waring, R. H. (1994). The normalized difference vegetation index of small Douglas-fir canopies with varying chlorophyll concentrations. *Remote Sensing Environ.* **49,** 81–91.

Yoder, B. J., Ryan, M. G., Waring, R. H., Schoettle, A. W., and Kaufmann, M. R. (1994). Evidence of reduced photosynthetic rates in old trees. *For. Sci.* **40,** 513–527.

Young, J. R., Ellis, E. C., and Hidy, G. M. (1988). Deposition of air-borne acidifers in the western environment. *J. Envir. Qual.* **17**, 1–18.

Zak, D. R., and Grigal, D. F. (1991). Nitrogen mineralization, nitrification and denitrification in upland and wetland ecosystems. *Oecologia* **88**, 189–196.

Zak, D. R., Host, G. E., and Pregitzer, K. S. (1989). Regional variability in nitrogen mineralization, nitrification, and overstory biomass in northern Lower Michigan. *Can. J. For. Res.* **19**, 1521–1526.

Zak, D. R., Groffman, P. M., Pregitzer, K. S., Christensen, S., and Tiedje, J. M. (1990). The vernal dam: Plant–microbe competition for nitrogen in northern hardwood forests. *Ecology* **71**, 651–656.

Zak, D. R., Tilman, D., Paramenter, R. P., Rice, C. W., Fisher, F. M., Vose, J., Milchunas, D., and Martin, C. W. (1994). Plant production and soil microorganisms in late-successional ecosystems: A continental-scale study. *Ecology* **75**, 2333–2347.

Zarin, D. J., and Johnson, A. H. (1995). Nutrient accumulation during primary succession in a montane tropical forest, Puerto Rico. *Soil Sci. Soc. Am. J.* **59**, 1444–1452.

Zhang, J., and Davies, W. J. (1991). Antitranspirant activity in xylem sap of maize plants. *J. Exp. Bot.* **42**, 317–321.

Zimmermann, M. H. (1983). "Xylem Structure and the Ascent of Sap." Springer-Verlag, New York.

Subject Index

Absorbed photosynthetically active radiation, *see also* Radiation
    canopy, 68–72
    relation to net primary production, 136
A-C_i relationships, *see* Photosynthesis
Aerodynamic conductance, 74–76
Aging, *see* Succession
Albedo, 65–66
Allocation, *see* Carbon allocation and accumulation
Allometry, 60
Architecture, *see* Canopy architecture
Assimilation, *see* Photosynthesis
Autotrophic respiration, *see* Respiration

Beer's Law, 68
Bowen ratio, 79–80

Canopy architecture
    clumping, 69
    extinction coefficient, 69
    foliage distribution, 57–58
    leaf area index, 59–61
    specific leaf area, 55–56
Carbon allocation
    allocation controls general, 145, 148
        defense compounds, 153–154
        fine roots, 148–151
        foliage, 145–150
        reproduction, 152–153
        respiration, 139–144
        storage, 151–152
    modeling, 145–148
    stand development, effects on, *see* Succession
Carbon budget, *see also* Carbon allocation, Photosynthesis, and Soil Surface CO_2 flux
    general, 126
    role of forests in global budget, 157–158
Catchment hydrology, *see* Hydrologic budget

Cavitation, due to
    freeze thaw, 24
    water stress, 61
Coarse woody debris, *see also* Decomposition
    carbon input, 167
    role of, 167
Conifers versus deciduous, *see* Deciduous versus evergreen

Deciduous versus evergreen
    detritus production, 167–169
    leaf area, 61
    nutrient use, 202–204, 206–210
    specific leaf area, 138–139
    water transport system, 24, 60–61
Decomposition
    constants (k), 173
    effects of
        environment, 172–174
        litter quality, 174–175
        surface area: volume, 175
    methods to estimate, 171
    role of organisms, 172
Defense compounds, *see* Carbon allocation
Detritus production
    chemical composition, 168–170
    nutrient inputs, 202–204
    sources of
        aboveground detritus, 166–167
        coarse woody debris, 167
        fine root turnover, 168
Drought stress, *see* Water stress

Embolism, *see* Cavitation
Energy balance, *see* Radiation budget
Epsilon, *see* Models
Exudates, root, *see* Carbon allocation

Fine roots, *see also* Mycorrhizae
    distribution, 103–105
    net primary production, 148–151
    turnover, 168–169

Forest biomes
  characteristics, general
    boreal, 35–37
    temperate broad-leaved evergreen, 41–43
    temperate conifers, 38–40
    temperate deciduous, 37–38
    temperate mixed, 40–41
    tropical deciduous, 46–48
    tropical evergreen, 43–46
  climate
    boreal, 26, 30–31
    temperate broad-leaved evergreen, 28–29, 32
    temperate coniferous, 27, 31
    temperate deciduous, 27, 31
    temperate mixed, 28, 31–32
    tropical deciduous, 30, 32–33
    tropical evergreen, 29, 32
  distribution, future, 48–49
  physiological adaptations, species, 22–24
Forest management
  decisions, 3–9
  sustainable, 2
Forest plantations, 21–22
Fungi, see also Mycorrhizae
  biomass, 172
  carbon:nutrient ratio, 191–192

Geographic information systems (GIS)
  applications in forestry
    land use change, 281–282
    coupling to remote sensing and process models, 289–299
    fragmentation, 281–282
  data layers, 289–299
  raster versus vector, 278–281
  use as a management tool, 289–299
Growth efficiency, 154–155

Heat transport, 72
Heterotrophic respiration, see Decomposition, Soil surface CO_2 flux
Hydrologic budget
  components of
    drainage, 101–103
    infiltration, 101–103
    interception, 94–97
    precipitation, 93–94
    runoff, 101–103
    transpiration, 105–111
  effects of forest management, 111–116
  measurements of soil moisture, 102
  soil hydraulic characteristics, 97–101
Hydraulic conductivity, effects of tree size

Ion exchange, soil
  anion exchange capacity, 191, 220
  cation exchange capacity, 166, 189–190

Leaf area index (LAI)
  biome averages, 61
  radiation attenuation, 68–72
Leaf area: sapwood area ratio, 60
Leaf life span, see Deciduous versus evergreen
Light, see Radiation
Light absorption, see Absorbed PAR
Light use efficiency, see Absorbed PAR and Models

Models
  application of, 269–271, 274–275
  empirical, 248–250
  process-based, 251–254
    BEX, 260–262
    BIOMASS, 257–259
    CENTURY, 265–266
    EPSILON, 266–269
    FOREST-BGC, 259–260
    LINKAGES, 264–265
    MAESTRO, 254–257
    PnET, 262–264
Mycorrhizae, see also Carbon allocation
  carbon costs, 126
  effects on nutrient uptake, 149–150, 205–206
  types, 148

Net ecosystem production, 155–157
Net primary production
  controls on allocation, see Carbon allocation
  measurement of, 142–145
Nitrification, 197–200, 241
Nitrogen, effects on
  carbon allocation, 147–151
  photosynthesis, 137–138
Nitrogen saturation, 225–226
Nutrient cycling
  atmospheric deposition, 195–196
  denitrification, 213–214
  effects of
    atmospheric deposition, 195–196

fire, 216–220
forest harvesting, 220–224
species, 214–216
stand age, *see* Succession
foliar leaching, 210
leaching losses, 210–213
litterfall, 202–204
immobilization
 microbial, 191–192, 221–222
 vegetation, 191–192, 221–222
mineralization, 197–200
nitrification, 197–200
nitrogen fixation
 asymbiotic, 200
 symbiotic, 200–202
nutrient use efficiency, 208–210
retranslocation, 206–210
root turnover, *see* Detritus
saturation, nitrogen, *see* nitrogen saturation
stemflow, 196
throughfall, 196
uptake
 canopy, 196
 roots, 204–206
weathering, 196–198
Nutrient distribution, 192–194
Nutrient retranslocation, 206–210
Nutrient stress, *see* Nitrogen
Nutrient use efficiency, 206–210
Nutrients
 essential, 187
 diagnosing nutrient deficiency, 187–188
 stoichiometry, 191–192

Photon flux density, *see* Radiation
Photosynthesis
 A-C_i curves, 128–131
 effects of
  carbon dioxide, 129–130
  leaf longevity, 138–139
  light, 137
  nitrogen, 137–138
  vapor pressure deficit, *see* Stomatal conductance
  water, *see* Stomatal conductance
 Farquhar and von Caemmerer model, 130–132
 modeling, 129–132, 135–136
 scaling, *see* Photosynthesis, modeling
Pollution, *see* Nitrogen saturation

Quantum yield, *see* Photosynthesis and absorbed photosynthetic active radiation

Radiation budget
 components of
  latent heat of vaporization, 67, 79–80
  net, 65–67
  sensible, 67
  storage, 67
 effects of topography, 84–87
 interception, *see* Absorbed PAR
 measurement, 64
 photosynthetic active (PAR), 63
 reflectivity, 65–66
 shortwave, 63
Remote sensing
 information obtainable from
  land cover, 284–285
  leaf area index (L*), 286–288
  stand structure, 285–286
  normalized difference vegetation index (NDVI), 286–288
 types of sensors, 282–284
 use as a management tool, 289–299
Respiration, *see also* Soil surface CO_2 flux
 construction, maintenance, and ion uptake, 139–140
 effects of
  nitrogen, 140
  temperature, 139–140
 measurement of, 141–142
 scaling, 140–144
Roots, *see* Fine roots, Mycorrhizae and Carbon allocation

Soil carbon
 accumulation rates, 163–164
 content, 163
 effects of forest management
  afforestation, 163–164
  fertilization, 181
  fire, 180
  harvesting, 179–180
  site preparation, 179–180
  species, 181–182
 humus, 166
 leaching losses, 178–180
 mean residence time 164–165
 role in global carbon budget 163, 179–183

Soil surface CO_2 flux
  biome averages, 176
  effects of
    nutrient availability, 177
    temperature, 176–177
    heterotrophic versus autotrophic, 178–179
  measurement of, 178
Soil water
  conductivity, 100
  measurement, 102
  potential, 97–99
Species diversity
  biome averages, 24–25
  succession, 233–234
Stand development, *see also* Succession
  linear versus non-linear, 231–233
Stomatal conductance, *see also* Transpiration
  controls by
    environmental conditions, 106–111
    hydraulic conductance, 107–109
    plant growth hormones, 107–108
  measurement of, 127
  modeling, 131–135
Succession
  autotrophic respiration, 237
  decomposition, 238
  gross primary production, 237
  leaf area index, 237
  net ecosystem production, 239–240
  net primary production, 236–241
  nitrogen mineralization, 241
  nutrient leaching, 244
  nutrient uptake and use, 241–243
  primary versus secondary, 230–231
  species composition, 233–234
  stomatal constraint, 238–239
  structural changes, 234–236
Sustainable management, *see* Forest management, sustainable

Temperature, effects on
  decomposition, 173–174
  respiration, 139–144
Transpiration, *see also* Stomatal conductance
  Bowen ratio, 79–80
  effects of forest management, 111–116
  modeling, 128–135

Water relations
  capacitance, 60–61
  leaf water potential, 117–119
  hysteresis, 117–119
  Ohm's law, 117–119
Water stress
  effects on growth, 120–122
  integral, 119–120
  plant growth regulators, 107–108
Water uptake
  environmental effects on root/shoot allocation, 150
  soil properties, 97–100

Physiological Ecology
A Series of Monographs, Texts, and Treatises

Series Editor
Harold A. Mooney
Stanford University, Stanford, California

Editorial Board
Fakhri A. Bazzaz   F. Stuart Chapin   James R. Ehleringer
Robert W. Pearcy   Martyn M. Caldwell   E.-D. Schulze

T. T. KOZLOWSKI (Ed.). Growth and Development of Trees, Volumes I and II, 1971
D. HILLEL (Ed.). Soil and Water: Physical Principles and Processes, 1971
V. B. YOUNGER and C. M. McKELL (Eds.). The Biology and Utilization of Grasses, 1972
J. B. MUDD and T. T. KOZLOWSKI (Eds.). Responses of Plants to Air Pollution, 1975
R. DAUBENMIRE (Ed.). Plant Geography, 1978
J. LEVITT (Ed.). Responses of Plants to Environmental Stresses, Second Edition Volume I: Chilling, Freezing, and High Temperature Stresses, 1980 Volume II: Water, Radiation, Salt, and Other Stresses, 1980
J. A. LARSEN (Ed.). The Boreal Ecosystem, 1980
S. A. GAUTHREAUX, JR. (Ed.). Animal Migration, Orientation, and Navigation, 1981
F. J. VERNBERG and W. B. VERNBERG (Eds.). Functional Adaptations of Marine Organisms, 1981
R. D. DURBIN (Ed.). Toxins in Plant Disease, 1981
C. P. LYMAN, J. S. WILLIS, A. MALAN, and L. C. H. WANG (Eds.). Hibernation and Torpor in Mammals and Birds, 1982
T. T. KOZLOWSKI (Ed.). Flooding and Plant Growth, 1984
E. L. RICE (Ed.). Allelopathy, Second Edition, 1984
M. L. CODY (Ed.). Habitat Selection in Birds, 1985
R. J. HAYNES, K. C. CAMERON, K. M. GOH, and R. R. SHERLOCK (Eds.). Mineral Nitrogen in the Plant-Soil System, 1986
T. T. KOZLOWSKI, P. J. KRAMER, and S. G. PALLARDY (Eds.). The Physiological Ecology of Woody Plants, 1991
H. A. MOONEY, W. E. WINNER, and E. J. PELL (Eds.). Response of Plants to Multiple Stresses, 1991
F. S. CHAPIN III, R. L. JEFFERIES, J. F. REYNOLDS, G. R. SHAVER, and J. SVOBODA (Eds.). Arctic Ecosystems in a Changing Climate: An Ecophysiological Perspective, 1991

T. D. SHARKEY, E. A. HOLLAND, and H. A. MOONEY (Eds.). Trace Gas Emissions by Plants, 1991

U. SEELIGER (Ed.). Coastal Plant Communities of Latin America, 1992

JAMES R. EHLERINGER and CHRISTOPHER B. FIELD (Eds.). Scaling Physiological Processes Leaf to Globe, 1993

JAMES R. EHLERINGER, ANTHONY E. HALL, and GRAHAM D. FARQUHAR (Eds.). Stable Isotopes and Plant Carbon-Water Relations, 1993

E.-D. SCHULZE (Ed.). Flux Control in Biological Systems, 1993

MARTYN M. CALDWELL and ROBERT W. PEARCY (Eds.). Exploitation of Environmental Heterogeneity by Plants: Ecophysiological Processes Above- and Belowground, 1994

WILLIAM K. SMITH and THOMAS M. HINCKLEY (Eds.). Resource Physiology of Conifers: Acquisition, Allocation, and Utilization, 1995

WILLIAM K. SMITH and THOMAS M. HINCKLEY (Eds.). Ecophysiology of Coniferous Forests, 1995

MARGARET D. LOWMAN and NALINI M. NADKARNI (Eds.). Forest Canopies, 1995

BARBARA L. GARTNER (Ed.). Plant Stems: Physiology and Functional Morphology, 1995

GEORGE W. KOCH and HAROLD A. MOONEY (Eds.). Carbon Dioxide and Terrestrial Ecosystems, 1996

CHRISTIAN KÖRNER and FAKHRI A. BAZZAZ (Eds.). Carbon Dioxide, Populations, and Communities, 1996

THEODORE T. KOZLOWSKI and STEPHEN G. PALLARDY. Growth Control in Woody Plants, 1997

J. J. LANDSBERG and S. T. GOWER. Applications of Physiological Ecology to Forest Management, 1997